THE THEORY OF ROTATING FLUIDS

BY

H. P. GREENSPAN

Professor of Applied Mathematics
Massachusetts Institute of Technology

BREUKELEN PRESS

BROOKLINE

Reprinted with corrections and a reference supplement by

Breukelen Press
P. O. Box 101
Brookline, MA

(First published in 1968 by Cambridge University Press as part of the Cambridge Monographs on Mechanics and Applied Mathematics.)

Library of Congress Catalog Card No. 68-12058

ISBN 0–9626998–0–2 (previously ISBN 0 521 05147 9)

CONTENTS

3. CONTAINED ROTATING FLUID MOTION: NON-LINEAR THEORIES

4. MOTION IN AN UNBOUNDED ROTATING FLUID

5. DEPTH-AVERAGED EQUATIONS: MODELS FOR OCEANIC CIRCULATION

6. STABILITY

ACKNOWLEDGMENTS

I am indebted to the following people for permitting the reproduction of certain drawings and photographs in this book: Drs A. B. Arons, D. J. Baker, W. H. Banks, E. R. Benton, F. P. Bowden, K. Bryan, D. Coles, A. J. Faller, P. Frenzen, D. Fultz, R. Hide, D. Lilly, R. R. Long, M. S. Longuet-Higgins, W. V. R. Malkus, E. Mollö-Christensen, C. E. Pearson, M. H. Rogers, N. Rott, K. Stewartson, A. Toomre, J. S. Turner, G. Veronis, B. Warren.

Permission to reproduce figures was received from the editors of the following periodicals:

Bulletin of American Meteorological Society, figures 5.10, 5.11.

Deep-Sea Research, figures 5.6, 5.7.

Icarus, figure 3.17.

Journal of Atmospheric Sciences, figures 2.11, 2.12, 4.9, 5.5, 5.12, 5.13, 6.6, 6.9, 6.12, 6.13.

Journal of Fluid Mechanics, figures 2.21, 3.2, 3.4, 3.8, 3.9, 6.1, 6.2, 6.5, 6.8, 6.10.

Journal of Marine Research, figure 5.2.

Proceedings of the Royal Society, figures 3.10, 5.4, 6.17.

Progress in Aeronautical Sciences, figure 3.14.

Quarterly Journal Math. & Applied Mech. figures 3.5, 3.6. 3.11, 3.18.

Tellus, figures 2.20, 5.8.

I especially wish to thank Dr V. Barcilon who did much to improve the presentation of this work.

This book is the outgrowth of a research programme sponsored by the Office of Scientific Research of the United States Air Force, Grants AF–AFOSR–492–64, 66.

PREFACE

The selection of subject matter and the manner of its presentation are not issues on which there can be exact agreement, much less so when the field in question is in the process of rapid and diverse growth. Perhaps, then, it would be of interest for me to describe the scope of this work, its rationale and some of the decisions and compromises that I have made or accepted.

It is my intention to provide a basic foundation for the support and promotion of research in rotating fluids. Because the subject has so many separate branches, I have tried to concentrate on those topics which I consider fundamental, of central importance to most, if not all, the areas of application. Practically speaking, this has been translated to mean the study of rotating fluids in quite ordinary circumstances, unembellished by very special and exotic effects—the motion of a contained, incompressible, viscous fluid such as water in a simple controlled environment. Furthermore, attention is focused almost exclusively upon primary phenomena, those that occur *only* in a *rotating* medium. To restrict the length of this monograph, I have also severely curtailed, and at times omitted, the discussion of material which receives extensive coverage in other books.

The amount of detail to be included in a theoretical development is a problem having no generally satisfactory solution. What is sufficient for one reader is inadequate or superfluous for another. My policy is as follows. If the material is new, the topic pregnant with possibility, or the methods basic, then the exposition is rather complete. On the other hand, I have not hesitated merely to highlight or to summarize, difficult, extended or inconclusive analyses, taking care to cite all the pertinent references for those interested. The intricate details of elaborate experimental and numerical programmes are also omitted from the text. However, it is strongly recommended that the serious student attempt, on his own, some of the experiments which can be performed with a modest outlay, and sufficient detail is provided for this purpose. These demonstrations really give the subject life and their role in developing intuition cannot be overestimated.

I have attempted to make as many sections as possible reasonably self-contained. The basic equations of motion are repeated often for this purpose as are certain important definitions and formulas. I regard this duplication of effort a minor penalty for the afforded convenience.

In all of this, there is the large undeniable element of personal preference. This is an author's prerogative even if it is, perhaps, a challenge to some to exercise their own literary judgments.

H. P. G.

Cambridge, Mass.
1968

CHAPTER 1

INTRODUCTION

PART A

1.1. A few simple experiments

All fluid phenomena on earth involve rotation to a greater or lesser extent. Those in which rotation is an absolutely essential factor include the large-scale circulations in the atmosphere and oceans, and so many small-scale flows that examples are unnecessary. Though the variety of motions is extensive, there are, however, common processes at work. It is this basic structure of rotating fluids that is the object of study here.

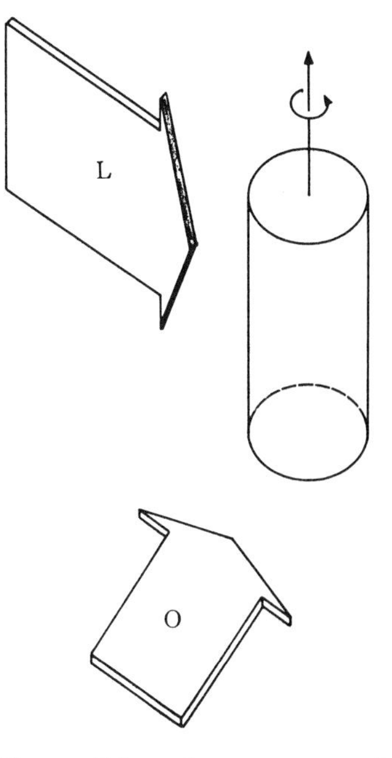

Fig. 1.1. Schematic arrangement of equipment: L, slit beam light source; O, observation direction.

The primary effects of rotation, those which do not exist otherwise, may be called remarkable without qualm; moreover, they can be observed in the very simplest of experiments. The ease with which important and extraordinary phenomena are reproduced and demonstrated engenders a special fascination for the subject.

Several interesting experiments can be performed with elementary apparatus consisting of a turntable, a light source, a transparent cylindrical container and minor subsidiary equipment. The closed container is *completely* filled with tap water into which a small amount of aluminium powder is suspended by mixing it with a little ordinary detergent. The tank is illuminated from the side with a vertical slit beam and it is best viewed from a direction perpendicular to the light source. The arrangement is shown schematically in fig. 1.1. The light reflected from the randomly oriented disk-like particles has a textured quality and the appearance of an undisturbed uniform mixture is shown by the central core region in fig. 1.4. A very slight shearing motion in any section of the tank is sufficient to align the metallic particles there, thus changing the light intensity seen by an observer. This provides an extremely sensitive indicator of relative fluid motion and is particularly suitable for visual or photographic observation.

In the first experiment, a small solid object is fixed securely to the bottom surface of the filled container and the entire system is put into a state of solid body rotation. The rotation rate is then changed a trifle, 0·5 % at 16 r.p.m., so that the shell and the contained fluid no longer have the same angular velocity. This velocity differential persists for a substantial time span, much longer than needed for the present demonstration. The essential feature here is the establishment of a fluid motion relative to the obstacle.

Fig. 1.2 (*a*), a photograph of the flow, clearly shows a column of fluid rigidly attached to the protuberance. All the fluid contained within this vertical cylinder, which circumscribes the object, is trapped there and must move with the body as a unit. The flow about this cylinder is then much the same as if the entire pillar were an impermeable solid, as seen in fig. 1.2 (*b*). The implication is that slow relative motion between a body and a rotating fluid produces a columnar, or two-dimensional, flow pattern. This was first predicted theoretically by Proudman[156] and confirmed experimentally shortly thereafter by Taylor[200, 201, 203]. (To produce a relative velocity, Taylor towed the obstacle slowly across the bottom of the rotating tank; of course, the net effect is basically the same.)

This most striking and unusual rotational phenomenon occurs frequently and in many equivalent forms during the course of this work. Other demonstrations of Taylor–Proudman columns will be described later.

The strong tendency towards two-dimensionality that is observed is really a special display of a more general property of rotating fluids, namely, that rotation allows an incompressible liquid, such as water, to support internal wave motion. This can be demonstrated with the equipment described, by oscillating a small disk that is

(*a*)

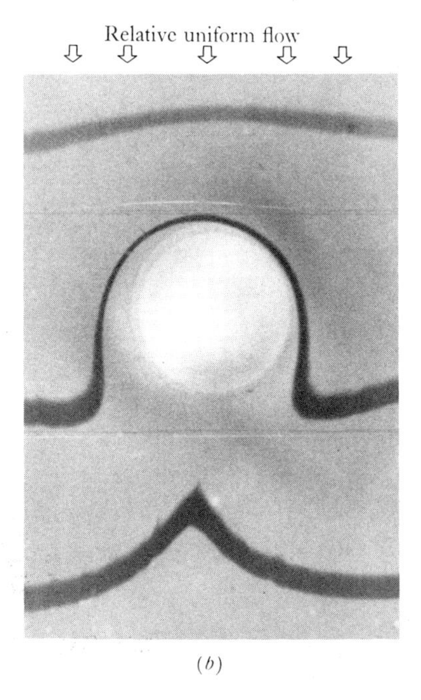

(*b*)

Fig. 1.2. (*a*) A Taylor–Proudman column. (*b*) This photo by D. J. Baker shows the flow past a Taylor–Proudman column. The dye lines lie in a level plane well above the cylindrical obstacle that produces the disturbance and the relative motion is uniform and steady. In the absence of rotation, a dye line remains straight (horizontal in the figure) as it passes over the body.

(*a*)

(*b*)

Fig. 1.3. (*a*) Waves produced by an oscillating disk with $\omega/\Omega = 1{\cdot}75$. The half apex angle is 59° and the theoretical value is 56°. (*b*) The apex angle increases for a larger value of ω/Ω.

placed in the midst of the cylindrical container, (Görtler[67], Oser[140]). The disk executes 'infinitesimal' oscillations, at frequency ω, in the direction normal to its own surface and parallel to the rotation axis. When the container is not rotating, the fluid motion produced in this manner is quite ordinary in appearance, a potential flow for all practical purposes. However, should the apparatus rotate uniformly at frequency Ω, extraordinary effects can be observed when $\omega \leqslant 2\Omega$ and these are shown in fig. 1.3. The pattern, made visible by the thin, free shear layers in the fluid, is certainly reminiscent of Mach cones in compressible aerodynamics, even to the extent of multiple reflections off the lateral wall of the cylinder. The internal boundary layers are actually viscosity modified characteristic surfaces typically found in hyperbolic or wave problems. Taylor–Proudman columns correspond to the characteristic surfaces at zero excitation frequency, $\omega = 0$. As ω increases towards 2Ω, the characteristic cones become more nearly horizontal planes. Beyond that critical value, the wave system disappears entirely and the flow resembles a potential motion once again. In mathematical terms, a changeover from hyperbolic to elliptic character must occur at the precise value $\omega = 2\Omega$.

Once the existence of wave motion caused solely by rotation is recognized, the subject must be accorded a status equalling that of surface wave theory in hydrodynamics, or the theory of stratified fluids. Indeed, these subjects are richly endowed structures for precisely the same reason—both involve wave propagation (a gravitational effect).

Another important process in rotating fluids concerns the manner in which a secondary flow can significantly alter the primary motion through a slight redistribution of angular momentum and vorticity. The secondary flow is often produced by viscous boundary layers and the control of the main motion is accomplished by vortex line stretching and the conservation of angular momentum. This is the interaction involved in vortex decay, whether it be a hurricane dissipating itself over land, or a stirred cup of tea settling to rest. Indeed, the latter is a rather trivial 'experiment', accessible to all, through which an understanding of this important mechanism can be attained.

Casual observation discloses that motion in an ordinary teacup ceases approximately one minute after stirring. Furthermore, the accumulation of impurities at the bottom centre is a clear indication of the presence of a secondary flow. The hypothesis that the vortex decay in this instance is caused by viscous processes diffusing across the breadth, $2L$, of the cup, implies a characteristic time scale, L^2/ν, ν being the kinematic viscosity of water. Any typical values for L and ν (say $L = 4$ cm, $\nu = 0{\cdot}01$ cm^2/sec at 100° C) lead to a dissipation time that is far too large—the hypothesis is clearly wrong. The correct time scale turns out to be $L/(\Omega\nu)^{\frac{1}{2}}$, ($\Omega$ is the initial rotation

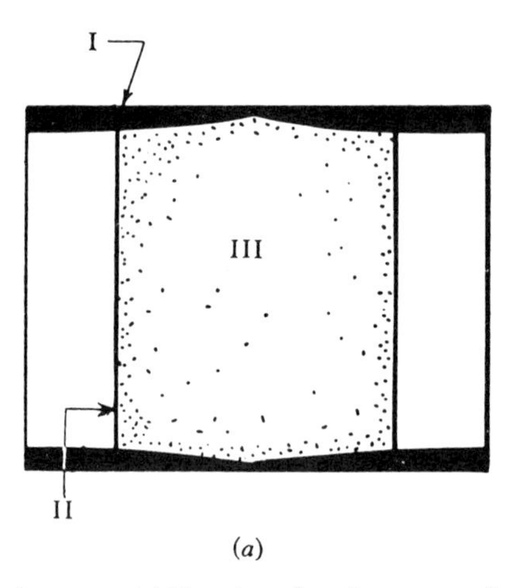

(*a*)

Fig. 1.4. Spin-up from rest: (*a*) Drawing of motion at an early time showing the Ekman boundary layer I (exaggerated), the light front II, and the almost quiescent core III.

rate), and represents the time required for the secondary flow to move a fluid particle into the bottom boundary layer where its excess angular momentum is eliminated.

The opposite experiment, spin-up from rest (Wedemeyer [223]), is very revealing when performed with the techniques used in the preceding demonstrations. Boundary layers are seen to form on the horizontal surfaces within the first few revolutions following the impulsive start. Non-rotating fluid is sucked into these layers, spun-up by viscous action, and returned to the interior at the vertical sidewall. Fig. 1.4 depicts the motion at two separate times; the boundary layers appear as the thin dark ribbons adjoining the horizontal surfaces. The original, undisturbed fluid occupies the

Fig. 1.4. Spin-up from rest: (*b*) Photograph of the flow at a later time than that for (*a*).

ever diminishing central core and is a much stronger source of reflected light than the spinning fluid returned from the boundary layers. The almost perfectly vertical light front separating the two interior regimes is convincing evidence that rotating flows have a strong disposition towards two-dimensionality—even in distinctly non-linear circumstances. This cylindrical front is the instantaneous position of those fluid particles initially at the sidewall which are all constrained by rotation to move inward as a vertical column. The time scale of the spin-up process is the same as in the teacup problem because the dominant mechanisms are identical.

This mechanism, involving the interplay of viscous boundary layers, secondary flows, vorticity, and angular momentum, is so fundamental that much of this book is devoted to its elucidation. The remainder deals with other processes really meriting equal consideration for in part they produce flows as interesting and remarkable. The atmospheric jet stream and the great ocean currents are in this category.

The demonstrations presented here are intended as a very brief survey of major effects. The complete analyses of these experiments, and others, appear in subsequent sections.

1.2. Equations of motion

We are principally concerned with the motion of an incompressible viscous liquid of constant material properties. The general problem of this type is formulated first and particular reductions of the theory are then discussed. Effects due to density stratification are examined sufficiently often in the course of this work to warrant the detailed development of an even more general theory. However, this is presented in §1.4 in order to defer the introduction of additional complexities for the time being.

The equations governing the motion of an incompressible viscous fluid are those stating the conservation of mass and momentum;

$$\nabla \cdot \mathbf{q} = 0, \tag{1.2.1}$$

$$\frac{\partial}{\partial t}\mathbf{q} + \tfrac{1}{2}\nabla(\mathbf{q}\cdot\mathbf{q}) + (\nabla \times \mathbf{q}) \times \mathbf{q} + 2\boldsymbol{\Omega} \times \mathbf{q} + \boldsymbol{\Omega} \times (\boldsymbol{\Omega} \times \mathbf{r})$$
$$= -\frac{1}{\rho}\nabla P + \mathfrak{F} - \nu\nabla \times (\nabla \times \mathbf{q}). \tag{1.2.2}$$

Here $\mathbf{q}$ is the particle velocity measured in a co-ordinate system *rotating* with *constant angular velocity* $\boldsymbol{\Omega}$ ($= \Omega\hat{\mathbf{k}}$)[1], and $\mathbf{r}$, t, P, ρ, ν and $\mathfrak{F}$ represent respectively, the position vector, time, pressure, density, kinematic viscosity and body force per unit mass. The body force is assumed to be conservative, $\mathfrak{F} = -\nabla\mathfrak{A}$, so that it and the centrifugal acceleration can be combined with P to form the *reduced pressure*

$$p = P + \rho\mathfrak{A} - \tfrac{1}{2}\rho(\boldsymbol{\Omega}\times\mathbf{r})\cdot(\boldsymbol{\Omega}\times\mathbf{r}), \qquad (1.2.3)$$

which is used to simplify (1.2.2):

$$\frac{\partial}{\partial t}\mathbf{q} + \mathbf{q}\cdot\nabla\mathbf{q} + 2\boldsymbol{\Omega}\times\mathbf{q} = -\frac{1}{\rho}\nabla p - \nu\nabla\times(\nabla\times\mathbf{q}). \qquad (1.2.4)$$

The common form of the convective acceleration $\mathbf{q}\cdot\nabla\mathbf{q}$ is employed more often than the invariant vector representation.

The equations of motion in an *inertial* coordinate system are obtained from the foregoing by setting $\boldsymbol{\Omega} = 0$. The formula

$$\mathbf{q}_{\text{inert}} = \boldsymbol{\Omega}\times\mathbf{r} + \mathbf{q}_{\text{rot}}$$

relates the fluid velocities in the inertial and rotating systems.

At solid impermeable surfaces, the viscous fluid must move with the boundary, since no slipping or penetration can occur. If the surface is permeable, the normal velocity component may be prescribed but the requirement of no slippage or relative tangential velocity stands. A common circumstance has a section of the bounding surface rotating uniformly with angular velocity $\boldsymbol{\Omega}_W$. With respect to the co-ordinate system rotating with angular velocity $\boldsymbol{\Omega}$, the boundary condition on this surface is

$$\mathbf{q} = (\boldsymbol{\Omega}_W - \boldsymbol{\Omega})\times\mathbf{r}. \qquad (1.2.5)$$

The formulation is completed by prescribing the initial velocity field

$$\mathbf{q}(\mathbf{r}, 0) = \mathbf{q}_*(\mathbf{r}). \qquad (1.2.6)$$

The problem is then to solve equations (1.2.1) and (1.2.4) in a specified domain subject to the boundary conditions of the type expressed in (1.2.5) and (1.2.6).

Let L, Ω^{-1}, U characterize the typical length, time, and relative

[1] If $\boldsymbol{\Omega}$ is time-dependent, the term $-\mathbf{r}\times\frac{d}{dt}\boldsymbol{\Omega}(t)$ must be added to the left-hand side of (1.2.2).

velocity of a particular motion. The replacement of the variables $\mathbf{r}$, t, $\mathbf{q}$, $\boldsymbol{\Omega}$, p by their scaled counterparts $L\mathbf{r}$, $\Omega^{-1}t$, $U\mathbf{q}$, $\Omega\hat{\mathbf{k}}$, $\rho\Omega ULp$, allows reduction of the problem to a *dimensionless* form:

$$\nabla\cdot\mathbf{q} = 0, \tag{1.2.7}$$

$$\frac{\partial}{\partial t}\mathbf{q}+\varepsilon\mathbf{q}\cdot\nabla\mathbf{q}+2\hat{\mathbf{k}}\times\mathbf{q} = -\nabla p - E\nabla\times\nabla\times\mathbf{q}, \tag{1.2.8}$$

with prescribed boundary conditions. (The caret, ^, denotes a unit vector.) Two important dimensionless parameters appear; the Ekman number

$$E = \frac{\nu}{\Omega L^2} \tag{1.2.9}$$

and the Rossby number

$$\varepsilon = \frac{U}{\Omega L}. \tag{1.2.10}$$

The former is a gross measure of how the typical viscous force compares to the Coriolis force and it is, in essence, the inverse Reynolds number for the flow. Likewise, the Rossby number, a ratio of the convective acceleration to the Coriolis force, provides an overall estimate of the relative importance of non-linear terms. The Ekman number is very small in most cases of interest where primary effects of rotation are displayed. Practical values of 10^{-5} are usual and henceforth the assumption $E \ll 1$ is made without further statement. The Rossby number, ε, is of unit magnitude or less; linear theories presume an infinitesimal value.

The *dimensionless* form of the vorticity equation in the *rotating frame* is

$$\frac{\partial}{\partial t}\mathfrak{B}+\nabla\times\{(\varepsilon\mathfrak{B}+2\hat{\mathbf{k}})\times\mathbf{q}\} = -E\nabla\times\nabla\times\mathfrak{B}, \tag{1.2.11}$$

where

$$\mathfrak{B} = \nabla\times\mathbf{q}. \tag{1.2.12}$$

The trivial solution of these equations, $\mathbf{q} = 0$, represents, of course, the non-trivial state of rigid rotation viewed from the rotating frame. In inertial co-ordinates, the corresponding scaled velocity is simply $\mathbf{q}_{\text{inert}} = \hat{\mathbf{k}}\times\mathbf{r}$. Obviously, a viscous fluid filling a closed uniformly rotating container tends, in the course of time, to this natural state of rigid rotation. A great deal can be learned from the manner in which this is accomplished and considerable effort is expended in this direction in chapter 2.

Perhaps some comments about notation are appropriate. *No special symbols are used to distinguish dimensional variables from those made dimensionless.* The context (i.e., the occurrence of parameters like ε and E) affords sufficient means to avoid confusion. Although the dimensionless formulation is favoured, it is not used exclusively. Hereafter, the characteristic values of length, time and velocity employed in any section will be indicated by a special notation, $[\![L, t, U]\!]$, which in the present case is $[\![L, \Omega^{-1}, \varepsilon\Omega L]\!]$. This should permit easy reference and conversion to different schemes of dimensionalization when necessary. Other scaled variables will be defined explicitly.

1.3. Theoretical survey

It can now be ascertained that this theoretical formulation is sufficiently general to account for the phenomena described in the first section. In fact, much can be inferred about the problems that confront us from an almost superficial examination of the governing equations.

Consider first the meaning of the very small Ekman number E which multiplies the most highly differentiated terms in (1.2.8). This is the formal criterion for the existence of boundary layers somewhere in the physical domain. It is well known that thin shear layers are located along the bounding surfaces and are the regions within which the tangential velocity is adjusted to its proper wall value by viscosity. However, viscous layers may also occur in the flow interior to counteract any effect that tends to produce sharp or even discontinuous velocity profiles. The structure of any boundary layer depends on the underlying force balance and in general it differs from case to case. It will be seen that boundary layer thicknesses $LE^{\frac{1}{2}}$, $LE^{\frac{1}{3}}$, $LE^{\frac{1}{4}}$ all occur in rotating fluid problems.

A concentration of viscous action into narrow layers means that elsewhere the fluid behaves in an essentially inviscid manner, as if $E = 0$. Classical boundary layer theory or the method of matched asymptotic expansions is surely designated as the analytical means of connecting the various flow regimes (see Van Dyke[213]).

Perhaps not so obvious or well known is the fact that there are also boundary layers associated with a small Rossby number ε, called inertial boundary layers. Once again the mathematical criterion is present: when E is zero, ε multiplies the remaining most highly differentiated terms. For $0 < E \ll \varepsilon \ll 1$ two boundary layers should develop with a 'wide' inertial layer containing a thinner viscous sub-stratum. The precise meaning of the inequality and the relationship of the two parameters and their associated boundary layers are matters for investigation. The exact nature of the rather subtle force balance in a non-linear inertial boundary layer is taken up in § 5.4.

Consider the simple special reduction of the general theory in which the motion is slow, $\varepsilon = 0$; steady, $(\partial/\partial t)\mathbf{q} = 0$; and inviscid, $E = 0$. The momentum equation is then

$$2\hat{\mathbf{k}} \times \mathbf{q} = -\nabla p. \qquad (1.3.1)$$

The curl of this expression and use of (1.2.7) easily yields a profound result known as the Taylor–Proudman theorem:

$$(\hat{\mathbf{k}} \cdot \nabla)\mathbf{q} = 0. \qquad (1.3.2)$$

Hence, the particle velocity must be independent of the co-ordinate measured along the rotation axis, z, under the prescribed conditions so that

$$\mathbf{q} = \mathbf{q}(x, y).$$

Moreover, since every particle in a vertical column of fluid has the same velocity, the entire pillar moves as a single elongated vertical fluid element. A column of given length, extending to solid boundaries at both ends, is constrained to maintain the same length as it moves about. For example, according to the theorem, there can be no flow over the small mound in the configuration shown in fig. 1.2 because such a motion would require a change in column height. (The boundary condition of no normal flow at impermeable walls, $\mathbf{q}\cdot\hat{\mathbf{n}} = 0$, implies that the vertical velocity component is zero at the top wall but non-zero on the protuberance. Since this component is independent of height, the contradiction is evident and no flow of this sort is possible.) One alternative is for the fluid to flow about the cylinder circumscribing the obstacle as if it were a solid. This agrees with observation. However, other possibilities exist because the original hypotheses are very restrictive. The contradiction need only mean that some assumptions are violated in real flows. This requires study but for introductory purposes it is sufficient to note that experiments do indeed reveal such two-dimensional flow patterns and that the theory can predict their occurrence. The complete linear theory is developed in the next chapter.

Finally, the linear, inviscid, time-dependent theory is shown to possess wave-like solutions. With $\varepsilon = 0$, $E = 0$ and

$$\left.\begin{aligned} \mathbf{q} &= \mathbf{Q}\, e^{i\lambda t}, \\ p &= \Phi\, e^{i\lambda t}, \end{aligned}\right\} \tag{1.3.3}$$

the problem reduces to solving

$$\nabla \cdot \mathbf{Q} = 0, \tag{1.3.4}$$

$$i\lambda \mathbf{Q} + 2\hat{\mathbf{k}} \times \mathbf{Q} = -\nabla\Phi, \tag{1.3.5}$$

subject to the *inviscid* boundary condition

$$\mathbf{Q} \cdot \hat{\mathbf{n}} = 0 \tag{1.3.6}$$

at the container wall (whose outwardly directed unit normal is $\hat{\mathbf{n}}$). In terms of the pressure alone, the boundary value problem is

$$\nabla^2\Phi - \frac{4}{\lambda^2}(\hat{\mathbf{k}} \cdot \nabla)^2\,\Phi = 0, \tag{1.3.7}$$

with $$-\lambda^2 \hat{\mathbf{n}} \cdot \nabla\Phi + 4(\hat{\mathbf{n}} \cdot \hat{\mathbf{k}})(\hat{\mathbf{k}} \cdot \nabla\Phi) + 2i\lambda(\hat{\mathbf{k}} \times \hat{\mathbf{n}}) \cdot \nabla\Phi = 0, \tag{1.3.8}$$

at the bounding surface. This complicated eigenvalue problem admits solutions in closed domains for real values of λ, $|\lambda| < 2$, but to prove this and many other interesting properties requires an extensive analysis. Only the fact that (1.3.7) is hyperbolic in its spatial dependence for $|\lambda| < 2$ is especially relevant now. A governing equation of hyperbolic type means that discontinuities can occur in the fluid across characteristic surfaces which are in this case the cones

$$(x^2 + y^2)^{\frac{1}{2}} \pm \lambda(4 - \lambda^2)^{-\frac{1}{2}} z = \text{const.} \tag{1.3.9}$$

These are the surfaces shown in fig. 1.3; the experiments of Görtler [66, 67], and Oser [140] confirm the predicted dependence of the apex angle of the cone on the oscillation frequency of the disk. Note that the 'sound speed' is a function of the oscillation frequency λ so that the wave system is dispersive.

This preliminary analysis indicates that some of the most important effects in rotating fluids can be accounted for by a simple theory. Furthermore, the balance between pressure gradient and Coriolis force, with corrections for viscous action at the boundaries, emerges as the backbone of the entire subject. The foundation for all further study is set once these processes are thoroughly understood. This is not to imply that other mechanisms are always of secondary importance but merely that they come second in any systematic exposition.

PART B

1.4. A formulation for stratified fluids

A more general formulation is needed to describe the effects of compressibility and density variation on rotating fluid motions. The combination of these two subjects, stratification and rotation, can be expected to raise serious difficulties especially since both individually are endowed with a particularly rich assortment of complex physical phenomena, (see Yih [227], for a recent account of stratified flows). In accordance with the aims set forth in the preface, only modifications of primary rotational processes are studied and the theory is developed to an extent consistent with these limited objectives.

Consider the motion of a slightly stratified *liquid* having constant material properties such as dynamic and bulk viscosities μ_*, β_*; specific heat c_P; and thermal conductivity κ. The equations governing the fluid motion are those pertaining to the conservation of mass, momentum and energy, supplemented by an equation of state. With respect to a *uniformly rotating* co-ordinate system, the dynamic equations are

$$\frac{\partial \rho}{\partial t} + \mathbf{q}\cdot\nabla\rho + \rho\nabla\cdot\mathbf{q} = 0, \tag{1.4.1}$$

$$\rho\left[\frac{\partial}{\partial t}\mathbf{q} + \tfrac{1}{2}\nabla(\mathbf{q}\cdot\mathbf{q}) + (\nabla\times\mathbf{q})\times\mathbf{q} + 2\boldsymbol{\Omega}\times\mathbf{q}\right]$$
$$= -\nabla P - \rho\boldsymbol{\Omega}\times(\boldsymbol{\Omega}\times\mathfrak{r}) + \rho\mathfrak{F} - \mu_*\nabla\times\nabla\times\mathbf{q}$$
$$+ (\beta_* + \tfrac{4}{3}\mu_*)\nabla\nabla\cdot\mathbf{q}, \tag{1.4.2}$$

$$\rho c_P\left(\frac{\partial T}{\partial t}+\mathbf{q}\cdot\nabla T\right) = \kappa\nabla^2 T$$
$$+\mu_*[\nabla^2\mathbf{q}\cdot\mathbf{q}+2\nabla\cdot(\nabla\times\mathbf{q})\times\mathbf{q}-2\mathbf{q}\cdot\nabla\nabla\cdot\mathbf{q}$$
$$+\nabla\times\mathbf{q}\cdot\nabla\times\mathbf{q}-\tfrac{2}{3}(\nabla\cdot\mathbf{q})^2]+\beta_*(\nabla\cdot\mathbf{q})^2, \quad (1.4.3)$$

where T is the temperature. The quantity

$$\frac{T}{\rho}\frac{\partial\rho}{\partial T}\left(\frac{\partial P}{\partial t}+\mathbf{q}\cdot\nabla P\right),$$

that ordinarily appears on the left-hand side of (1.4.3), is neglected from the onset because for liquids it is very small compared to the other terms in the expression. This permits some simplification in procedure at no real penalty. As a matter of fact, the viscous dissipation term can also be neglected for the same reason but this may not yet be obvious and it will be carried along for a while.

The equation of state

$$\rho = \rho(P, T) \quad (1.4.4)$$

is generally quite complex but for many liquids an excellent approximation over a wide range is

$$\rho = \rho_m - \alpha_*(T - T_m) \quad (1.4.5)$$

where subscript m denotes an average value and the coefficient of thermal expansion α_* is very small. For water, the density variation is about one percent from 0 to 50° C and another three percent from 50 to 100° C.

The class of motions of most interest to us concerns slight departures from a state of equilibrium, or near equilibrium, in which the body force is gravitational, $\mathfrak{F} = -g\hat{\mathbf{k}}$, and parallel to the angular velocity vector, $\boldsymbol{\Omega} = \Omega\hat{\mathbf{k}}$.

Static equilibrium, a state of no relative fluid motion in the rotating frame, is not possible under the postulated conditions. (See also Zeipel [228].) To prove this, assume that $\mathbf{q}_e = 0$, where the subscript e signifies an equilibrium variable. It follows from the momentum equation (1.4.2) that

$$\nabla P_e = -\rho_e\,\Omega^2\hat{\mathbf{k}}\times(\hat{\mathbf{k}}\times\mathbf{r})-\rho_e g\hat{\mathbf{k}}. \quad (1.4.6)$$

The curl of this expression results in the following equation (in cylindrical co-ordinates) for the axially symmetric equilibrium density distribution,

$$g\frac{\partial \rho_e}{\partial r} + r\Omega^2 \frac{\partial \rho_e}{\partial z} = 0,$$

which has the general solution

$$\rho_e = \rho_e\left(z - \frac{\Omega^2 r^2}{2g}\right). \tag{1.4.7}$$

The pressure and the temperature T_e, which is obtained from the state law (1.4.4), must also be functions of $\left(z - \frac{\Omega^2 r^2}{2g}\right)$. However, under all these constraints, the energy equation reduces to

$$\nabla^2 T_e = 0 \tag{1.4.8}$$

and this has no non-trivial solution of the required form

$$T_e = T_e\left(z - \frac{\Omega^2 r^2}{2g}\right).$$

This contradiction proves the assertion—some convection is necessary in the steady state. However, states of near static equilibrium can be established that persist almost unaltered for much longer than the lifetimes of many rotational phenomena to be studied. For example, if gravity is much larger than the centrifugal acceleration and the width and height of the container are comparable, then

$$\frac{\Omega^2 r^2}{2g} \ll z$$

and ρ_e, T_e, etc. are essentially functions of height only. In this case,

$$T_e \simeq T_0 + z\Delta T, \tag{1.4.9}$$

is an exact solution of (1.4.8) and represents a state of approximate equilibrium for small Froude number, $F_R = \Omega^2 L/g$, with an incurred error of order F_R.

Media possessing extremely small values of thermal (or saline) diffusivity also exhibit near static equilibrium conditions. Diffusion effects are often completely negligible because the underlying stratification changes little in the duration of the main action. There may even be a substantial time interval when a 'steady' state is almost achieved before the neglected processes contribute significantly. In other words, the diffusion time scale for temperature (or salinity) can be an order of magnitude larger than that characterizing the motion. The near equilibrium distributions which are actually slowly varying functions of this longer time scale appear almost steady over shorter intervals. An exact analysis and expansion procedure incorporating all the small variations and multiple time scales would be unnecessarily complicated in view of our limited purposes. The development here deals with perturbations from the near equilibrium linear temperature field given by (1.4.9). This distribution seems sometimes to develop in the presence of steady convection too. However, the complete analytical description of an equilibrium field involving fluid motion can be a most difficult problem (see § 6.7).

Fluid motion in a rotating system is often induced by moving sections of the bounding surface at slightly different angular velocities. Although other means are available, the dimensional analysis of the fundamental equations is based on this kind of excitation and on the supposition that density stratification has a major effect only in so far as the buoyancy force is important.

As before, the velocity excess over rigid rotation is characterized by $\varepsilon\Omega L$ to make the variation of the dimensionless velocity of unit magnitude. The scaling rule is $[\![L, \Omega^{-1}, \varepsilon\Omega L]\!]$. Three different quantities are needed to describe the density structure: ρ_0, the average value over the whole field; $\Delta\rho$, indicative of the amount of stratification in the near equilibrium state; $\varepsilon\Omega^2 L\rho_0/g$, which is characteristic of the deviation produced by rotational processes. The last scale is arrived at by equating the size of the buoyancy and Coriolis forces. Therefore, we write

$$\frac{\rho}{\rho_0} = 1 + \frac{\Delta\rho}{\rho_0}\rho_e(z) + \frac{\varepsilon\Omega^2 L}{g}\rho, \tag{1.4.10}$$

where ρ is the *dimensionless* density perturbation. The first two terms, with the scale factor, constitute ρ_e. The temperature distribution is resolved in a similar fashion as

$$T = T_0 + \Delta T\left(T_e(z) + \frac{\varepsilon\Omega^2 L}{\frac{\Delta\rho}{\rho_0}g}T\right). \tag{1.4.11}$$

Usually, ΔT is given and the value of $\Delta \rho$ determined from the relationship involving the known expansivity:

$$\alpha_* = \frac{\Delta \rho}{\Delta T} = \left(\frac{d\rho}{dT}\right)_m.$$

Finally, the pressure is

$$P = \rho_0 g L P_e + \varepsilon \rho_0 \Omega^2 L^2 p. \qquad (1.4.12)$$

Density variations introduce a number of new parameters in addition to ε and E. These are:

the Froude number, $F_R = \dfrac{\Omega^2 L}{g}$;

a density ratio, $\varkappa = \dfrac{\Delta \rho}{\rho_0}$;

the internal Froude number, $f_R = \dfrac{\Omega^2 L}{\dfrac{\Delta\rho}{\rho_0} g} = \dfrac{F_R}{\varkappa}$;

the Prandtl number, $\sigma_P = \dfrac{\mu_* c_P}{\kappa}$;

a diffusivity coefficient, $D = \nu\Omega/c_P \Delta T$.

Of these, ε, E, F_R, D, $\varkappa$ are all assumed to be small compared to unity (though in varying degrees), f_R is order one, and σ_P can be large. These assumptions restrict the discussion to a very small range of the parameter space but one that covers a wide variety of interesting and important physical processes.

The specific values for water are reasonably representative. At 20° C, $\mu_* = 0{\cdot}01002$ gm cm^{-1} sec^{-1}, $\kappa = 6{\cdot}10^4$ gm cm sec^{-3} deg^{-1}, $c_P = 4{\cdot}10^7$ cm^2 sec^{-2} deg^{-1}, and $\rho = 0{\cdot}9982$ gm cm^{-3}. Moreover, $\rho = 1$ gm cm^{-3} at 4° C, 0·988 gm cm^{-3} at 50° C and 0·958 gm cm^{-3} at 100° C and this shows that $\alpha_* \cong 10^{-4}$ gm cm^{-3} deg^{-1}. Therefore, the value $\varkappa = 0{\cdot}001$ requires a temperature differential of several degrees. If $\alpha_* = 2{\cdot}10^{-4}$ gm cm^{-3} deg^{-1} in particular, then $\varkappa = 0{\cdot}001$, and 0·01 correspond to $\Delta T = 5$ and 50° respectively. A typical experimental setting might be $\varepsilon = 0{\cdot}01$, $\varkappa = 0{\cdot}01$, $L = 20$ cm, $\Omega = 1$ rad/sec so that $E = 2{\cdot}5 \times 10^{-5}$, $F_R \cong 0{\cdot}02$, $f_R = 2{\cdot}0$, $\sigma_P = 6{\cdot}67$ and $D = 5 \times 10^{-12}$. The large Prandtl number means that little change in the near equilibrium temperature field occurs before motion is dissipated by viscous action. The minute value of D implies that the dissipation of mechanical energy by friction is an unimportant term in the energy equation. The reader can also check that the term dropped from the energy equation is truly small.

These scaled variables are used to cast the theory into a *dimensionless* form. The perturbation equations for mass, momentum, and energy become the following:

$$\mathrm{F_R}\frac{\partial\rho}{\partial t}+\varkappa\mathbf{q}\cdot\nabla\rho_e+\varepsilon\mathrm{F_R}\mathbf{q}\cdot\nabla\rho+(1+\varkappa\rho_e+\varepsilon\mathrm{F_R}\rho)\nabla\cdot\mathbf{q}=0, \tag{1.4.13}$$

$$(1+\varkappa\rho_e+\varepsilon\mathrm{F_R}\rho)\left(\frac{\partial}{\partial t}\mathbf{q}+\varepsilon\mathbf{q}\cdot\nabla\mathbf{q}+2\hat{\mathbf{k}}\times\mathbf{q}\right)$$
$$=-\nabla p-\mathrm{F_R}\rho\hat{\mathbf{k}}\times(\hat{\mathbf{k}}\times\mathbf{r})-\rho\hat{\mathbf{k}}$$
$$+\mathrm{E}\left(\nabla^2\mathbf{q}+\left(\frac{\beta_*}{\mu_*}+\frac{4}{3}\right)\nabla\nabla\cdot\mathbf{q}\right), \tag{1.4.14}$$

$$(1+\varkappa\rho_e+\varepsilon\mathrm{F_R}\rho)\left(\mathrm{f_R}\frac{\partial T}{\partial t}+\mathbf{q}\cdot\nabla T_e+\varepsilon\mathrm{f_R}\mathbf{q}\cdot\nabla T\right)$$
$$=\frac{\mathrm{Ef_R}}{\sigma_P}\nabla^2 T+\varepsilon\mathrm{D}\Big(\nabla^2\mathbf{q}\cdot\mathbf{q}-2\nabla\cdot\mathbf{q}\times\nabla\times\mathbf{q}$$
$$-2\mathbf{q}\cdot\nabla\nabla\cdot\mathbf{q}+\nabla\times\mathbf{q}\cdot\nabla\times\mathbf{q}+\left(\frac{\beta_*}{\mu_*}-\frac{2}{3}\right)(\nabla\cdot\mathbf{q})^2\Big). \tag{1.4.15}$$

A general state law, (1.4.4), supplements this set.

The fundamental linear theory is obtained by setting $\varepsilon=0$ in the foregoing equations, adding the requisite number of boundary conditions and replacing the general equation of state (1.4.4) with the Boussinesq approximation, (1.4.5). (See Spiegel and Veronis [182].) This still leaves a formidable system to deal with and further approximations are inevitable.

If *all* terms multiplied by small parameters are discarded, then the theory reduces to its simplest form:

$$\nabla\cdot\mathbf{q}=0, \tag{1.4.16}$$

$$\frac{\partial}{\partial t}\mathbf{q}+2\hat{\mathbf{k}}\times\mathbf{q}=-\nabla p-\rho\hat{\mathbf{k}}\quad\{+\mathrm{E}\nabla^2\mathbf{q}\}, \tag{1.4.17}$$

$$\mathrm{f_R}\frac{\partial T}{\partial t}+\mathbf{q}\cdot\nabla T_e=0\qquad\left\{+\frac{\mathrm{Ef_R}}{\sigma_P}\nabla^2 T\right\}, \tag{1.4.18}$$

$$\rho=-T. \tag{1.4.19}$$

Although we expect this system to govern the motion in most of the fluid domain, it does not allow for the existence of boundary layers at the container walls which are of crucial importance in rotating flow phenomena. Perhaps the easiest way of rectifying this omission is to retain terms multiplied by the Ekman number and marked by brackets in (1.4.17) and (1.4.18). The problem would then be well posed upon completely specifying the boundary conditions.

Of course, the physical processes that *must* be included in an analytical study depend on the particular problem. At a later time (see §2.21), we will consider stratified fluid motions in which boundary layer thermal convection remains unimportant and in these cases the simplest viscous theory is quite satisfactory.

Often, a salinity distribution, $\mathfrak{S}$, is responsible for the density stratification instead of, or in addition to, a temperature field. This can be incorporated within the theory by replacing (1.4.5) with a more general approximation,

$$\rho = \rho_m - \alpha_*(T - T_m) + \alpha_S(\mathfrak{S} - \mathfrak{S}_m), \qquad (1.4.20)$$

and by adding a salt conservation equation to the system:

$$\frac{\partial \mathfrak{S}}{\partial t} + \mathbf{q} \cdot \nabla \mathfrak{S} = \kappa_S \nabla^2 \mathfrak{S}. \qquad (1.4.21)$$

A linearized version of (1.4.21) is derived in a manner similar to that used in establishing (1.4.18).

An interesting and important analogy exists between rotating fluid motions (non-stratified) and stratified flows (non-rotating). Many effects found in one subject have counterparts in the other and to facilitate the comparison, the linear theory governing non-rotating stratified motions is developed now.

A new dimensionalization scheme must be introduced to characterize properly the scale of motion in a stratified *non-rotating* flow. Choose the scales $\left[L, \left(L \Big/ \frac{\Delta\rho}{\rho_0} g \right)^{\frac{1}{2}}, \varepsilon_S \left(L \frac{\Delta\rho}{\rho_0} g \right)^{\frac{1}{2}} \right]$ and let

$$\left.\begin{aligned} \rho &= \rho_0 + \Delta\rho(\rho_e(z) + \varepsilon_S \rho), \\ T &= T_0 + \Delta T(T_e(z) + \varepsilon_S T), \\ P &= \rho_0 g L P_e + \varepsilon_S \Delta\rho g L p, \end{aligned}\right\} \qquad (1.4.22)$$

where ε_S is a small number analogous to ε in the rotating system. The time scale is just that appropriate for pendulum oscillations in the reduced gravitational field, $(\Delta\rho/\rho_0)g$. The lowest order approximation is derived in a manner identical with that just completed and the procedure is rather straightforward. Thus, the linear theory is found to be

$$\left.\begin{aligned} \nabla\cdot\mathbf{q} &= 0, \\ \frac{\partial}{\partial t}\mathbf{q} &= -\nabla p - \rho\hat{\mathbf{k}} + E_S\nabla^2\mathbf{q}, \\ \frac{\partial T}{\partial t} + \mathbf{q}\cdot\nabla T_e &= \frac{E_S}{\sigma_P}\nabla^2 T, \\ \rho &= -T, \end{aligned}\right\} \tag{1.4.23}$$

with $E_S = \nu((\Delta\rho/\rho_0)gL)^{-\frac{1}{2}}L^{-1}$ taking the role of the Ekman number. For simplicity, let $\mathbf{q}\cdot\nabla T_e = w = \mathbf{q}\cdot\hat{\mathbf{k}}$. A single equation for the disturbance pressure, obtainable from these, is

$$\left(\frac{\partial}{\partial t} - \frac{E_S}{\sigma_P}\nabla^2\right)\left(\frac{\partial}{\partial t} - E_S\nabla^2\right)\nabla^2 p + \left(\frac{\partial^2}{\partial x^2} + \frac{\partial^2}{\partial y^2}\right)p = 0. \tag{1.4.24}$$

In the special case of two-dimensional motion, $p = p(x, z, t)$, with $\sigma_P = 1$, this becomes

$$\left(\frac{\partial}{\partial t} - E_S\nabla^2\right)^2\nabla^2 p + \frac{\partial^2}{\partial x^2}p = 0. \tag{1.4.25}$$

This reduction is important because (1.4.25) is identical with the pressure equation arising in the theory of rotating flows, (2.2.2). The implications stemming from this observation will be examined more closely in § 2.2. Further comparison of the two subjects is undertaken periodically in subsequent chapters.

1.5. Rudiments of vorticity theory

The theory of rotating fluids forms only a part of the general theory of vorticity. However, since our objectives are limited, this brief discussion of fundamentals is restricted to those aspects of the general theory that are of direct and immediate relevance. Surely, knowledge of the basic concepts of vorticity is an assumed prerequisite and this has already been taken for granted in previous sections.

Comprehensive discussions of vorticity appear in many treatises on fluid dynamics and the following references are particularly noteworthy: Goldstein[65], Lighthill[109], Truesdell[209].

It will be necessary to distinguish absolute vorticity, $\mathfrak{B}_a$, measured in an *inertial* frame from the relative vorticity, $\mathfrak{B}$, measured in a *uniformly rotating* co-ordinate system. The two are related by

$$\mathfrak{B}_a = \nabla\times\mathbf{q}_{inert} = 2\mathbf{\Omega}+\nabla\times\mathbf{q}_{rot} = 2\mathbf{\Omega}+\mathfrak{B}. \qquad (1.5.1)$$

The same distinction is also made for the circulation about a closed circuit Υ:

$$\Gamma_a = \oint_{\Upsilon}\mathbf{q}_{inert}\cdot\mathbf{ds} = \oint_{\Upsilon}\mathbf{q}_{rot}\cdot\mathbf{ds}+\oint_{\Upsilon}\mathbf{\Omega}\times\mathbf{r}\cdot\mathbf{ds} = \Gamma+\oint_{\Upsilon}\mathbf{\Omega}\times\mathbf{r}\cdot\mathbf{ds}.$$

Some manipulation of the last integral leads to the formula

$$\Gamma_a = \Gamma+2\Omega\int\hat{\mathbf{n}}\cdot\hat{\mathbf{k}}\,d\Sigma = \Gamma+2\Omega\Sigma_P. \qquad (1.5.2)$$

Here Σ_P is the projection of the area Σ enclosed by contour Υ onto a plane perpendicular to $\mathbf{\Omega} = \Omega\hat{\mathbf{k}}$. $\hat{\mathbf{n}}$ is the unit normal vector to Σ.

Application of Stokes' theorem yields the equivalent definition of circulation

$$\Gamma_a = \int\mathfrak{B}_a\cdot\hat{\mathbf{n}}\,d\Sigma. \qquad (1.5.3)$$

The vorticity at any point is proportional to the instantaneous angular momentum of a (rigid) spherical fluid element at that position. In fact, the instantaneous angular velocity of the particle is just $\frac{1}{2}\mathfrak{B}_a$. A line in the fluid that is everywhere tangent to $\mathfrak{B}_a$ is called a vortex line; the vortex lines through every point of a small closed curve make up a vortex tube. If the cross-sectional area $\Delta\Sigma$ of the tube is small then $\mathfrak{B}_a\cdot\hat{\mathbf{n}}\,\Delta\Sigma$ has the same value anywhere along the tube and is called the tube strength. This follows from the application of the divergence theorem,

$$\nabla\cdot\mathfrak{B}_a = 0, \qquad (1.5.4)$$

to the volume consisting of a section of tube capped by two surfaces $\Delta\Sigma_1$ and $\Delta\Sigma_2$. According to (1.5.3), this result is equivalent to the statement that the circulation about any contour girdling the lateral surface of the tube is a constant. Another conclusion derived from

(1.5.4) is that vortex lines cannot begin or end in the fluid; they are either closed curves or terminate on the solid boundaries.

The structure of the vorticity field can be visualized by means of vortex tubes of equal strength threading throughout the medium. The location of each tube is given by a central vortex line and the density of tubes in any region is then proportional to the magnitude of the vorticity, increasing as the tubes crowd together. The vorticity increases when a vortex tube of constant strength is stretched to compensate for a reduction in cross-sectional area.

The vorticity equation in *inertial* space for a stratified incompressible fluid is accurately approximated by

$$\frac{D}{Dt}\mathfrak{B}_a = \frac{\partial}{\partial t}\mathfrak{B}_a + \mathbf{q}\cdot\nabla\mathfrak{B}_a = \mathfrak{B}_a\cdot\nabla\mathbf{q} - \nabla\frac{1}{\rho}\times\nabla P + \nu\nabla^2\mathfrak{B}_a. \quad (1.5.5)$$

This equation relates the convective rate of change of the absolute vorticity of a particle (left-hand side) with three processes producing change (right-hand side). These processes, in the order they are written, are the stretching and tilting of vortex lines, the creation of vorticity by density variation and the diffusive transport of vorticity from surrounding elements.

In the case of constant density, (1.5.5) is a homogeneous partial differential equation of parabolic type. The steady solution must be $\mathfrak{B}_a = 0$ if vorticity is not generated at the bounding surfaces as a result of the viscous condition there that requires the relative tangential velocity to be zero. Vorticity diffusion is related to momentum transfer among fluid particles through the momentum gradient and this gradient is largest at the container wall. The situation is similar to the transfer of heat to a fluid at a hot boundary; the parallel is actually between the total flow of vorticity at the surface and the heat flux. In the computation of the vorticity field, solid boundaries are in effect replaced by equivalent distributions of vorticity sources.

Vorticity can be made a primary variable in flow calculations and this is desirable for several reasons. First, vorticity really is the principal physical quantity in many flows. (Lighthill [109] cites as an example the problem of blowing out a candle with a lip generated ring vortex.) Secondly, vorticity transport is accomplished by convection or diffusion whereas changes in pressure and velocity propagate with an infinite sound speed in an incompressible fluid. Hence, the range and time scale of significant vorticity variation is more restricted and slower than that of the pressure or velocity either of which can experience large changes at infinity, instantaneously. Thus, it is often much easier to compute the vorticity field and this approach has led to significant progress in the numerical solution of viscous flow problems, Payne [141], Pearson [142, 143]. The results of Pearson's calculations of certain rotating fluid motions is recounted in chapter 3.

Several classic vorticity theorems which concern inviscid fluids are now stated briefly. Reference is made to Goldstein [65] for a complete discussion of these fundamental results.

Lagrange's theorem asserts that in an inviscid, constant density medium, a vorticity field initially zero everywhere must remain so, $\mathfrak{V}_a = 0$. (The conditions obviously eliminate all the means of producing vorticity.) The same result holds in a stratified fluid when $\nabla\rho \times \nabla P = 0$.

Kelvin's theorem is derivable from the formula for the total rate of change of circulation about a closed contour that moves with the fluid:

$$\frac{D}{Dt}\Gamma_a = -\oint_{\Upsilon} \frac{1}{\rho} \nabla P \cdot \mathbf{ds}. \qquad (1.5.6)$$

Therefore, if the fluid is homogeneous, or barotropic, $P = P(\rho)$, and dissipative processes are neglected then *the circulation around a circuit moving with the fluid is a constant for all time.* Helmholtz's theorem follows directly from this and states that under the same conditions a *vortex tube moves with the fluid and its strength remains constant.*

These important theorems when written in a form appropriate for a rotating co-ordinate system show how changes in relative vorticity and circulation are effected by the basic rotation field. For example, the relationship (1.5.6),

$$\frac{D}{Dt}\Gamma = -\oint_{\Upsilon} \frac{1}{\rho} \nabla P \cdot \mathbf{ds} - 2\frac{D}{Dt}\int \mathbf{\Omega} \cdot \hat{\mathbf{n}}\, d\Sigma,$$

implies that the relative circulation is reduced when the cross-sectional area normal to $\mathbf{\Omega}$ is enlarged.

A theorem due to Ertel[43], of particular value in the study of rotating fluid motions, concerns a non-dissipative flow in a *uniformly rotating* co-ordinate system in which some special quantity such as the salinity, $\mathfrak{S}$, or the temperature, T, or anything else, is conserved as the particle moves about. To be specific, let

$$\frac{D\mathfrak{S}}{Dt} = \frac{\partial \mathfrak{S}}{\partial t} + \mathbf{q}\cdot\nabla\mathfrak{S} = 0. \qquad (1.5.7)$$

The relative vorticity equation, written in terms of the *rotating* system of axes, is

$$\frac{\partial}{\partial t}\mathfrak{B} + \nabla\times((\mathfrak{B}+2\boldsymbol{\Omega})\times\mathbf{q}) = \nabla P\times\nabla\frac{1}{\rho}, \qquad (1.5.8)$$

but it is advantageous to proceed using the absolute vorticity (measured in the same frame). Since $\boldsymbol{\Omega}$ is a constant, (1.5.8) becomes

$$\frac{\partial}{\partial t}\mathfrak{B}_a + \nabla\times(\mathfrak{B}_a\times\mathbf{q}) = \nabla P\times\nabla\frac{1}{\rho}. \qquad (1.5.9)$$

After multiplying this by $\nabla\mathfrak{S}\cdot$, the resultant equation can be manipulated into the following form

$$\nabla\mathfrak{S}\cdot\frac{\partial}{\partial t}\mathfrak{B}_a - \nabla\cdot(\nabla\mathfrak{S}\times(\mathfrak{B}_a\times\mathbf{q})) = \nabla\mathfrak{S}\cdot\nabla P\times\nabla\frac{1}{\rho}. \qquad (1.5.10)$$

Equation (1.5.7) is used to show that

$$\nabla\mathfrak{S}\times(\mathfrak{B}_a\times\mathbf{q}) = -\frac{\partial\mathfrak{S}}{\partial t}\mathfrak{B}_a - \mathbf{q}(\mathfrak{B}_a\cdot\nabla\mathfrak{S})$$

and the replacement of this in (1.5.10) with a suitable rearrangement of terms leads to Ertel's equation:

$$\left(\frac{\partial}{\partial t}+\mathbf{q}\cdot\nabla\right)\left[\frac{(2\boldsymbol{\Omega}+\mathfrak{B})\cdot\nabla\mathfrak{S}}{\rho}\right] = \frac{1}{\rho}\nabla\mathfrak{S}\cdot\nabla P\times\nabla\frac{1}{\rho}. \qquad (1.5.11)$$

The bracketed quantity in this expression must be conserved in the motion of a particle whenever the right-hand side is identically zero. This can occur in any one of several ways. The thermodynamic state law, which has not been used thus far, may relate ρ to P or $\mathfrak{S}$ alone; ρ may be a constant; the three vectors, $\nabla\mathfrak{S}$, ∇P and $\nabla\rho$ may be coplanar. It should be noted once again that the theorem holds for any quantity satisfying (1.5.7). Generalizations of this theorem have been given by Ertel and Rossby[44], and Truesdell[210].

If the flow is incompressible but stratified, then the density ρ itself satisfies (1.5.7) and may be used in the theorem to obtain the result

$$\frac{D}{Dt}\left[\frac{(2\boldsymbol{\Omega}+\boldsymbol{\mathfrak{B}})\cdot\nabla\rho}{\rho}\right] = 0.$$

Motion in a thin fluid layer, discussed in chapter 5, is similar to this but the density ρ is replaced there by the total layer depth H, which is a function of position. It turns out that the quantity

$$\frac{1}{H}(2\boldsymbol{\Omega}+\boldsymbol{\mathfrak{B}})\cdot\hat{\mathbf{k}},$$

called the potential vorticity, is conserved under these conditions. But these are matters for another time.

1.6. Rudiments of viscous boundary layer theory

The basic purpose here, as in the preceding section, is to extract from an old and fully developed subject a few of those results most fundamental to the study of rotating fluids. Aside from the matter of completeness, this will at least establish the minimal prerequisite upon which subsequent work is based. Several comprehensive accounts of boundary layer theory are available (Schlichting [178], Rosenhead [173], Moore [129], to name some) and an elementary introduction to the subject is given by Carrier [25].

The observation that viscosity in many flow problems is important only in certain thin layers becomes the basis of mathematical approximations which simplify the equations of motion. (Inertial boundary layers will be considered in § 5.4.) The usual circumstance is one in which viscous processes force the fluid velocity to satisfy the no-slip boundary condition on a solid wall. This adjustment occurs within a very thin viscous layer and may involve changes in the flow variables that are comparable to their original magnitudes.

As we mentioned previously, the occurrence of such rapid transition regimes is to be anticipated whenever the most highly differentiated terms in a partial differential equation are multiplied by a small parameter, as for example $E\nabla\times\nabla\times\mathbf{q}$ in (1.2.8). This is certainly not a sufficient condition and too often questions of existence and location must be resolved on the basis of intuition, experience or solid experimental evidence.

Suppose now that a particular viscous, incompressible fluid motion is known to have a boundary layer at a surface Σ. The exact *dimensional* equations of motion in the *rotating* frame are, of course, (1.2.1) and (1.2.2) but the object is to use the rapid variation of flow components within the boundary layer to simplify this system. This end is achieved by consistently neglecting the tangential derivatives along Σ of any flow variable when they compare directly with the normal derivative of the same function. The formal development can proceed in several equivalent ways and the presentation here is based in part on that of Crabtree, Küchemann and Sowerby [36].

The co-ordinate system that is used to form the boundary layer equations is constructed in the following manner. Let (ξ_1, ξ_2) be orthogonal curvilinear co-ordinates on the surface Σ of the container and let ξ_3 measure distance along $\hat{\mathbf{n}}$ the unit normal to Σ. The surfaces, $\xi_3(x, y, z) = \text{const.}$, are then all parallel to Σ, which is taken to be $\xi_3 = 0$, and, by construction, the unit normal vector to each of these parallel surfaces is $\hat{\mathbf{n}}(\xi_1, \xi_2)$, a continuous function of the surface co-ordinates only. (This extension of the normal $\hat{\mathbf{n}}$, defined on Σ, into a field of normal vectors to the family of parallel surfaces, is one for which $\nabla \times \hat{\mathbf{n}} = 0$.) Finally, it should be observed that the three co-ordinates, (ξ_1, ξ_2, ξ_3), form a mutually orthogonal set on Σ but not necessarily elsewhere.

The position vector to any point P within the boundary layer can be written as

$$\mathbf{r}_P = \mathbf{r}_\Sigma(\xi_1, \xi_2) + \xi_3 \hat{\mathbf{n}}(\xi_1, \xi_2),$$

where $\mathbf{r}_\Sigma$ is the position vector to Σ. In this vicinity, ξ_3 is very small and the arc length is approximately given by

$$ds^2 = \mathbf{dr}_P \cdot \mathbf{dr}_P \cong \mathbf{dr}_\Sigma \cdot \mathbf{dr}_\Sigma + d\xi_3^2.$$

This is equivalent to

$$ds^2 = h_1^2\, d\xi_1^2 + h_2^2\, d\xi_2^2 + d\xi_3^2,$$

where the metric coefficients are defined by

$$h_j = \left| \frac{\partial}{\partial \xi_j} \mathbf{r}_P \right|.$$

Indeed, for $j = 1, 2$,

$$h_j \cong \left| \frac{\partial}{\partial \xi_j} \mathbf{r}_\Sigma \right| \quad \text{and} \quad h_3 \cong 1.$$

The unit vectors on Σ are

$$\hat{\mathbf{i}}_j = \frac{1}{h_j} \frac{\partial}{\partial \xi_j} \mathbf{r}_\Sigma,$$

and with $\hat{\mathbf{i}}_3 = \hat{\mathbf{n}}$ form an orthogonal triad.

The particle velocity can be resolved into tangential and normal components near Σ as

$$\mathbf{q} = \hat{\mathbf{n}} \cdot \mathbf{q} \hat{\mathbf{n}} - \hat{\mathbf{n}} \times (\hat{\mathbf{n}} \times \mathbf{q}) \tag{1.6.1}$$

and the substitution of this expression into the equation for mass conservation results in

$$(\hat{\mathbf{n}} \cdot \nabla)(\hat{\mathbf{n}} \cdot \mathbf{q}) + \hat{\mathbf{n}} \cdot \nabla \times (\hat{\mathbf{n}} \times \mathbf{q}) + \hat{\mathbf{n}} \cdot \mathbf{q} \nabla \cdot \hat{\mathbf{n}} - \hat{\mathbf{n}} \times \mathbf{q} \cdot \nabla \times \hat{\mathbf{n}} = 0.$$

At impermeable bounding surfaces, the normal flux $\hat{\mathbf{n}} \cdot \mathbf{q}$ is zero and accordingly it must remain relatively small across the entire boundary layer. However, there is a rapid variation with ξ_3, the normal distance, and the derivative in this direction, $(\hat{\mathbf{n}} \cdot \nabla)(\hat{\mathbf{n}} \cdot \mathbf{q})$, is an order of magnitude larger than the quantity $\hat{\mathbf{n}} \cdot \mathbf{q}$ itself. Hence, the third term in the preceding formula is negligible compared to the first. The last term is negligible because the boundary layer co-ordinate system is constructed from a family of surfaces which are parallel to Σ and $\hat{\mathbf{n}}$ belongs to the field of normal vectors to these surfaces. Thus, the reduced form of the mass conservation equation in the boundary layer is

$$(\hat{\mathbf{n}} \cdot \nabla)(\hat{\mathbf{n}} \cdot \mathbf{q}) + \hat{\mathbf{n}} \cdot \nabla \times (\hat{\mathbf{n}} \times \mathbf{q}) = 0. \tag{1.6.2}$$

Expressed in terms of curvilinear components

$$\mathbf{q} = q_1 \hat{\mathbf{i}}_1 + q_2 \hat{\mathbf{i}}_2 + q_3 \hat{\mathbf{i}}_3,$$

this becomes

$$\frac{1}{h_1 h_2} \left(\frac{\partial}{\partial \xi_1} (h_2 q_1) + \frac{\partial}{\partial \xi_2} (h_1 q_2) \right) + \frac{\partial q_3}{\partial \xi_3} = 0. \tag{1.6.3}$$

The elimination of obviously small terms from the momentum equations leads to the following boundary layer equations:

$$\frac{\partial q_1}{\partial t} + \frac{q_1}{h_1} \frac{\partial q_1}{\partial \xi_1} + \frac{q_2}{h_2} \frac{\partial q_1}{\partial \xi_2} + q_3 \frac{\partial q_1}{\partial \xi_3} + \frac{q_1 q_2}{h_1 h_2} \frac{\partial h_1}{\partial \xi_2}$$

$$- \frac{q_2^2}{h_1 h_2} \frac{\partial h_2}{\partial \xi_1} + 2(\Omega_2 q_3 - \Omega_3 q_2) = -\frac{1}{\rho h_1} \frac{\partial p}{\partial \xi_1} + \nu \frac{\partial^2}{\partial \xi_3^2} q_1, \tag{1.6.4}$$

$$\frac{\partial q_2}{\partial t}+\frac{q_1}{h_1}\frac{\partial q_2}{\partial \xi_1}+\frac{q_2}{h_2}\frac{\partial q_2}{\partial \xi_2}+q_3\frac{\partial q_2}{\partial \xi_3}-\frac{q_1^2}{h_1 h_2}\frac{\partial h_1}{\partial \xi_2}$$

$$+\frac{q_1 q_2}{h_1 h_2}\frac{\partial h_2}{\partial \xi_1}+2(\Omega_3 q_1-\Omega_1 q_3) = -\frac{1}{\rho h_2}\frac{\partial p}{\partial \xi_2}+\nu\frac{\partial^2}{\partial \xi_3^2}q_2, \quad (1.6.5)$$

$$0 = \frac{\partial p}{\partial \xi_3}. \quad (1.6.6)$$

The last equation implies that the pressure is constant through the boundary layer and equals the value assumed just outside this transition zone in the inviscid domain. Although this is an entirely correct statement, it is sometimes insufficient in certain problems where the next order of approximation is crucial. It would be safer to replace (1.6.6) by a more complete form of this component of the momentum equation in order to prepare for this possibility. However, when this is necessary, the appropriate boundary layer equations are usually rederived from the exact theory by a formal procedure.[1] The extension of (1.6.6) appears then as part of that process and, at the same time, other negligible terms in the above equations are discarded (removed to a higher order of approximation).

At the surface Σ, the fluid velocity and the wall velocity must be identical. Conditions at the outer edge of the boundary layer, ξ_3 large, require that the flow variables attain prescribed values appropriate to the *inviscid* domain.

It is often convenient to express the dependent variables in a form that clearly identifies the rapid variation near the surface Σ by a separate function, as for example

$$\mathbf{q} = \mathbf{q}_I + \tilde{\mathbf{q}}.$$

The function $\mathbf{q}_I$ is then the solution of the inviscid equations of

[1] The order symbol, O, which occurs as part of the boundary layer perturbation formalism is used frequently throughout this work. It has the following definition: $F(x, E) = O(E^a)$ if for sufficiently small E there exists a constant A such that $|F(x, E)| \leqslant AE^a$. This is equivalent to $\lim_{E\to 0} F(x, E)/E^a < \infty$. The ordering is uniform if A does not depend on x.

The approximate value of a quantity is sometimes indicated by $\doteq$. For example, $t_c \doteq E^{-\frac{1}{2}}$ is to be interpreted as equality in a loose sense. The exponent of E, which determines the scale and essential magnitude, is given precisely by this relationship but for exact equality a multiplicative factor, of say $\frac{1}{2}$ or 2, might be necessary. The symbol $\cong$ implies equality in a more exact sense say to at least a few figures; $\sim$ means an asymptotic equality.

motion whereas $\tilde{\mathbf{q}}$ denotes the boundary layer contribution, which decays exponentially fast with normal distance from the surface Σ. The replacement of $\mathbf{q}$ and p in the foregoing theory by such decompositions leads to the modified system of equations for the perturbation boundary layer functions. Many problems to be studied will be formulated in this manner.

CHAPTER 2

CONTAINED ROTATING FLUID MOTION: LINEAR THEORIES

2.1. Classification

A somewhat arbitrary division of rotating fluid motions into two classes will be made. The distinction is based largely on whether the fluid is considered to be enclosed within a body or vice versa. Belonging to the first class, are all flows of an essentially bounded or contained character, where the outer bounding walls are the significant cause of the fluid motion. This category is itself subdivided into a study of linear and non-linear theories. The second class consists of problems involving motion of a body or object in an essentially unbounded rotating fluid. In these cases, the influence of the outer walls is only secondary compared to the effects produced by the motion of the object and for most purposes the concept of an unbounded fluid (moving the outer walls to infinity) is appropriate.

2.2. Almost rigid rotation

Most linear problems concern motions that represent slight deviations from an established state of rigid rotation. The velocity magnitude characteristic of the motion is then small compared to the basic rotation speed and $\varepsilon \ll 1$. Non-linear terms are negligible as a consequence.

The linear theory for the motion of a viscous incompressible fluid is obtained by setting $\varepsilon = 0$ in (1.2.7) and (1.2.8):

$$\left.\begin{aligned} \nabla\cdot\mathbf{q} &= 0, \\ \frac{\partial}{\partial t}\mathbf{q}+2\hat{\mathbf{k}}\times\mathbf{q} &= -\nabla p - E\nabla\times\nabla\times\mathbf{q}, \end{aligned}\right\} \qquad (2.2.1)$$

with appropriate boundary conditions.

A formal derivation of (2.2.1) requires each dependent variable to be expanded as a perturbation series in powers of ε; for example,

$$\mathbf{q}(\mathbf{r}, t, E, \varepsilon) = \sum_{n=0}^{\infty} \mathbf{q}^{(n)}(\mathbf{r}, t, E)\,\varepsilon^n.$$

The substitution of these expansions into the basic equations leads to a sequence of problems, each corresponding to a different power of ε. The first of the sequence (ε^0 terms) is the linear theory. This programme is employed in the next chapter to discuss non-linear effects but the simpler notation in (2.2.1), omitting superscript numerals, is retained for present work.

Some idea of the type of problem confronting us may be had by eliminating $\mathbf{q}$ from the theory to obtain a single sixth order partial differential equation for the pressure alone,

$$\left(\frac{\partial}{\partial t} - E\nabla^2\right)^2 \nabla^2 p + 4\frac{\partial^2}{\partial z^2} p = 0. \tag{2.2.2}$$

(The boundary conditions cannot be written in terms of just one dependent scalar variable.) Obviously, solving this equation is a formidable task, one that necessitates further approximations if progress is to be made.

There is a class of important problems dealing with the uniform motion of a body through a rotating fluid or the streaming flow past it. Usually in these cases, both body and fluid rotate with the same angular velocity and the disturbance is produced by a relative difference in the forward speeds of the obstacle and the distant fluid. The Rossby number, ε, need not be small but a linear problem is still obtainable in special circumstances. For example, when there is a uniform flow, $\mathfrak{U}$, past a small obstacle (in the rotating system) the velocity deviations from this state are usually quite small except, perhaps, in the vicinity of the body. A formal procedure based on this description involves the matching of inner and outer expansions in an overlap domain to obtain a uniformly valid representation (see Van Dyke[213]). To lowest order, the net effect is the replacement of the non-linear convective term $\mathbf{q}\cdot\nabla\mathbf{q}$ by $\mathfrak{U}\cdot\nabla\mathbf{q}$. This is just the classical Oseen approximation and we shall use this 'replacement concept' without resorting to a more elaborate procedure for justification.

Therefore, a linear theory in this case with ε moderate is

$$\left.\begin{aligned} \nabla\cdot\mathbf{q} &= 0, \\ \frac{\partial}{\partial t}\mathbf{q} + \varepsilon\hat{\mathfrak{U}}\cdot\nabla\mathbf{q} + 2\hat{\mathbf{k}}\times\mathbf{q} &= -\nabla p + E\nabla^2\mathbf{q}. \end{aligned}\right\} \tag{2.2.3}$$

Solutions of this system should be excellent approximations over most of the fluid domain. The substitution is certainly valid in the far field and it may be of no real consequence in the immediate vicinity of the body where other terms in the equations are predominant. This is, in part, confirmed by exact analyses of other problems where the Oseen linearization has been used.

Once again, a single equation for the pressure can be determined from (2.2.3) and it is

$$\left(\frac{\partial}{\partial t}+\epsilon\hat{\mathfrak{U}}\cdot\nabla-E\nabla^2\right)^2\nabla^2 p+4(\hat{\mathbf{k}}\cdot\nabla)^2 p = 0. \qquad (2.2.4)$$

The close relationship between non-rotating stratified flows and almost rigid rotations of an incompressible fluid now becomes evident. Equations (1.4.24) and (2.2.2) are obviously quite similar and the comparison with (1.4.25) is even closer. The co-ordinate z, measuring distance along the rotation axis, plays the same role as co-ordinate x, measuring length along a surface of constant density. The phenomenon of Taylor columns should then have its counterpart in the form of horizontal plugging in the stratified medium and indeed it does, (see Yih[227]). Many other phenomena are found to be roughly equivalent including the manifested wave systems (Trustrum[211]). In fact, rotating flows, which are easier to produce experimentally, can be used effectively in certain instances to simulate stratification. The literature in one subject is often applicable to the other and this is an important connection to bear in mind, even though it will *not* be emphasized at every conceivable opportunity in subsequent sections.

2.3. The Ekman layer

Consider an idealized situation comprising a viscous fluid confined to the upper half plane, $z > 0$, by an impermeable plate at $z = 0$. Both the fluid and the disk rotate rigidly with the same constant angular velocity and, at some initial instant, the rotation rate of the disk is increased slightly. Viewed from a *rotating* co-ordinate system in which the initial state is one of relative rest, the dimensionless boundary value problem consists of equation (2.2.1) complemented by the boundary conditions $\mathbf{q}(\mathbf{r},0) = 0$ and $\mathbf{q} = \hat{\mathbf{k}}\times\mathbf{r}$ on the plane $z = 0$. (The characteristic velocity is a length multiplied by the change in rotation speed of the disk.) It is also supposed that $\mathbf{q} \to 0$ as $z \to \infty$, a condition that in effect provides for the maintenance of rigid rotation of the main body of fluid by the remote boundaries.

The inherent axial symmetry of the resultant flow allows considerable simplification of the equations of motion when written in terms of a cylindrical co-ordinate system, (r, θ, z). The velocity, $\mathbf{q} = (u, v, w)$, and the pressure are then independent of the azimuthal angle θ. A further reduction is attained in this simple case by eliminating the radial co-ordinate entirely through the substitution

$$\mathbf{q} = -\nabla \times (r\chi(z,t)\,\hat{\boldsymbol{\theta}}) + rV(z,t)\,\hat{\boldsymbol{\theta}}. \tag{2.3.1}$$

The boundary value problem for the 'stream function' χ and the azimuthal velocity V becomes

$$\left(E\frac{\partial^2}{\partial z^2} - \frac{\partial}{\partial t}\right) V - 2\frac{\partial}{\partial z}\chi = 0, \tag{2.3.2}$$

$$\left(E\frac{\partial^2}{\partial z^2} - \frac{\partial}{\partial t}\right)\frac{\partial^2}{\partial z^2}\chi + 2\frac{\partial}{\partial z}V = 0, \tag{2.3.3}$$

with $V - 1 = \chi = (\partial/\partial z)\chi = 0$ on $z = 0$, $V = \chi = 0$ for $t = 0$, and both functions to vanish at $z = \infty$. The solution may be determined using Laplace transforms (see Campbell and Foster[24] for the necessary inversion integrals) and it is

$$V = \mathscr{Re}[\exp(-(2i)^{\frac{1}{2}}E^{-\frac{1}{2}}z) + F(z,t)], \tag{2.3.4}$$

$$\chi = E^{\frac{1}{2}}\mathscr{Im}[(2i)^{-\frac{1}{2}}\exp(-(2i)^{\frac{1}{2}}E^{-\frac{1}{2}}z) - (2i)^{-\frac{1}{2}}\operatorname{erf}(2it)^{\frac{1}{2}} - (2i)^{-\frac{1}{2}}F(z,t)], \tag{2.3.5}$$

with

$$F(z,t) = \tfrac{1}{2}[\exp((2i)^{\frac{1}{2}}E^{-\frac{1}{2}}z)\operatorname{erfc}\{(2i)^{\frac{1}{2}}t^{\frac{1}{2}} + \tfrac{1}{2}(Et)^{-\frac{1}{2}}z\} - \exp(-(2i)^{\frac{1}{2}}E^{-\frac{1}{2}}z)\operatorname{erfc}\{(2i)^{\frac{1}{2}}t^{\frac{1}{2}} - \tfrac{1}{2}(Et)^{-\frac{1}{2}}z\}]. \tag{2.3.6}$$

Of particular interest is the final steady motion when the individual velocity components are

$$\frac{u}{r} = \exp(-E^{-\frac{1}{2}}z)\,\sin E^{-\frac{1}{2}}z, \tag{2.3.7}$$

$$\frac{v}{r} = \exp(-E^{-\frac{1}{2}}z)\,\cos E^{-\frac{1}{2}}z, \tag{2.3.8}$$

$$w = E^{\frac{1}{2}}[-1 + \exp(-E^{-\frac{1}{2}}z)\,(\sin E^{-\frac{1}{2}}z + \cos E^{-\frac{1}{2}}z)], \tag{2.3.9}$$

and the final reduced pressure is

$$p = -2E\exp(-E^{-\frac{1}{2}}z)\sin E^{-\frac{1}{2}}z. \qquad (2.3.10)$$

The elementary form of solution for steady motion in its dependence solely on the variable $\zeta = E^{-\frac{1}{2}}z$, fig. 2.1, reveals that the motion of the plate is communicated to the fluid through a viscous boundary layer of thickness $E^{\frac{1}{2}}$ ($(\nu/\Omega)^{\frac{1}{2}}$ in dimensional units). (This was discovered by Ekman[41] who studied a closely related problem involving wind-stress on the ocean surface.) Within this thin layer, the Coriolis force is balanced by the viscous shear and the pressure

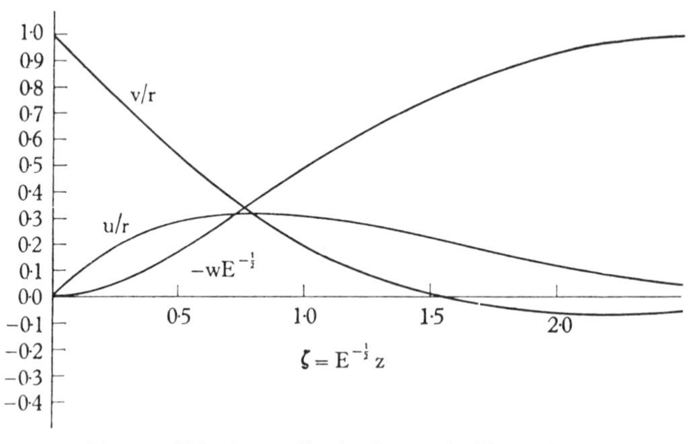

Fig. 2.1. Velocity profiles in the steady Ekman layer.

is a constant to very small order, O(E). Thus, in the steady flow, fluid particles near the plate are spun-up to a larger angular velocity by direct viscous action. The increased Coriolis force overcomes the pressure gradient along the plate and the fluid is propelled radially outwards as in a centrifugal fan. To compensate for the mass flow in the boundary layer, a small normal flux from the inviscid interior of order $E^{\frac{1}{2}}$ is required. In the absence of other nearby boundaries, a purely vertical secondary flow is produced throughout the main body of fluid by the boundary layer suction. The flow completes a closed circuit at infinity in an, as yet, indefinite manner.

The velocities within the boundary layer are of unit magnitude but subsist only in this thin layer adjacent to the wall whereas, a small interior circulation, of magnitude $O(E^{\frac{1}{2}})$, is produced in the entire fluid domain. The coiled pattern, fig. 2.2, that is formed by plotting the horizontal component of velocity versus position in the boundary layer is known as the Ekman spiral.

The steady vorticity equation

$$-2\frac{\partial}{\partial z}\mathbf{q} = E\nabla^2\boldsymbol{\mathfrak{B}} \tag{2.3.11}$$

shows that a stationary boundary layer of finite thickness can exist because the diffusion of vorticity from the plate is counteracted by the distortion of vortex lines. Most often in fluid problems, convection is the process that balances diffusion to establish a steady viscous or thermal boundary layer. In this respect, rotating fluids

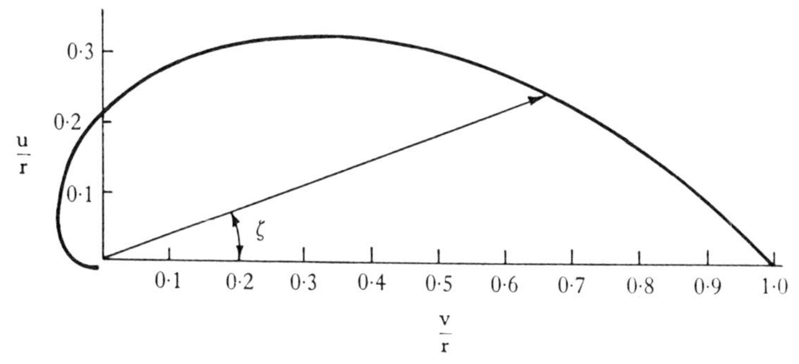

Fig. 2.2. The Ekman spiral. The boundary layer co-ordinate is plotted as a polar angle.

present a different and intrinsically more complex means of accomplishing the same end.

The examination of the transient solution shows that the Ekman layer forms in a relatively short time, just a few revolutions of the system. Fig. 2.3 illustrates the development of the azimuthal velocity component. The process is complicated, however, by continual viscous diffusion and minor oscillations at the non-dimensional frequency 2 (i.e., 2Ω). The asymptotic form of the typical complementary error function of complex argument in (2.3.6) makes this clearly evident,

$$\operatorname{erfc}(2it)^{\frac{1}{2}} \sim (2\pi i t)^{-\frac{1}{2}}\exp(-2it).$$

Vibrations at this frequency are due to the excitation of the natural inviscid modes. By setting E equal to zero in (2.3.2) and (2.3.3), it is a simple matter to show that solutions proportional to $\exp(i\lambda t)$ exist only if $\lambda = \pm 2$, and $\lambda = 0$.

The description in the case of plate spin-down to a lower rotation rate is entirely analogous to the foregoing except for a reversal of sign.

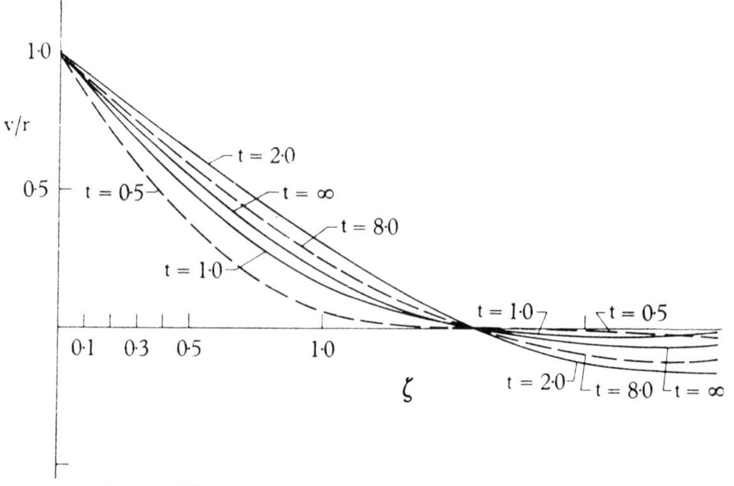

Fig. 2.3. The azimuthal velocity versus time showing the approach to the steady Ekman layer.

2.4. Spin-up

Suppose the configuration of the last section is modified by confining the fluid between two parallel concentric planes, the entire system to rotate rigidly as before. The rotation speeds of both disks are then simultaneously increased a like amount. In due course, the interior fluid rotates rigidly once again and this transition is called spin-up. Spin-up in any arbitrary closed container has the same meaning and involves an abrupt change in angular speed followed by a transient adjustment of the fluid to a final state of rigid rotation.

Although the fluid is still laterally unbounded in the parallel plate geometry, the combination of two infinite planes rotating in unison produces a motion that typifies all contained spin-up flows.

The mathematical problem is almost identical with equations (2.3.2), (2.3.3) et seq., except that the boundary conditions at infinity are replaced by conditions on the second plate; i.e.,

$$\chi = (\partial/\partial z)\chi = V - 1 = 0$$

at $z = 1$. (The plate separation distance is the characteristic length L.)

The solution of this problem was determined by Greenspan and Howard[75] using transform methods. However, the detailed analysis is complicated and is not, for this reason, presented here. Fortunately, the entire motion can be depicted in relatively simple terms. The complete description necessarily involves some repetition with the last section, especially in relation to the early phases of motion, but this is perhaps justified by the very fundamental nature of the process.

The initial impulsive change in the angular velocity immediately produces a Rayleigh shear layer at each disk which then starts to thicken by viscous diffusion. Within a few revolutions, $t \doteq 1$, the effects of rotation have made themselves felt, and a quasi-steady Ekman boundary layer develops from the vorticity diffusion. The analytical description of the evolution involves functions like those in equation (2.3.6). In addition, there are inertial oscillations at twice the rotation frequency but of very small amplitude. The Ekman layer is characterized by an outward radial flow of unit dimensionless magnitude caused by centrifugal action and this transport is balanced by a small flux into the boundary layer from the interior. However, in the presence of the other disk, this vertical flow into the boundary layer can be maintained only through the establishment of an equally small interior flow that is radially inward. In other words, the convergence of fluid into the Ekman layer, together with the constraints of the geometrical configuration, produce a small radial convection in the interior in order to conserve mass.

Since the interior flow is practically inviscid, the angular momentum of a ring of fluid moving inward to replace the fluid entering the Ekman layer is conserved, and the ring must acquire an increased angular velocity. The Ekman layer acts as a sink for low angular momentum fluid in the interior, this fluid being replaced by higher angular momentum fluid drawn from larger radii. As the conditions in the interior approach the values appropriate to the final steady state, the Ekman layer decays. This happens in a dimensionless time of order $E^{-\frac{1}{2}}$, a time scale not long enough for the boundary layers to thicken appreciably. In the meantime, the small oscillations set up by the initial impulse have been modified very slightly in the interior by viscosity and more markedly near the boundaries. They persist until they are finally destroyed by viscosity at a dimensionless time of the order of E^{-1}. (In containers of *finite* volume, the dissipation time is approximately $E^{-\frac{1}{2}}$.) At this late time, the viscous boundary layers at each wall have been so extended by the diffusion process alone that they overlap and there is no longer any interior inviscid domain. Viscous forces are then important at all interior positions and act to eliminate the residual modal oscillations. Thus, the transient motion between *infinite* disks consists of three distinct phases: the development of viscous boundary layers for $t \doteq 1$; spin-up, $t \doteq E^{-\frac{1}{2}}$; viscous decay of residual effects, $t \doteq E^{-1}$.

Given this basic physical picture, one can derive the characteristic spin-up time, t_c, by elementary physical arguments. If the increase in rotation rate is $\varepsilon\Omega$, then $\varepsilon\Omega L = U$ is the characteristic transport velocity within the Ekman layer of thickness $\delta = (\nu/\Omega)^{\frac{1}{2}}$. Mass conservation requires a mass influx into the viscous layer from the interior with a velocity of magnitude

$$w_I = 2U\delta/L.$$

Here L is the characteristic vertical length of the container so that w_I is also the typical radial velocity of the interior circulation. An annular ring of interior fluid of mass M and angular momentum $ML^2\Omega$ acquires an angular velocity $(1+\varepsilon)\,\Omega$ by moving radially inward a distance $\frac{1}{2}\varepsilon L$. Angular momentum is conserved because the interior flow is inviscid. The time required for the fluid ring to traverse this distance and thereby to attain the angular velocity of the new steady state is

$$t_c \doteq \frac{\varepsilon L}{w_I} \doteq \left(\frac{L^2}{\nu\Omega}\right)^{\frac{1}{2}} \doteq E^{-\frac{1}{2}}\Omega^{-1}.$$

The Ekman layer acts as a sink (or source), of strength proportional in fact to $E^{\frac{1}{2}}$ times the local difference in the angular velocities of the boundary and the interior flow. This equivalence was deduced by Charney and Eliassen[33], who gave the characteristic time t_c in a meteorological context. Bondi and Lyttleton[11], in a discussion of the secular retardation of the earth's core, determined that t_c is the time-lag of the angular velocity near the axis of the core behind the angular velocity possessed by the shell at any instant.

Another approach, which also clarifies a different aspect of the physical picture, is to consider the vorticity. The curl of the inviscid momentum equation (2.2.1) yields

$$\frac{\partial}{\partial t}\nabla\times\mathbf{q}-2\frac{\partial}{\partial z}\mathbf{q}=0,$$

the time-dependent form of the Taylor–Proudman theorem for the motion outside the boundary layers. The vertical velocity induced by the Ekman convergence is of order $E^{\frac{1}{2}}$ and of opposite signs at the two boundaries. Its vertical gradient is then also $E^{\frac{1}{2}}$; hence, the relative vertical vorticity can only be increased (by 'stretching') from zero to its final value of 2 in an approximate dimensionless time $E^{-\frac{1}{2}}$.

The following table illustrates the characteristic times involved in a case of practical interest, $L = 4$ cm, $\Omega = 200\pi\ \text{sec}^{-1}$.

Material	ν (cm² sec⁻¹)	$L^2\nu^{-1}$ (sec)	$\left(\frac{L^2}{\Omega\nu}\right)^{\frac{1}{2}}$ (sec)	E
Lubricating oil (40° C)	1·00	16	0·16	$1{\cdot}0\times10^{-4}$
Water	0·01	1,600	1·60	$1{\cdot}0\times10^{-6}$
Mercury	0·001	16,000	5·05	$1{\cdot}0\times10^{-7}$

An analytical approximation for the interior flow between the disks valid through the spin-up time is

$$V_I = 2S(2t)(1-\exp(-2E^{\frac{1}{2}}t)), \tag{2.4.1}$$

$$\chi_I = -2(z-\tfrac{1}{2})\,E^{\frac{1}{2}}S(2t)\exp(-2E^{\frac{1}{2}}t), \tag{2.4.2}$$

where $S(x)$ is the Fresnel integral:

$$S(x) = \int_0^x (2\pi t)^{-\frac{1}{2}}\sin t\,dt. \tag{2.4.3}$$

Experiments performed to determine the decay rate in spin-up resulted in very close agreement, [75], with the predicted factor $2E^{\frac{1}{2}}$. It is also shown in this reference that the essential quantitative features of spin-up in arbitrary axially symmetric containers can be determined, given the above description as a guide, by the methods of boundary layer theory. Aspects of this procedure were developed by Bondi and Lyttleton[11], and also by Stern[183] in connection with a study of Ekman layer instabilities. The complete analytical solution, in the case of the parallel disk configuration, provides a vertification of the basic physical picture and a convincing mathematical justification for the use of boundary layer methods. These methods appear to be the only feasible ones to use in the general case of motion in a container of arbitrary shape which is taken up in the next section. In addition, the 'exact' solution illuminates the role of the inertial oscillations and the manner in which all three time scales—the rotation period, the viscous decay time, and the spin-up time enter into the problem; these finer details are suppressed by boundary layer theory.

2.5. The initial value problem: formulation

An arbitrarily shaped container is filled with fluid and rotated at a constant angular velocity. At some definite time, $t = 0$, a general, but physically acceptable, initial state of fluid motion is prescribed. The object is to analyse the ensuing transient motion and to describe accurately the ultimate approach to solid-body rotation.

If the initial conditions differ only slightly from rigid rotation then the linear theory is applicable. This consists of (2.2.1) and the boundary conditions $\mathbf{q}(\mathbf{r}, 0) = \mathbf{q}_*(\mathbf{r})$, $\mathbf{q}(\mathbf{r}, t) = 0$ on Σ, the container surface. In the *rotating reference frame*, the final flow is $\mathbf{q} = 0$. (The spin-up problem of the last section is formulated in just this way using a co-ordinate system fixed in the body.)

Let the top and bottom surfaces of the container, $\Sigma = \Sigma_T + \Sigma_B$, be represented by

$$z = f(x, y) \quad (\Sigma_T), \qquad z = -g(x, y) \quad (\Sigma_B), \tag{2.5.1}$$

and let $\mathbf{n}$ *always* denote the outwardly directed normal vector on Σ. The normal is assumed to be a *continuous* vector function on Σ; in particular

$$\left.\begin{aligned} \mathbf{n}_T &= \hat{\mathbf{k}} - \nabla f = (1 + (\nabla f)^2)^{\frac{1}{2}}\, \hat{\mathbf{n}}_T, \\ \mathbf{n}_B &= -\hat{\mathbf{k}} - \nabla g = (1 + (\nabla g)^2)^{\frac{1}{2}}\, \hat{\mathbf{n}}_B, \end{aligned}\right\} \tag{2.5.2}$$

where $\hat{\mathbf{n}}_T$ and $\hat{\mathbf{n}}_B$ are unit vectors.

The motion in this general configuration has essentially the same underlying structure as in the special case of the parallel disk geometry. Since the Ekman number E is small, direct viscous action is confined to a thin boundary layer at the container wall throughout the principal phase of motion. Furthermore, this Ekman layer produces a secondary interior circulation that exerts a major influence by means of vortex line stretching and the transport of angular momentum. Viscosity, in this fashion, eliminates the initial velocity distribution in the spin-up time scale, $E^{-\frac{1}{2}}$, and $E^{\frac{1}{2}}$ emerges as a significant expansion parameter. Consequently, the analysis of the general motion is based upon the concepts of boundary layer theory. An approximate solution is sought by this approach that consists of two parts: an almost inviscid motion throughout the interior of the container is matched to a motion in the viscous boundary layer in order to satisfy the boundary conditions. Moreover, the representation must be uniformly valid in its spatial dependence and for a sufficiently long period of time, $t \doteq E^{-\frac{1}{2}}$ at least, to ensure that all the important phenomena are included and described. These are important but difficult requirements to meet because there are several sources of non-uniformity: three different time scales, viscous boundary layers and, as it turns out, boundary layer resonances and free shear layers. The methods employed must be flexible enough to permit some control of difficulties as they arise.

Briefly, the solution procedure is as follows: Expansions in half powers of the Ekman number E are introduced into the governing equations and a problem sequence resolved. The first of these is for the primary or zeroth order solution of the inviscid interior motion corresponding to $E = 0$. In the second problem, the internal motion is corrected for viscous effects to make the basic tangential velocity zero at the boundary. However, the boundary layers induce further interior motion by establishing a small normal mass flux and this sets up a third problem. Once the secondary interior circulation is determined, it too must be corrected at the boundary and the procedure continues until the mutual interactions of the interior and boundary layer flows are determined to the desired accuracy. Practically speaking, and for good mathematical reasons, the analysis ends with the secondary motion of order $E^{\frac{1}{2}}$. The goal then is to achieve an approximate solution for the motion that is uniformly valid to $O(E^{\frac{1}{2}})$ through the spin-up time $t \doteq E^{-\frac{1}{2}}$.

In many cases, it is possible to develop the interior solution as a superposition of all the inviscid modes, which are essentially of two types. There are an infinite number of **inertial modes** typically represented by

$$\left.\begin{aligned} \mathbf{q} &= \mathbf{Q}_m(\mathbf{r}) \exp \lambda_m t, \\ p &= \Phi_m(\mathbf{r}) \exp \lambda_m t, \end{aligned}\right\} \qquad (2.5.3)$$

($\mathbf{Q}$ and Φ are complex-valued functions) and a single **geostrophic mode**, a mode corresponding to zero frequency, of the form

$$\left.\begin{aligned} \mathbf{q} &= \mathbf{q}_0(\mathbf{r}, E^{\frac{1}{2}}t), \\ p &= \varphi_0(\mathbf{r}, E^{\frac{1}{2}}t). \end{aligned}\right\} \qquad (2.5.4)$$

A geostrophic flow is one in which the pressure gradient and Coriolis force are in exact balance over many periods of revolution. This motion is slowly varying and depends only on the long time

$$\tau = E^{\frac{1}{2}}t. \qquad (2.5.5)$$

Each of these modes must be corrected for viscous action in the manner outlined above. For example, $\mathbf{q}_0$ requires a boundary layer correction, $\tilde{\mathbf{q}}_0$, which in turn induces further interior motion, $\mathbf{q}_1$. This too, must be adjusted at the walls by another viscous correction, $\tilde{\mathbf{q}}_1$, and the iteration continues indefinitely. Therefore, an approximate solution of the following form is sought, (Greenspan[74]):

$$\mathbf{q} = \mathbf{q}_0(\mathbf{r}, \tau) + \sum_m A_m \mathbf{Q}_m(\mathbf{r}) e^{\lambda_m t} + E^{\frac{1}{2}}\{\mathbf{q}_1(\mathbf{r}, \tau) + \sum_m \mathbf{q}_{m1}(\mathbf{r}, t, E^{\frac{1}{2}})\} + \dots$$
$$+ \tilde{\mathbf{q}}_0 + \sum_m \tilde{\mathbf{q}}_{m0} + E^{\frac{1}{2}}\{\tilde{\mathbf{q}}_1 + \sum_m \tilde{\mathbf{q}}_{m1}\} + \dots, \qquad (2.5.6)$$

$$p = \varphi_0(\mathbf{r}, \tau) + \sum_m A_m \Phi_m(\mathbf{r}) e^{\lambda_m t} + E^{\frac{1}{2}}\{\varphi_1(\mathbf{r}, \tau) + \sum_m \varphi_{m1}(\mathbf{r}, t, E^{\frac{1}{2}})\} + \dots$$
$$+ \tilde{\varphi}_0 + \sum_m \tilde{\varphi}_{m0} + E^{\frac{1}{2}}\{\tilde{\varphi}_1 + \sum_m \tilde{\varphi}_{m1}\} + \dots \qquad (2.5.7)$$

The tilde symbol denotes a boundary layer function of a stretched boundary layer co-ordinate, ζ. These functions approach zero exponentially fast as $\zeta \to \infty$. The outer edge of the boundary layer, $\zeta = \infty$, is, of course, still located at the wall in terms of the ordinary spatial variables. On the container surface, ζ is zero, and in the viscous layer the normal derivative is

$$\hat{\mathbf{n}} \cdot \nabla \cong -E^{-\frac{1}{2}} \frac{\partial}{\partial \zeta}$$

when applied to a boundary layer function. The tangential derivative, $\hat{\mathbf{n}} \times \nabla$, of any function is no more than an $O(1)$ quantity. Interior functions are independent of ζ and do not vary across the width of the boundary layer.

The effects of spatial and temporal non-uniformities obscure the correct form of the expansions beyond the terms indicated. Thus, despite appearances, we are limited to a discussion of the primary flow field, its boundary layer and the secondary circulation induced.

The Ekman layers produced by the geostrophic motion change rapidly with time during the period of formation but thereafter, the variation is exceedingly slow. As far as the basic inviscid motion, $\mathbf{q}_0$, is concerned, the boundary layers can be considered essentially steady because the actual time-dependence in $t \leqslant E^{-\frac{1}{2}}$, has an insignificant effect. The results of §2.2 support this assertion and in order to simplify the analysis we take

$$\frac{\partial}{\partial t}\tilde{\mathbf{q}}_0 = O(E^{\frac{1}{2}}),$$

which accounts for the assumed form of the solution. The approximate velocity $\mathbf{q}_0$ obtained on this basis is very accurate in the entire interval of interest, $0 \leqslant t \leqslant t_c \doteq E^{-\frac{1}{2}}$. However, we do lose the capability of exactly describing the boundary layers and secondary flow at the very earliest times, but only then. The initial phases can be studied with minor modifications (see §2.9) but the extended transient analysis leads to almost the same approximation for $\mathbf{q}_0$ derived on the basis of a steady boundary layer. The boundary layers and secondary flows are also substantially the same as those determined from the simpler theory, at least during the spin-up time. Hence, there is little to be gained at present by this more elaborate approach and very much to be lost in the way of clarity and simplicity. For these reasons, we continue upon the present course.

The third time scale, $t \doteq E^{-1}$, characterizing the time required for viscous diffusion to permeate into the interior, is of little importance in problems of transient motions in containers of finite volume. (It acquires renewed prominence when the fluid is stratified, §2.21.)

To achieve uniform validity for t large, the frequency parameter

s_m in (2.5.3) is also expanded as

$$s_m = s_{m,0} + E^{\frac{1}{2}} s_{m,1} + \dots \tag{2.5.8}$$

where

$$s_{m,0} = i\lambda_m \tag{2.5.9}$$

represents the inviscid eigenvalue. The parameter $s_{m,1}$ will be chosen to eliminate secular terms possessing unacceptable growth rates that arise in the formal development. The procedure is very similar to the classical method of Poincaré [153].

The substitution of these expansions into the basic equations (2.2.1) and the requisite boundary conditions leads to a sequence of problems for the inviscid modes, the boundary layer flows and their mutual interactions, [74].

For the geostrophic mode, the first three problems of this sequence are the following:

$\mathscr{A}_1$: $2\hat{\mathbf{k}} \times \mathbf{q}_0 = -\nabla\varphi_0, \quad \nabla \cdot \mathbf{q}_0 = 0;$

boundary conditions: $\mathbf{q}_0 \cdot \hat{\mathbf{n}} = 0$ on Σ.

$\mathscr{A}_2$: $2\hat{\mathbf{k}} \times \tilde{\mathbf{q}}_0 - \hat{\mathbf{n}} \dfrac{\partial}{\partial \zeta} \tilde{\varphi}_1 = \dfrac{\partial^2}{\partial \zeta^2} \tilde{\mathbf{q}}_0,$

$-\dfrac{\partial}{\partial \zeta}(\hat{\mathbf{n}} \cdot \tilde{\mathbf{q}}_1) + \hat{\mathbf{n}} \cdot \nabla \times (\hat{\mathbf{n}} \times \tilde{\mathbf{q}}_0) = 0$ (see (1.6.2));

boundary conditions: $\mathbf{q}_0 + \tilde{\mathbf{q}}_0 = 0$ on Σ, $\zeta = 0$.

$\mathscr{A}_3$: $\dfrac{\partial}{\partial \tau} \mathbf{q}_0 + 2\hat{\mathbf{k}} \times \mathbf{q}_1 = -\nabla\varphi_1, \quad \nabla \cdot \mathbf{q}_1 = 0;$

boundary conditions: $\hat{\mathbf{n}} \cdot (\mathbf{q}_1 + \tilde{\mathbf{q}}_1) = 0$ on Σ, $\mathbf{q}_0(\mathbf{r}, 0)$ prescribed.

The problem sequence for the typical inertial mode begins as follows:

$\mathscr{B}_1$: $i\lambda_m \mathbf{Q}_m + 2\hat{\mathbf{k}} \times \mathbf{Q}_m + \nabla\Phi_m = 0, \quad \nabla \cdot \mathbf{Q}_m = 0;$

boundary conditions: $\mathbf{Q}_m \cdot \hat{\mathbf{n}} = 0$ on Σ.

$\mathscr{B}_2$: $\dfrac{\partial}{\partial t} \tilde{\mathbf{q}}_{m0} + 2\hat{\mathbf{k}} \times \tilde{\mathbf{q}}_{m0} - \hat{\mathbf{n}} \dfrac{\partial}{\partial \zeta} \tilde{\varphi}_{m1} = \dfrac{\partial^2}{\partial \zeta^2} \tilde{\mathbf{q}}_{m0},$

$-\dfrac{\partial}{\partial \zeta}(\hat{\mathbf{n}} \cdot \tilde{\mathbf{q}}_{m1}) + \hat{\mathbf{n}} \cdot \nabla \times (\hat{\mathbf{n}} \times \tilde{\mathbf{q}}_{m0}) = 0;$

boundary conditions: $\tilde{\mathbf{q}}_{m0} = -\mathbf{Q}_m e^{s_m t}$ on Σ, $\zeta = 0$,
$\tilde{\mathbf{q}}_{m0} = 0$ at $t = 0$.

$$\mathscr{B}_3: \frac{\partial}{\partial t}\mathbf{q}_{m1} + 2\hat{\mathbf{k}} \times \mathbf{q}_{m1} = -\nabla\varphi_{m1} - s_{m,1}\mathbf{Q}_m e^{s_m t},$$
$$\nabla\cdot\mathbf{q}_{m1} = 0;$$

boundary conditions: $\hat{\mathbf{n}}\cdot(\mathbf{q}_{m1} + \tilde{\mathbf{q}}_{m1}) = 0$ on Σ, $\mathbf{q}_{m1} = 0$ at $t = 0$.

It would be simplest to assume that all functions appearing in sequence ($\mathscr{B}$) have the same exponential time behaviour, but severe difficulties are encountered in this classical approach that are none too easy to surmount. Trouble arises from the interchange of limit processes involving E and ζ at positions of resonance on the container surface. The present method is a little more satisfactory because it avoids these snags, produces a better approximation for the actual time-dependent boundary layers, and leads to the same calculation for the decay factor $s_{m,1}$ as the classical analysis. None of these matters are considered herein and the parameter $s_{m,1}$ is determined in the easiest way available, even if an appeal must be made to another method for justification.

2.6. The geostrophic mode

Problem $\mathscr{A}_1$ for φ_0 and $\mathbf{q}_0$,

$$\left.\begin{aligned} 2\hat{\mathbf{k}} \times \mathbf{q}_0 &= -\nabla\varphi_0, \\ \nabla\cdot\mathbf{q}_0 &= 0, \end{aligned}\right\} \tag{2.6.1}$$

with $\mathbf{q}_0\cdot\hat{\mathbf{n}} = 0$ on the boundary Σ, is a special case of the general inertial mode problem $\mathscr{B}_1$ corresponding to $\lambda = 0$. Clearly then, these functions do not vary in a period of revolution and this is the basis for the assumed form of solution involving the longer time scale $\tau = E^{\frac{1}{2}}t$.

The curl of the momentum equation in (2.6.1) yields

$$(\hat{\mathbf{k}}\cdot\nabla)\mathbf{q}_0 = 0 \tag{2.6.2}$$

showing, once more, that $\mathbf{q}_0$ is a three-dimensional vector independent of the height z. The motion is columnar; the entire vertical pillar of fluid from the lower surface, $z = -g$, to the upper, $z = f$, moves as a unit.

It is not difficult to solve (2.6.1) for the velocity in terms of the pressure and the result is

$$\mathbf{q}_0 = \tfrac{1}{4}(\mathbf{n}_T - \mathbf{n}_B) \times \nabla\varphi_0. \tag{2.6.3}$$

Here, $\mathbf{n}_T$ and $\mathbf{n}_B$ are the normal vectors to Σ at the points where the top and bottom surfaces are pierced by the same vertical line. An equivalent statement of the boundary condition, $\hat{\mathbf{n}}\cdot\mathbf{q}_0 = 0$, is then

$$\hat{\mathbf{n}}_T\cdot(\hat{\mathbf{n}}_B \times \nabla\varphi_0) = (\hat{\mathbf{n}}_T \times \hat{\mathbf{n}}_B)\cdot\nabla\varphi_0 = 0 \tag{2.6.4}$$

and, since φ_0 is independent of z, this can be rewritten as

$$\hat{\mathbf{k}}\cdot\nabla\varphi_0 \times \nabla(f+g) = 0. \tag{2.6.5}$$

This implies that the *pressure is a function only of the total height*, $h = f+g$, with time acting as a parameter:

$$\varphi_0 = \varphi_0(f+g, \tau) = \varphi_0(h, \tau). \tag{2.6.6}$$

The special case in which h is a constant, so that $\nabla h = 0$ and (2.6.5) is automatically satisfied, is considered separately after a development of the general theory. (h is also designated as the total depth when this seems more fitting.)

Having established the dependence of pressure on h, the velocity is found to be

$$\mathbf{q}_0 = -\frac{1}{2}\left[\frac{\partial}{\partial h}\varphi_0(h, \tau)\right]\mathbf{n}_T \times \mathbf{n}_B. \tag{2.6.7}$$

Hence, in geostrophic flow, a column of fluid of height h moves about the interior of the container as a unit, always maintaining a constant length. The pressure plays the role of a stream function.

Obviously, geostrophy can exist only when the container has closed contours, $\mathfrak{C}$, of constant total height h, as that shown in fig. 2.4. This does not always occur and a simple container consisting of a hemisphere rotating about a diameter on its flat surface is one for which no geostrophic flow is possible. Constant-height contours in this geometry, as for the sliced cylinder configuration depicted in fig. 2.14, do not form closed stream lines. It turns out that in these configurations an infinite number of rather special low frequency inertial waves, called Rossby waves, arise to replace geostrophy. Rossby waves are studied in §§ 2.16 and 5.5.

The line contour $\mathfrak{C}$ traced out on the bounding surface plays a crucial role in the general theory and it is well to note that its unit tangent vector is

$$\frac{\hat{\mathbf{n}}_T \times \hat{\mathbf{n}}_B}{|\hat{\mathbf{n}}_T \times \hat{\mathbf{n}}_B|} = \frac{\mathbf{n}_T \times \mathbf{n}_B}{|\mathbf{n}_T \times \mathbf{n}_B|}.$$

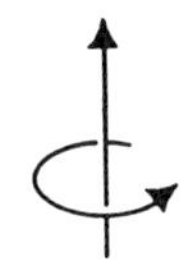

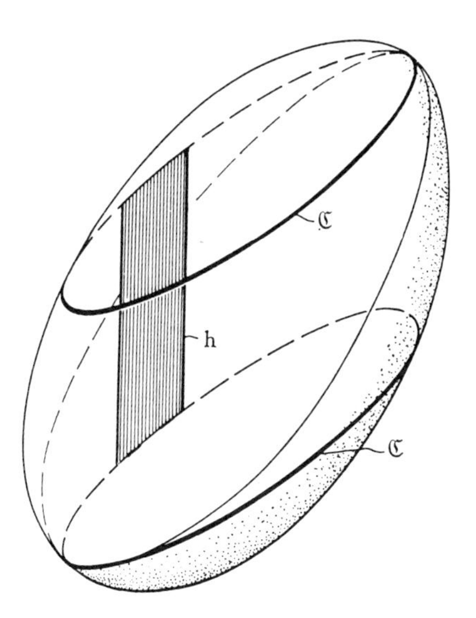

Fig. 2.4. An arbitrary container configuration showing a column of constant height and geostrophic contours.

Furthermore, the circulation of the geostrophic velocity about $\mathfrak{C}$, is

$$\Gamma(h, \tau) = \oint_{\mathfrak{C}} \mathbf{q}_0 \cdot \mathbf{ds} = -\frac{1}{2}\left[\frac{\partial}{\partial h}\varphi_0(h, \tau)\right]\oint_{\mathfrak{C}} |\mathbf{n}_T \times \mathbf{n}_B|\, ds, \tag{2.6.8}$$

a non-zero quantity. *The geostrophic mode possesses circulation!*

Problems $\mathscr{A}_2$ and $\mathscr{A}_3$ must be solved in order to study the viscous effects on the basic geostrophic flow. Consider the former first:

$$2\hat{\mathbf{k}}\times\tilde{\mathbf{q}}_0-\hat{\mathbf{n}}\frac{\partial}{\partial\zeta}\tilde{\phi}_1=\frac{\partial^2}{\partial\zeta^2}\tilde{\mathbf{q}}_0, \tag{2.6.9}$$

$$-\frac{\partial}{\partial\zeta}(\hat{\mathbf{n}}\cdot\tilde{\mathbf{q}}_1)+\hat{\mathbf{n}}\cdot\nabla\times(\hat{\mathbf{n}}\times\tilde{\mathbf{q}}_0)=0, \tag{2.6.10}$$

with $\mathbf{q}_0+\tilde{\mathbf{q}}_0=0$ on $\zeta=0$. This is the usual formulation for the steady Ekman boundary layer, a special case of which appeared in §2.3. The first equation arises from the requirement that the tangential component of the interior velocity be reduced to zero at the wall. The second equation determines the normal flux, $\hat{\mathbf{n}}\cdot\tilde{\mathbf{q}}_1$, induced by this viscous layer which produces further interior motion (problem $\mathscr{A}_3$).

Simple vector manipulations of (2.6.9) (and the fact that $\hat{\mathbf{n}}\cdot\tilde{\mathbf{q}}_0=0$ in the layer) lead to the boundary layer equation governing the tangential components of velocity:

$$\frac{\partial^2}{\partial\zeta^2}(\hat{\mathbf{n}}\times\tilde{\mathbf{q}}_0+i\tilde{\mathbf{q}}_0)=2i(\hat{\mathbf{n}}\cdot\hat{\mathbf{k}})(\hat{\mathbf{n}}\times\tilde{\mathbf{q}}_0+i\tilde{\mathbf{q}}_0). \tag{2.6.11}$$

This has the solution

$$\hat{\mathbf{n}}\times\tilde{\mathbf{q}}_0+i\tilde{\mathbf{q}}_0=-\{\hat{\mathbf{n}}\times\mathbf{q}_0+i\mathbf{q}_0\}_\Sigma\exp\left[-(2i\hat{\mathbf{n}}\cdot\hat{\mathbf{k}})^{\frac{1}{2}}\zeta\right], \tag{2.6.12}$$

where the positive real-valued root is always implied and $\mathbf{q}_0$ *is evaluated on* Σ. The normal flux into the boundary layer is found by integrating (2.6.10) and the following relationship holds *at the wall*, $\zeta=0$:

$$\hat{\mathbf{n}}\cdot\tilde{\mathbf{q}}_1=\tfrac{1}{2}\hat{\mathbf{n}}\cdot\nabla\times\left\{\left[\hat{\mathbf{n}}\times\mathbf{q}_0+\frac{\hat{\mathbf{n}}\cdot\hat{\mathbf{k}}}{|\hat{\mathbf{n}}\cdot\hat{\mathbf{k}}|}\mathbf{q}_0\right]|\hat{\mathbf{n}}\cdot\hat{\mathbf{k}}|^{-\frac{1}{2}}\right\}_{\zeta=0}. \tag{2.6.13}$$

If the relative tangential wall velocity is $\mathscr{U}$ instead of zero, then the value of the interior velocity $\mathbf{q}_0$ on Σ should be replaced by the corresponding value of $\mathbf{q}_0-\mathscr{U}$ in the two preceding formulas.

Although (2.6.13) was derived using the boundary layer coordinate system (see p. 24), the vector formula is not restricted by this choice and is of general validity. In other words, the final result no longer depends on a particular method of extending a vector defined only on Σ into a vector function of position $\mathbf{r}$. To prove this, assume that $\mathbf{F}$ is a vector defined on Σ alone and let $\mathscr{F}(x,y,z)$ be any continuous vector function which has the property that $\mathscr{F}=\mathbf{F}$

on Σ. Furthermore, let ξ_1, ξ_2, ξ_3 be orthogonal curvilinear co-ordinates, with corresponding unit vectors $\hat{\mathbf{i}}_1$, $\hat{\mathbf{i}}_2$, $\hat{\mathbf{i}}_3$, so that ξ_3 = constant is the surface Σ. From the definition of the curl,

$$\nabla \times \mathscr{F} = \hat{\mathbf{i}}_1 \times \frac{1}{h_1} \frac{\partial}{\partial \xi_1} \mathscr{F} + \hat{\mathbf{i}}_2 \times \frac{1}{h_2} \frac{\partial}{\partial \xi_2} \mathscr{F} + \hat{\mathbf{i}}_3 \times \frac{1}{h_3} \frac{\partial}{\partial \xi_3} \mathscr{F},$$

it follows that on Σ, where $\hat{\mathbf{i}}_3 = \hat{\mathbf{n}}$,

$$\hat{\mathbf{n}} \cdot \nabla \times \mathscr{F} = \hat{\mathbf{i}}_2 \cdot \frac{1}{h_1} \frac{\partial}{\partial \xi_1} \mathscr{F} - \hat{\mathbf{i}}_1 \cdot \frac{1}{h_2} \frac{\partial}{\partial \xi_2} \mathscr{F}.$$

Since normal derivatives do not appear in this expression, $\mathscr{F}$ can be replaced by its surface value so that

$$\hat{\mathbf{n}} \cdot \nabla \times \mathscr{F} = \hat{\mathbf{i}}_2 \cdot \frac{1}{h_1} \frac{\partial}{\partial \xi_1} \mathbf{F} - \hat{\mathbf{i}}_1 \cdot \frac{1}{h_2} \frac{\partial}{\partial \xi_2} \mathbf{F}.$$

Therefore, the value of $\hat{\mathbf{n}} \cdot \nabla \times \mathscr{F}$ on Σ depends only on the surface derivatives of $\mathbf{F}$ and not at all on the extended field. Hence, $\hat{\mathbf{n}} \cdot \nabla \times \mathbf{F}$, the right-hand side of the last equation, is a definite and uniquely defined quantity. As a practical matter, the explicit calculation of (2.6.13) is simple and certainly need not involve surface gradients on the boundary. For example, $\mathbf{q}_0$ and $\hat{\mathbf{n}}$ on Σ are often given as functions of x, y, z. In this case, $\mathbf{q}_0(x, y, z)$ and $\hat{\mathbf{n}}(x, y)$ are also proper vector functions of position (fields which obviously have the correct form on Σ) and the calculation is then one of conventional vector manipulation in cartesian co-ordinates.

The next step is to solve problem $\mathscr{A}_3$:

$$2\hat{\mathbf{k}} \times \mathbf{q}_1 = -\nabla \varphi_1 - \frac{\partial}{\partial \tau} \mathbf{q}_0, \tag{2.6.14}$$

$$\nabla \cdot \mathbf{q}_1 = 0, \tag{2.6.15}$$

with $\hat{\mathbf{n}} \cdot \mathbf{q}_1 = -\hat{\mathbf{n}} \cdot \tilde{\mathbf{q}}_1$ at the boundary $\zeta = 0$. The curl of (2.6.14) reduces it to

$$\frac{\partial}{\partial z} \mathbf{q}_1 = \frac{1}{2} \frac{\partial}{\partial \tau} \nabla \times \mathbf{q}_0. \tag{2.6.16}$$

Upon integrating this, noting that (2.6.15) must be satisfied, we find that

$$\mathbf{q}_1 = \nabla \times \left(\frac{1}{2} z \frac{\partial}{\partial \tau} \mathbf{q}_0 \right) + \nabla \times \mathbf{B}(x, y, \tau). \tag{2.6.17}$$

A single equation relating $\mathbf{q}_0$ and its time derivative is obtained by applying Stokes' theorem to the surface sections of the top and bottom boundaries ($z = f$, $z = -g$) enclosed by contour $\mathfrak{C}$ shown in fig. 2.4. Let Σ'_T, Σ'_B denote these parts of Σ respectively, and let

$$\mathbf{ds} = \frac{\hat{\mathbf{n}}_T \times \hat{\mathbf{n}}_B}{|\hat{\mathbf{n}}_T \times \hat{\mathbf{n}}_B|}\, ds$$

be the directed arc length along $\mathfrak{C}$. Since all vectors are independent of z, then from (2.6.17) (exercising proper care for signs)

$$\left.\begin{aligned} \int \hat{\mathbf{n}}_T \cdot \mathbf{q}_1 \, d\Sigma'_T &= \frac{1}{2}\oint_{\mathfrak{C}} f \frac{\partial \mathbf{q}_0}{\partial \tau} \cdot \mathbf{ds} + \oint_{\mathfrak{C}} \mathbf{B} \cdot \mathbf{ds}, \\ \int \hat{\mathbf{n}}_B \cdot \mathbf{q}_1 \, d\Sigma'_B &= \frac{1}{2}\oint_{\mathfrak{C}} g \frac{\partial \mathbf{q}_0}{\partial \tau} \cdot \mathbf{ds} - \oint_{\mathfrak{C}} \mathbf{B} \cdot \mathbf{ds}. \end{aligned}\right\} \tag{2.6.18}$$

However, use of (2.6.13) implies that

$$\left.\begin{aligned} \int \hat{\mathbf{n}}_T \cdot \mathbf{q}_1 \, d\Sigma'_T &= -\frac{1}{2}\oint_{\mathfrak{C}} |\hat{\mathbf{n}}_T \cdot \hat{\mathbf{k}}|^{-\frac{1}{2}} (\hat{\mathbf{n}}_T \times \mathbf{q}_0 + \mathbf{q}_0) \cdot \mathbf{ds}, \\ \int \hat{\mathbf{n}}_B \cdot \mathbf{q}_1 \, d\Sigma'_B &= \frac{1}{2}\oint_{\mathfrak{C}} |\hat{\mathbf{n}}_B \cdot \hat{\mathbf{k}}|^{-\frac{1}{2}} (\hat{\mathbf{n}}_B \times \mathbf{q}_0 - \mathbf{q}_0) \cdot \mathbf{ds}. \end{aligned}\right\} \tag{2.6.19}$$

The replacement of these expressions in the preceding set and the elimination of the integral involving $\mathbf{B}$, yields

$$-\oint_{\mathfrak{C}} \mathbf{ds} \cdot [(\mathbf{q}_0 + \hat{\mathbf{n}}_T \times \mathbf{q}_0) \,|\, \hat{\mathbf{n}}_T \cdot \hat{\mathbf{k}}|^{-\frac{1}{2}} + (\mathbf{q}_0 - \hat{\mathbf{n}}_B \times \mathbf{q}_0) \,|\, \hat{\mathbf{n}}_B \cdot \hat{\mathbf{k}}|^{-\frac{1}{2}}]$$
$$= \oint_{\mathfrak{C}} (f+g) \frac{\partial \mathbf{q}_0}{\partial \tau} \cdot \mathbf{ds}. \tag{2.6.20}$$

Since $\mathbf{ds} \times \mathbf{q}_0 = 0$ and $f+g = h$, which is a constant on contour $\mathfrak{C}$, (2.6.20) simplifies to

$$-\oint_{\mathfrak{C}} \mathbf{ds} \cdot \mathbf{q}_0 (|\hat{\mathbf{n}}_T \cdot \hat{\mathbf{k}}|^{-\frac{1}{2}} + |\hat{\mathbf{n}}_B \cdot \hat{\mathbf{k}}|^{-\frac{1}{2}}) = h \oint_{\mathfrak{C}} \frac{\partial \mathbf{q}_0}{\partial \tau} \cdot \mathbf{ds}. \tag{2.6.21}$$

Finally, upon substituting equation (2.6.7) for the geostrophic velocity $\mathbf{q}_0$, this in turn becomes

$$-\left[\frac{\partial}{\partial h} \varphi_0(h, \tau)\right] \oint_{\mathfrak{C}} (|\hat{\mathbf{n}}_T \cdot \hat{\mathbf{k}}|^{-\frac{1}{2}} + |\hat{\mathbf{n}}_B \cdot \hat{\mathbf{k}}|^{-\frac{1}{2}}) |\mathbf{n}_T \times \mathbf{n}_B| \, ds$$
$$= h \left[\frac{\partial^2}{\partial h \, \partial \tau} \varphi_0(h, \tau)\right] \oint_{\mathfrak{C}} |\mathbf{n}_T \times \mathbf{n}_B| \, ds. \tag{2.6.22}$$

from which $(\partial/\partial h)\,\varphi_0$ can be determined by integration. Thus

$$\frac{\partial}{\partial h}\varphi_0(h,\tau) = \left(\frac{\partial}{\partial h}\varphi_0(h,0)\right)\exp\left(-\frac{I(h)}{J(h)}\tau\right), \qquad (2.6.23)$$

where

$$\left.\begin{aligned} I(h) &= \oint_{\mathfrak{C}} |\mathbf{n}_T\times\mathbf{n}_B|\,(|\hat{\mathbf{n}}_T\cdot\hat{\mathbf{k}}|^{-\frac{1}{2}}+|\hat{\mathbf{n}}_B\cdot\hat{\mathbf{k}}|^{-\frac{1}{2}})\,ds,\\ J(h) &= h\oint_{\mathfrak{C}} |\mathbf{n}_T\times\mathbf{n}_B|\,ds,\end{aligned}\right\} \qquad (2.6.24)$$

and $\tau = E^{\frac{1}{2}}t$.

The fundamental solution is completed by calculating the geostrophic velocity from (2.6.7). The function $\varphi_0(h,0)$ remains unspecified and must be determined from the initial conditions. Before this can be done, the properties of the inertial modes must be set out. The determination of $\nabla\times\mathbf{B}$, and hence the final form of the secondary motion $\mathbf{q}_1$, is obtained from (2.6.17) and the boundary conditions at the top and bottom surfaces. However, the theory, at this stage of development, provides only a formula for the velocity components in the direction normal to the geostrophic contours. The next order approximation is required to calculate the tangential component. If

$$\mathbf{q}_1 = \mathbf{v}_\parallel + \mathbf{v}_\perp$$

with $\mathbf{v}_\parallel\times(\hat{\mathbf{n}}_T\times\hat{\mathbf{n}}_B) = 0$, then it follows that

$$\hat{\mathbf{n}}_T\cdot\mathbf{v}_\perp = \hat{\mathbf{n}}_T\cdot\mathbf{q}_1|_{z=f} + \tfrac{1}{2}(z-f)\left(\hat{\mathbf{n}}_T\cdot\nabla\times\frac{\partial}{\partial\tau}\mathbf{q}_0\right),$$

$$\hat{\mathbf{n}}_B\cdot\mathbf{v}_\perp = \hat{\mathbf{n}}_B\cdot\mathbf{q}_1|_{z=-g} + \tfrac{1}{2}(z+g)\left(\hat{\mathbf{n}}_B\cdot\nabla\times\frac{\partial}{\partial\tau}\mathbf{q}_0\right).$$

The two components of $\mathbf{v}_\perp$ are determinable from this pair of equations but this is left as an exercise, (see also §2.17).

The exponential factor in (2.6.23) is a function only of the total height h, implying that the decay rate of a geostrophic flow varies from contour to contour. This can be made the basis for a number of elementary experiments. Greenspan and Howard[75] compared theory versus experiment for a container of variable depth consisting of a cylindrical tank whose base is a cone of vertex angle 110°. A geostrophic motion was produced by spin-up of the tank and measurements were made of the time required for a float to come to rest as a function of radial position. (Actually the top surface was free and this necessitates a slight modification of the theory, but see the cited reference for details.) The results of this experiment are shown in fig. 2.5; the solid line is the theoretical value (for h = g) of the e-folding time, i.e., $\exp -t/t_c$;

$$t_c = E^{-\frac{1}{2}}\Omega^{-1}g\left(1+\left[1+\left(\frac{dg}{dr}\right)^2\right]^{\frac{1}{2}}\right)^{-1}$$

which is linear in r for the conical bottom. There does not seem to be any significant difference between theory and experiment.

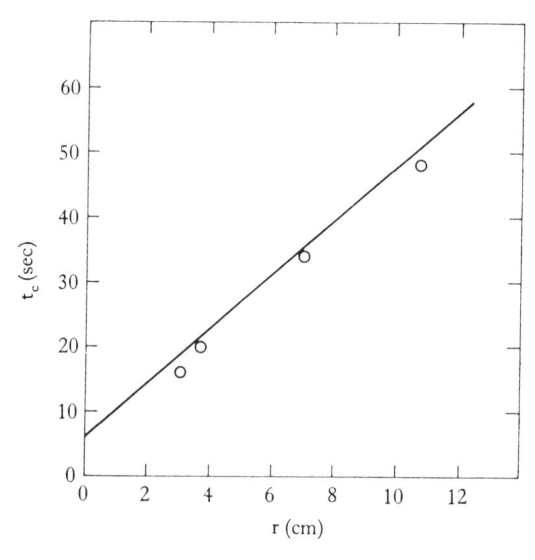

Fig. 2.5. The e-folding time versus radius for spin-up in a cylinder with a conical bottom of vertex angle 110°, [75].

To demonstrate geostrophic motion in an arbitrary configuration, a more qualitative experiment can be performed. A non-symmetrical football-shaped container similar to that in fig. 2.4 is rotated about the vertical axis as shown in fig. 2.6(*a*). When the interior fluid rotation is rigid, ink is introduced into the container and the rotation rate is then increased slightly. This initial state resolves itself mainly into the geostrophic mode and fig. 2.6(*b*) shows the ink traversing that geostrophic contour, $\mathfrak{C}$, which passes through the entry orifice. The theoretical curve, taped on the container, appears as the solid line. (No attempt has been made in this case to determine the decay

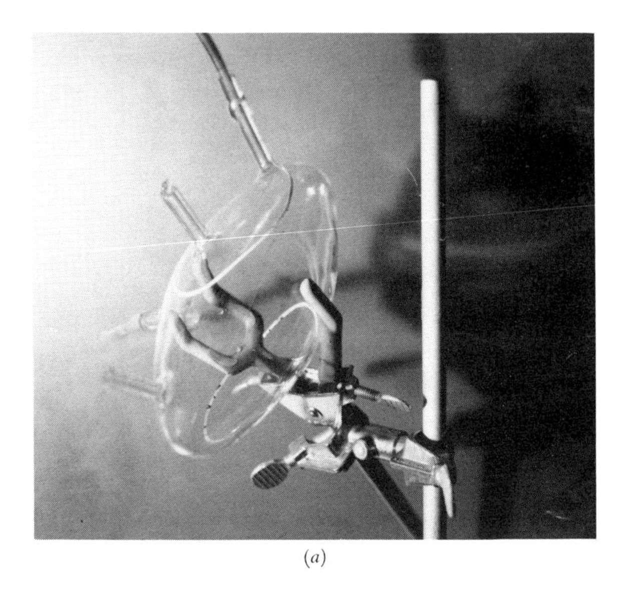

(*a*)

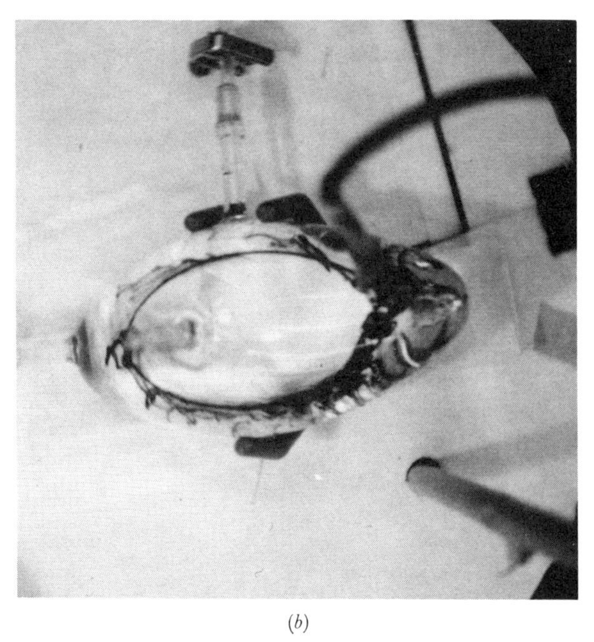

(*b*)

Fig. 2.6. (*a*) An ovular container, similar to that in fig. 2.4, with geostrophic contours marked on the surface. (*b*) Ink injected in one portal during spin-up travels about the geostrophic contour. Three circuits were completed when the photograph was made.

rate along $\mathfrak{C}$.) Looked at in another way, this experiment is seen to be a demonstration of the Taylor–Proudman theorem because the relative motion is small and essentially steady (for a long time period) in the rotating co-ordinate system.

The general theory provides an explanation for the column experiment shown in fig. 1.1. There, constant-height contours above the protuberance form a separate and distinct set of closed curves. In spin-up, fluid particles can neither penetrate this region from the outside nor escape from it and the two regions are separated by a shear surface. The aluminium particles in suspension align in this shear thereby causing a variation in the diffused light pattern which makes the column visible.

2.7. Inertial waves

Consider now problem $\mathscr{B}_1$ for the inviscid inertial oscillations:

$$\left.\begin{aligned} i\lambda\mathbf{Q}+2\hat{\mathbf{k}}\times\mathbf{Q}+\nabla\Phi &= 0,\\ \nabla\cdot\mathbf{Q} &= 0,\end{aligned}\right\} \tag{2.7.1}$$

with $\mathbf{Q}\cdot\hat{\mathbf{n}} = 0$ on the boundary. (The subscript notation is omitted for the present.) The complex velocity vector is expressible in terms of the pressure as

$$(1-\tfrac{1}{4}\lambda^2)\,\mathbf{Q} = \tfrac{1}{2}\hat{\mathbf{k}}\times\nabla\Phi-\tfrac{1}{4}i\lambda\nabla\Phi-\frac{1}{i\lambda}\,\hat{\mathbf{k}}(\hat{\mathbf{k}}\cdot\nabla\Phi), \tag{2.7.2}$$

if $\lambda \neq 0,\ \pm 2$. The first exception, $\lambda = 0$, corresponds to the geostrophic mode discussed in the last section. The second exception will be considered separately.

The problem for the pressure alone (studied by Poincaré [154] and named after him by Cartan [29]) is

$$\nabla^2\Phi-\frac{4}{\lambda^2}(\hat{\mathbf{k}}\cdot\nabla)^2\,\Phi = 0, \tag{2.7.3}$$

with
$$-\lambda^2\hat{\mathbf{n}}\cdot\nabla\Phi+4(\hat{\mathbf{n}}\cdot\hat{\mathbf{k}})\,(\hat{\mathbf{k}}\cdot\nabla\Phi)+2i\lambda(\hat{\mathbf{k}}\times\hat{\mathbf{n}})\cdot\nabla\Phi = 0 \tag{2.7.4}$$

on the boundary. Although the explicit determination of modes and frequencies for a particular configuration (e.g., the cylinder or sphere) necessitates solving this boundary value problem for Φ, the theoretical properties of the system are often more clearly discerned and proved by retaining the velocity vector intact and using the original formulation, (2.7.1) et seq. No exception for $|\lambda| = 2$ need then be made.

Several properties of this eigenvalue problem are now established. In what follows, the *complex conjugate* of a function ψ is denoted by $\psi^\dagger$.

The eigenvalues λ are real and $|\lambda| \leqslant 2$

If (2.7.1) is multiplied by $\mathbf{Q}^\dagger\cdot$ and integrated over the volume of the container, the result is

$$\mathrm{i}\lambda\int \mathbf{Q}^\dagger\cdot\mathbf{Q}\,\mathrm{d}V + 2\int \mathbf{Q}^\dagger\cdot\hat{\mathbf{k}}\times\mathbf{Q}\,\mathrm{d}V = -\int \mathbf{Q}^\dagger\cdot\nabla\Phi\,\mathrm{d}V.$$

Since $\mathbf{Q}^\dagger$ satisfies the divergence equation and $\mathbf{Q}^\dagger\cdot\hat{\mathbf{n}} = 0$ on the boundary, the right-hand side is easily shown to be zero. (Note in particular that if $(\mathbf{Q}, \lambda)$ is an eigenfunction–eigenvalue pair then so is $(\mathbf{Q}^\dagger, -\lambda^\dagger)$.) Therefore

$$\lambda = 2\mathrm{i}\frac{\int \mathbf{Q}^\dagger\cdot\hat{\mathbf{k}}\times\mathbf{Q}\,\mathrm{d}V}{\int \mathbf{Q}^\dagger\cdot\mathbf{Q}\,\mathrm{d}V} = -2\frac{\int \mathscr{I}m[\hat{\mathbf{k}}\cdot\mathbf{Q}\times\mathbf{Q}^\dagger]\,\mathrm{d}V}{\int \mathbf{Q}^\dagger\cdot\mathbf{Q}\,\mathrm{d}V}, \tag{2.7.5}$$

proving the first part of the theorem.

Let the velocity be expressed in terms of its real and imaginary parts

$$\mathbf{Q} = \mathbf{Q}_{\mathscr{R}} + \mathrm{i}\mathbf{Q}_{\mathscr{I}}.$$

A simple bound for λ, obtainable directly from (2.7.5), is then

$$|\lambda| \leqslant 4\frac{\int |\hat{\mathbf{k}}\cdot\mathbf{Q}_{\mathscr{R}}\times\mathbf{Q}_{\mathscr{I}}|\,\mathrm{d}V}{\int (\mathbf{Q}_{\mathscr{R}}\cdot\mathbf{Q}_{\mathscr{R}} + \mathbf{Q}_{\mathscr{I}}\cdot\mathbf{Q}_{\mathscr{I}})\,\mathrm{d}V}. \tag{2.7.6}$$

However,

$$2\int |\hat{\mathbf{k}}\cdot\mathbf{Q}_{\mathscr{R}}\times\mathbf{Q}_{\mathscr{I}}|\,\mathrm{d}V \leqslant 2\int |\mathbf{Q}_{\mathscr{R}}|\,|\mathbf{Q}_{\mathscr{I}}|\,\mathrm{d}V \leqslant \int (|\mathbf{Q}_{\mathscr{R}}|^2 + |\mathbf{Q}_{\mathscr{I}}|^2)\,\mathrm{d}V$$

which upon replacement in (2.7.6) proves that

$$|\lambda| \leqslant 2. \tag{2.7.7}$$

Orthogonality

Let $(\mathbf{Q}_n, \lambda_n)$, $(\mathbf{Q}_m, \lambda_m)$ be any two eigenfunction–eigenvalue pairs satisfying (2.7.1) for which $\lambda_n \neq \lambda_m$. From the basic equations, the following expressions can be arranged:

$$\left.\begin{aligned} i\lambda_n \mathbf{Q}_m^\dagger \cdot \mathbf{Q}_n + 2\mathbf{Q}_m^\dagger \cdot \hat{\mathbf{k}} \times \mathbf{Q}_n &= -\mathbf{Q}_m^\dagger \cdot \nabla \Phi_n, \\ -i\lambda_m \mathbf{Q}_n \cdot \mathbf{Q}_m^\dagger + 2\mathbf{Q}_n \cdot \hat{\mathbf{k}} \times \mathbf{Q}_m^\dagger &= -\mathbf{Q}_n \cdot \nabla \Phi_m^\dagger. \end{aligned}\right\} \tag{2.7.8}$$

If these are added and integrated over the volume V, then since $\mathbf{Q}_m^\dagger \cdot (\hat{\mathbf{k}} \times \mathbf{Q}_n) = -\mathbf{Q}_n \cdot (\hat{\mathbf{k}} \times \mathbf{Q}_m^\dagger)$, it follows that

$$(\lambda_n - \lambda_m) \int \mathbf{Q}_m^\dagger \cdot \mathbf{Q}_n \, dV = 0.$$

But by assumption, $\lambda_n \neq \lambda_m$; therefore

$$\int \mathbf{Q}_m^\dagger \cdot \mathbf{Q}_n \, dV = 0. \tag{2.7.9}$$

In terms of the pressure alone, this orthogonality relationship can be written as

$$\int dV \left[\nabla \Phi_n \cdot \nabla \Phi_m^\dagger + \frac{4}{\lambda_n \lambda_m} (\hat{\mathbf{k}} \cdot \nabla \Phi_n)(\hat{\mathbf{k}} \cdot \nabla \Phi_m^\dagger) \right] = 0, \tag{2.7.10}$$

an expression that is simpler to use in any specific computation because the solution procedure invariably involves the determination of Φ. On the other hand, (2.7.9) is more desirable from the theoretical viewpoint. The presence of eigenvalues in the last equation is a feature that makes the mathematical problem interesting but also difficult.

If a velocity distribution consists only of a superposition of inertial modes,

$$\mathbf{Q}_* = \sum_m A_m \mathbf{Q}_m, \tag{2.7.11}$$

then the Fourier coefficients can be evaluated with the aid of (2.7.9):

$$A_m = \frac{\int \mathbf{Q}_* \cdot \mathbf{Q}_m^\dagger \, dV}{\int \mathbf{Q}_m^\dagger \cdot \mathbf{Q}_m \, dV}. \tag{2.7.12}$$

In general, an arbitrary velocity distribution must involve geostrophic motion as well as inertial oscillations. The geostrophic mode, which corresponds to the totality of eigenfunctions having zero as an eigenvalue, is certainly orthogonal to any other inertial mode. But this is not sufficient! There must be some other distinguishing property that differentiates geostrophy from the inertial oscillations if the complete synthesis of any initial distribution is to be accomplished. In other words, how is the arbitrary function $\varphi_0(h, 0)$ in equations (2.6.7) and (2.6.23) to be determined? The answer is contained in the mean circulation theorem.

2.8. Mean circulation theorem

Define the depth-averaged velocity vector

$$\langle \mathbf{Q} \rangle = \int_{-g}^{f} \mathbf{Q}\, dz, \tag{2.8.1}$$

which is a three component vector independent of the z co-ordinate. The basic equations can then be rewritten in terms of $\langle \mathbf{Q} \rangle$. For example,

$$\int_{-g}^{f} \nabla \cdot \mathbf{Q}\, dz = \nabla \cdot \langle \mathbf{Q} \rangle - \mathbf{Q}_T \cdot (\nabla f - \hat{\mathbf{k}}) - \mathbf{Q}_B \cdot (\nabla g + \hat{\mathbf{k}}) = 0, \tag{2.8.2}$$

where subscripts T or B indicate that the attached function is evaluated at the top or bottom surface of the container.

The boundary conditions

$$\left.\begin{aligned} \mathbf{Q} \cdot \hat{\mathbf{n}} = 0 &= \mathbf{Q}_T \cdot (\nabla f - \hat{\mathbf{k}}), \\ = 0 &= \mathbf{Q}_B \cdot (\nabla g + \hat{\mathbf{k}}), \end{aligned}\right\} \tag{2.8.3}$$

allow the last equation to be simplified to

$$\nabla \cdot \langle \mathbf{Q} \rangle = 0. \tag{2.8.4}$$

The averaging of the momentum equations takes the form

$$i\lambda \langle \mathbf{Q} \rangle + 2\hat{\mathbf{k}} \times \langle \mathbf{Q} \rangle = -\nabla \left(\int_{-g}^{f} \Phi\, dz \right) - \Phi_T\, \mathbf{n}_T - \Phi_B\, \mathbf{n}_B, \tag{2.8.5}$$

and in particular

$$-i\lambda\, \nabla \times \langle \mathbf{Q} \rangle = \nabla \times (\Phi_T\, \mathbf{n}_T) + \nabla \times (\Phi_B\, \mathbf{n}_B); \tag{2.8.6}$$

the normal vectors are those defined in (2.5.2). Application of Stokes' theorem to *any* section of the container surface Σ' bounded by a contour Υ implies that

$$-\mathrm{i}\lambda \oint_{\Upsilon} \langle \mathbf{Q} \rangle \cdot \mathbf{ds} = \oint_{\Upsilon} \Phi_{\mathrm{T}} \mathbf{n}_{\mathrm{T}} \cdot \mathbf{ds} + \oint_{\Upsilon} \Phi_{\mathrm{B}} \mathbf{n}_{\mathrm{B}} \cdot \mathbf{ds}.$$

Let Υ be the geostrophic contour $\mathfrak{C}$ defined in §2.6. This is the surface contour corresponding to constant total height, $\mathrm{f}+\mathrm{g} = \mathrm{h}$, for which $\mathbf{n}_{\mathrm{T}} \cdot \mathbf{ds} = 0 = \mathbf{n}_{\mathrm{B}} \cdot \mathbf{ds}$. The right-hand side of the last equation is then identically zero and

$$\oint_{\mathfrak{C}} \langle \mathbf{Q} \rangle \cdot \mathbf{ds} = 0 \qquad (2.8.7)$$

for $\lambda \neq 0$. *The mean circulation about the geostrophic contour* $\mathfrak{C}$ *is zero for all inertial modes*. Only the geostrophic mode can possess mean circulation in the rotating co-ordinate system and this property allows us to complete the synthesis of an arbitrary initial state of motion. Note that since h is constant on $\mathfrak{C}$, the conventional average, $\mathrm{h}^{-1} \langle \mathbf{Q} \rangle$, may be used in (2.8.7).

If the container is one for which no closed, constant-height contours exist, then there is, of course, no simple geostrophic mode. In this case, the theorem does not hold and the inertial modes can possess circulation. Indeed special inertial modes arise to replace the geostrophic flow, but the study of these Rossby waves is deferred for the present.

Let $\mathbf{Q}_*(\mathfrak{r})$ now represent a possible velocity distribution inside a container, and assume it to be a synthesis of all the natural modes, inertial and geostrophic;

$$\mathbf{Q}_*(\mathfrak{r}) = \mathbf{q}_0 + \sum_m \mathrm{A}_m \mathbf{Q}_m. \qquad (2.8.8)$$

The Fourier coefficients A_m are obtained from the orthogonality integral and have already been given in (2.7.12).

The mean circulation theorem is used to determine $\mathbf{q}_0$ as follows: Integrate over the depth to obtain

$$\int_{-\mathrm{g}}^{\mathrm{f}} \mathbf{Q}_*(\mathfrak{r})\, \mathrm{dz} = \mathrm{h}\mathbf{q}_0 + \sum_m \mathrm{A}_m \langle \mathbf{Q}_m \rangle$$

and compute the mean circulation about $\mathfrak{C}$,

$$\frac{1}{\mathrm{h}} \oint_{\mathfrak{C}} \langle \mathbf{Q}_* \rangle \cdot \mathbf{ds} = \oint_{\mathfrak{C}} \mathbf{q}_0 \cdot \mathbf{ds}. \qquad (2.8.9)$$

The right-hand side of the last equation is the circulation of the geostrophic flow recorded in (2.6.8) and it follows that

$$\frac{\partial \varphi_0}{\partial h} = -\frac{2}{J(h)} \oint_{\mathfrak{C}} \langle \mathbf{Q}_* \rangle \cdot \mathbf{ds}, \tag{2.8.10}$$

where $\mathbf{q}_0$ and J are given by (2.6.7) and (2.6.24) respectively. The synthesis of an arbitrary distribution is thereby completed.

2.9. Viscous dissipation

The full solution, uniformly valid to $O(E^{\frac{1}{2}})$ for a sufficiently long time span to include spin-up, requires a determination of the decay factor $s_{m,1}$ appearing in (2.5.8). Accordingly, problems $\mathscr{B}_2$ and $\mathscr{B}_3$, p. 42, must be solved next, a laborious task only sketched here. The approach follows that of Greenspan[73] and Kudlick[100].

Problem $\mathscr{B}_2$ is a typical calculation of an unsteady Ekman layer. The boundary layer equation for the tangential velocity vector is

$$\frac{\partial}{\partial t}\tilde{\mathbf{q}}_{m0} + 2\hat{\mathbf{n}}\cdot\hat{\mathbf{k}}\,\hat{\mathbf{n}}\times\tilde{\mathbf{q}}_{m0} = \frac{\partial^2}{\partial\zeta^2}\tilde{\mathbf{q}}_{m0}, \tag{2.9.1}$$

with

$$\tilde{\mathbf{q}}_{m0} + \mathbf{Q}_m\, e^{s_m t} = \mathscr{U} \tag{2.9.2}$$

on $\zeta = 0$ and the initial condition, $\tilde{\mathbf{q}}_{m0} = 0$, at $t = 0$. Here $\mathscr{U}$ is the relative wall velocity in the general case (when sections of the boundary have different speeds) although at present $\mathscr{U} \equiv 0$. The solution is obtained by Laplace transforms in a straightforward manner. If the Laplace transform of a function ψ is

$$\mathscr{L}\{\psi\} = \int_0^\infty e^{-st}\,\psi\, dt, \tag{2.9.3}$$

then

$$\begin{aligned}\tilde{\mathbf{q}}_{m0} = \tfrac{1}{2}\mathscr{L}^{-1}\{&\mathscr{L}\{\mathscr{U}^* - i\hat{\mathbf{n}}\times\mathscr{U}^*\}\exp[-(s+2i\hat{\mathbf{n}}\cdot\hat{\mathbf{k}})^{\frac{1}{2}}\zeta] \\ &+\mathscr{L}\{\mathscr{U}^* + i\hat{\mathbf{n}}\times\mathscr{U}^*\}\exp[-(s-2i\hat{\mathbf{n}}\cdot\hat{\mathbf{k}})^{\frac{1}{2}}\zeta]\},\end{aligned} \tag{2.9.4}$$

where

$$\mathscr{U}^* = \mathscr{U} - \mathbf{Q}_m\, e^{s_m t}$$

and is evaluated on Σ, $\zeta = 0$, and the square root has a positive real value.

Once the boundary layer velocity has been found, the normal flux is obtained by integrating

$$\frac{\partial}{\partial\zeta}(\hat{\mathbf{n}}\cdot\tilde{\mathbf{q}}_{m1}) = \hat{\mathbf{n}}\cdot\nabla\times(\hat{\mathbf{n}}\times\tilde{\mathbf{q}}_{m0}). \tag{2.9.5}$$

With $\mathscr{U} = 0$, $\hat{\mathbf{n}}\cdot\mathbf{q}_{m1}$ at $\zeta = 0$ is an almost purely oscillatory function of time except in the neighbourhood of certain critical curves on Σ. This calculation makes known the boundary condition for the next problem and the secondary internal motion can now be determined. (The time-dependent boundary layer for geostrophic motion is a special case of this analysis and the examination of this particular solution substantiates the statements made earlier, p. 41.)

We remarked previously that the factor $s_{m,1}$ can be evaluated by classical methods if the inherent mathematical difficulties are dismissed or, more accurately, ignored. The answer is correct and may be justified by other means.

Consider then problem $\mathscr{B}_3$ and let

$$\mathbf{q}_{m1} = \mathbf{Q}_{m1}\,\mathrm{e}^{s_m t}. \tag{2.9.6}$$

If E is zero, the boundary value problem is

$$\left.\begin{aligned} \nabla\cdot\mathbf{Q}_{m1} &= 0,\\ \mathrm{i}\lambda_m\,\mathbf{Q}_{m1} + 2\hat{\mathbf{k}}\times\mathbf{Q}_{m1} &= -\nabla\Phi_{m1} - s_{m,1}\,\mathbf{Q}_m \end{aligned}\right\} \tag{2.9.7}$$

with $\hat{\mathbf{n}}\cdot\mathbf{Q}_{m1}$ a known function on the boundary. A formula for the flux, given in [100], is

$$\hat{\mathbf{n}}\cdot\mathbf{Q}_{m1} = \mathrm{F}_m; \tag{2.9.8}$$

$$\mathrm{F}_m = \lim_{\mathrm{E}\to 0}\ \lim_{s\to s_m}\left\{-\int_0^\infty \hat{\mathbf{n}}\cdot\nabla\times(\hat{\mathbf{n}}\times\mathfrak{W}_m)\,\mathrm{d}\zeta\right\}, \tag{2.9.9}$$

with
$$\begin{aligned} \mathfrak{W}_m = {} & \tfrac{1}{2}(\mathbf{Q}_m - \mathrm{i}\hat{\mathbf{n}}\times\mathbf{Q}_m)|_{\zeta=0}\exp\left(-\sqrt{\mathfrak{x}_+}\,\zeta\right)\\ & + \tfrac{1}{2}(\mathbf{Q}_m + \mathrm{i}\hat{\mathbf{n}}\times\mathbf{Q}_m)|_{\zeta=0}\exp\left(-\sqrt{\mathfrak{x}_-}\,\zeta\right), \end{aligned} \tag{2.9.10}$$

and
$$\mathfrak{x}_\pm = s \pm 2\mathrm{i}\hat{\mathbf{n}}\cdot\hat{\mathbf{k}}.$$

If (2.9.7) is multiplied by $\mathbf{Q}_m^\dagger\cdot$ and the conjugate of (2.7.1) multiplied by $\mathbf{Q}_{m1}\cdot$ (in the manner used to establish the orthogonality relationship), the two may be added and integrated over the volume to obtain

$$\int \mathbf{Q}_m^\dagger\cdot\nabla\Phi_{m1}\,\mathrm{d}V + \int \mathbf{Q}_{m1}\cdot\nabla\Phi_m^\dagger\,\mathrm{d}V + s_{m,1}\int \mathbf{Q}_m\cdot\mathbf{Q}_m^\dagger\,\mathrm{d}V = 0. \tag{2.9.11}$$

Further simplification, using (2.9.8), leads to the result

$$s_{m,1} = \frac{-\int \Phi_m^\dagger\,\mathrm{F}_m\,\mathrm{d}\Sigma}{\int \mathbf{Q}_m\cdot\mathbf{Q}_m^\dagger\,\mathrm{d}V}. \tag{2.9.12}$$

This is a necessary condition for the existence of a solution of the inhomogeneous boundary value problem, (2.9.7) et seq., when there is a non-trivial homogeneous solution, in this case the eigenfunction $\mathbf{Q}_m$. It is a familiar solvability requirement in all perturbed eigenvalue problems.

It can be determined after much algebraic reduction that

$$\int \Phi_m^\dagger F_m \, d\Sigma = \frac{1}{2^{\frac{3}{2}}} \int d\Sigma \{1-(\hat{\mathbf{n}}\cdot\hat{\mathbf{k}})^2\}^{-1}$$
$$\times \left\{ |\hat{\mathbf{n}}\cdot\hat{\mathbf{k}}\times\mathbf{Q}_m - i\hat{\mathbf{k}}\cdot\mathbf{Q}_m|^2 \, |\sigma_+|^{\frac{1}{2}} \left(1+\frac{i\sigma_+}{|\sigma_+|}\right)\right.$$
$$\left. + |\hat{\mathbf{n}}\cdot\hat{\mathbf{k}}\times\mathbf{Q}_m + i\hat{\mathbf{k}}\cdot\mathbf{Q}_m|^2 \, |\sigma_-|^{\frac{1}{2}} \left(1+i\frac{\sigma_-}{|\sigma_-|}\right)\right\}, \quad (2.9.13)$$

where $\sigma_\pm = \lambda_m \pm 2\hat{\mathbf{k}}\cdot\hat{\mathbf{n}}.$

In particular, this proves an important result:

$$\mathcal{R}e\, s_{m,1} < 0. \quad (2.9.14)$$

All contained inertial modes decay in the spin-up time scale, $t \doteq E^{-\frac{1}{2}}$.

This formula for $s_{m,1}$ can be checked experimentally for a number of special container shapes and agreement, in general, is good. A report of this confirmation appears in §2.12.

2.10. The initial value problem: solution and critique

The interior solution of the initial value problem, accurate to $O(E^{\frac{1}{2}})$ through the spin-up time, may now be determined in many cases from the results of the preceding sections. The general time-dependent solution for the velocity vector is

$$\mathbf{q}(\mathbf{r},t) = \mathbf{q}_0(x,y,E^{\frac{1}{2}}t) + \sum_m A_m \mathbf{Q}_m(\mathbf{r}) \exp(i\lambda_m + s_{m,1}E^{\frac{1}{2}})t. \quad (2.10.1)$$

If $\mathbf{q}(\mathbf{r},0) = \mathbf{q}_*(\mathbf{r})$, then

$$\mathbf{q}_*(\mathbf{r}) = \mathbf{q}_0(x,y,0) + \sum_m A_m \mathbf{Q}_m(\mathbf{r}). \quad (2.10.2)$$

The mean circulation theorem and orthogonality relationships are used to evaluate the unknown quantities with the result that

$$\mathbf{q}_0(x,y,\tau) = \frac{1}{J(h)} \left(\oint_{\mathfrak{C}} \langle \mathbf{q}_* \rangle \cdot \mathbf{ds}\right) \left(\exp - \frac{I(h)}{J(h)}\tau\right) \mathbf{n}_T \times \mathbf{n}_B; \quad (2.10.3)$$

(I and J are defined in (2.6.24)) and

$$A_m = \frac{\int \mathbf{q}_* \cdot \mathbf{Q}_m^\dagger \, dV}{\int \mathbf{Q}_m^\dagger \cdot \mathbf{Q}_m \, dV}. \tag{2.10.4}$$

Equations (2.6.23), (2.7.2) and (2.8.10) provide the relationship of modal velocity and pressure functions.

The geostrophic mode, in a container whose surface is comprised entirely of closed constant-height contours, is excited whenever the initial velocity distribution possesses mean circulation. The component of the initial flow that has no mean circulation stimulates inertial modes, all of which decay in the spin-up time as does the geostrophic mode. Viscosity, by means of the Ekman boundary layer, strongly affects the interior regime through the processes of momentum transport and vortex line stretching and not by the diffusion of vorticity. This accounts for the comparatively short time scale for the restoration of a disturbed flow to rigid rotation.

In view of the rather peculiar features of Poincaré's problem, attention must be directed to some of the important unanswered mathematical questions. The most obvious of these concern the completeness of the inviscid eigenfunctions and the nature of the inviscid eigenvalue spectrum. If it were practical, the solution of the general interior problem by Laplace transform methods (see [75] and [192]) would automatically resolve these issues. Unfortunately, this is a very difficult task even though the transformed boundary value problem is well-set and a classic type, i.e., an elliptic differential equation with mixed boundary conditions. It is for this reason that the initial value problem was solved by a modal synthesis and the uncertainties of this approach accepted knowingly. The extraordinary complexity of the inviscid limit process is also the root of much difficulty. The extent to which there is a one-to-one correspondence between the inviscid and viscous modes is not clear and, in fact, the very existence of continuous inviscid modes in certain configurations demands study (see Høiland [86], and Wood [226]). The boundary layer approximation may require modification in these cases when sharp corners exist and the effects of internal shear layers associated with characteristics and eruptions from the surface should be determined. Subsequent analysis will provide a few of the answers and many clues but the theory remains incomplete at present.

2.11. Special cases

The restrictions on the motion are relaxed greatly when the total height, h, is everywhere a constant since equation (2.6.5) is then automatically satisfied. In this situation, the upper and lower surfaces of the vessel are of identical shapes but they are placed a fixed vertical distance apart. The lateral surface of the container is a vertical sidewall. The normal vectors to Σ, at the points on the top and bottom surfaces pierced by the same straight line, are anti-parallel:

$$\mathbf{n}_T(x, y) = -\mathbf{n}_B(x, y). \tag{2.11.1}$$

These vectors are functions of the horizontal co-ordinates only. Hence, (2.6.3) becomes

$$\mathbf{q}_0 = \tfrac{1}{2}\mathbf{n}_T \times \nabla\varphi_0. \tag{2.11.2}$$

Geostrophic motion remains independent of the vertical height, z, but every contour on the surface of the container is now one of constant total height.

Equation (2.8.6) for the vorticity of the depth-averaged inertial velocity is

$$i\lambda\nabla \times \langle \mathbf{Q} \rangle = \nabla \times [(\Phi_T - \Phi_B)\,\mathbf{n}_T]$$

and it follows that

$$\mathbf{n}_T \cdot \nabla \times \langle \mathbf{Q} \rangle = 0. \tag{2.11.3}$$

If viscous corrections are ignored for the moment, then the general velocity solution is

$$\mathbf{q}(\mathbf{r}, t) = \mathbf{q}_0(\mathbf{r}) + \sum_m A_m\, \mathbf{Q}_m(\mathbf{r})\, e^{i\lambda_m t}.$$

The Fourier coefficients are determined as before, from the orthogonality integral, and $\mathbf{q}_0(\mathbf{r})$ is obtained from the initial condition $\mathbf{q}(\mathbf{r}, 0) = \mathbf{q}_*(\mathbf{r})$. Since

$$\langle \mathbf{q}_* \rangle = h\mathbf{q}_0 + \sum_m A_m \langle \mathbf{Q}_m \rangle,$$

equation (2.11.3) is used to show that

$$\mathbf{n}_T \cdot \nabla \times \langle \mathbf{q}_* \rangle = h\mathbf{n}_T \cdot \nabla \times \mathbf{q}_0 \tag{2.11.4}$$

and this is the fundamental equation governing $\mathbf{q}_0$ when h is constant.

The boundary condition at the sidewall is

$$\mathbf{n}_L \cdot \mathbf{q}_0 = 0,$$

where $\mathbf{n}_L$ is the normal to this vertical surface. Thus, the complete boundary value problem for the geostrophic pressure, $\varphi_0(x, y)$, is

$$\frac{h}{2}\,\mathbf{n}_T\cdot\nabla\times(\mathbf{n}_T\times\nabla\varphi_0) = \mathbf{n}_T\cdot\nabla\times\langle\mathbf{q}_*\rangle, \qquad (2.11.5)$$

with

$$\mathbf{n}_L\cdot(\mathbf{n}_T\times\nabla\varphi_0) = 0 \qquad (2.11.6)$$

at the sidewall. The system consists of an inhomogeneous elliptic partial differential equation in two dimensions and a proper boundary condition. In other words, the problem is well-set but non-trivial. Motion in a cylindrical tank is one important example of a flow of this type that is easy to solve, (see § 2.15).

Equation (2.11.3) is really a special statement of the mean circulation theorem. The theorem in this instance asserts that

$$\oint_{\Upsilon}\langle\mathbf{Q}\rangle\cdot\mathbf{ds} = \int\hat{\mathbf{n}}_T\cdot\nabla\times\langle\mathbf{Q}\rangle\,d\Sigma' = 0, \qquad (2.11.7)$$

for *any contour* Υ on the container surface and the usual arguments for integrals of this form lead directly to (2.11.3).

If only part of the container is of constant height, it is quite possible that internal shear layers would be required to join the solutions in the different regions. Containers possessing no closed geostrophic contours receive attention in §§ 2.16 and 2.20.

Turning to another exception made in the analysis, we outline the proof given in [100] that $|\lambda| = 2$ is not an eigenvalue for a container of finite volume. Let $(\mathbf{Q}_n, \lambda_n)$, $(\mathbf{Q}_m, \lambda_m)$ be any two different eigenfunction–eigenvalue pairs. From the basic equations for these modes, it may be established that

$$\nabla\Phi_n\cdot\nabla\Phi_m^\dagger = (4-\lambda_n\lambda_m)\,\mathbf{Q}_n\cdot\mathbf{Q}_m^\dagger - \frac{4}{\lambda_n\lambda_m}(\hat{\mathbf{k}}\cdot\nabla\Phi_n)(\hat{\mathbf{k}}\cdot\nabla\Phi_m^\dagger)$$
$$+\,i\lambda_m\,\mathbf{Q}_m^\dagger\cdot\nabla\Phi_n - i\lambda_n\,\mathbf{Q}_n\cdot\nabla\Phi_m^\dagger.$$

The result of integrating this expression over the container volume is

$$(4-\lambda_n\lambda_m)\int\mathbf{Q}_n\cdot\mathbf{Q}_m^\dagger\,dV$$
$$= \int\left\{\nabla\Phi_n\cdot\nabla\Phi_m^\dagger + \frac{4}{\lambda_n\lambda_m}(\hat{\mathbf{k}}\cdot\nabla\Phi_n)(\hat{\mathbf{k}}\cdot\nabla\Phi_m^\dagger)\right\}dV,$$

whence, in particular,

$$(4-\lambda_m^2)\int \mathbf{Q}_m\cdot\mathbf{Q}_m^\dagger\,\mathrm{d}V = \int\left\{\nabla\Phi_m\cdot\nabla\Phi_m^\dagger + \frac{4}{\lambda_m^2}(\hat{\mathbf{k}}\cdot\nabla\Phi_m)(\hat{\mathbf{k}}\cdot\nabla\Phi_m^\dagger)\right\}\mathrm{d}V. \tag{2.11.8}$$

If $\lambda_m^2 = 4$ and the velocity integral is finite, then the right-hand side is equated identically to zero which means that $\Phi_m \equiv$ constant. It is relatively easy to show next that a constant pressure implies a zero velocity field.

An important effect rendered obscure by the omission of the analytical details in previous sections concerns boundary layer eruptions or resonances at certain critical curves on Σ. It has been noted by many authors that the ordinary Ekman boundary layer for inertial oscillations becomes singular wherever the oscillation frequency is related to the angular velocity by

$$\lambda = 2\hat{\mathbf{n}}\cdot\hat{\mathbf{\Omega}}.$$

Roberts and Stewartson[165] analysed the steady flow in the neighbourhood of a critical position in a special problem. Their analysis included lateral shear terms and indicated that the boundary layer structure changed locally from a characteristic thickness $E^{\frac{1}{2}}$ to $E^{\frac{2}{5}}$. This change would appear as a singularity in the linear theory. However, the total effect of these eruptions on the interior flow is evidently negligible compared to the Ekman layer flux. Of course, the form of the asymptotic expansions beyond the lowest order terms is probably dictated by the nature of this non-uniformity. The analysis of Greenspan[73] avoids some of the difficulty by retaining the time-dependence and by taking limits judiciously. Both analyses leave much to be desired (are non-linear terms needed?) although it seems probable that eruptions are not significant to the primary flow. The likely effect of the critical zones might be to establish weak internal shear layers along the characteristic direction, although surely not of the magnitude proposed by Bondi and Lyttleton[12]. However, even weak shear layers could provide a source of flow instability and in that way become significant. In any event, non-uniformities do make uncertain the formal method of continuing the analysis to terms of all orders.

Another mathematical aspect of interest concerns the inviscid eigenvalue problem for the inertial waves. The eigenvalue spectrum for the interior problem will be shown in two cases, to be denumerable but dense in the interval $|\lambda| \leqslant 2$ (see section 2.15). This is solely a manifestation of non-uniform limit processes; the viscous spectrum is really discrete. Actually, the entire concept upon which the inviscid analysis is based, that each mode is separable into an inviscid component and a boundary layer correction, breaks down when the effective modal wave length is of the order of the boundary layer thickness. The conclusion that the spectrum is dense results from the failure of the asymptotic method to locate the eigenvalues precisely. The approximation yields only the projection of the correct eigenvalue position onto the real λ axis; the projected positions are dense.

2.12. Motion in a sphere

The spherical container provides an excellent illustration of the general theory because the eigenmodes, eigenfrequencies, and their viscous corrections may all be determined explicitly. Furthermore, this is also a convenient configuration for experimental purposes as well as the natural geometry in many applications.

In this case, geostrophic contours are the circles on the sphere of constant cylindrical radius, r, so that the geostrophic velocity has only an azimuthal component. Since

$$f = g = (1-r^2)^{\frac{1}{2}}, \tag{2.12.1}$$

(2.6.7) and (2.6.23) for the geostrophic velocity and pressure become

$$\mathbf{q}_0 = \hat{\boldsymbol{\theta}}\left(\frac{1}{2}\frac{\partial}{\partial r}\varphi_0(r,0)\right)\exp\left(-E^{\frac{1}{2}}(1-r^2)^{-\frac{3}{4}}t\right), \tag{2.12.2}$$

and

$$\frac{\partial}{\partial r}\varphi_0(r,t) = \left(\frac{\partial}{\partial r}\varphi_0(r,0)\right)\exp\left(-E^{\frac{1}{2}}(1-r^2)^{-\frac{3}{4}}t\right). \tag{2.12.3}$$

Poincaré's problem for the modes, in cylindrical co-ordinates, is

$$\frac{1}{r}\frac{\partial}{\partial r}r\frac{\partial\Phi}{\partial r}+\frac{1}{r^2}\frac{\partial^2\Phi}{\partial\theta^2}+\left(1-\frac{4}{\lambda^2}\right)\frac{\partial^2\Phi}{\partial z^2} = 0, \tag{2.12.4}$$

with

$$r\frac{\partial\Phi}{\partial r}+\frac{2}{i\lambda}\frac{\partial\Phi}{\partial\theta}+\left(1-\frac{4}{\lambda^2}\right)z\frac{\partial\Phi}{\partial z} = 0, \tag{2.12.5}$$

on $r^2+z^2 = 1$. Separable solutions can be found in the modified 'oblate spheroidal' co-ordinate system introduced by Bryan[19]

$$\left.\begin{aligned} r &= \left(\frac{4}{4-\lambda^2}-\eta^2\right)^{\frac{1}{2}}(1-\mu^2)^{\frac{1}{2}}, \\ z &= \left(\frac{4}{\lambda^2}-1\right)^{\frac{1}{2}}\eta\mu; \end{aligned}\right\} \tag{2.12.6}$$

they are

$$\Phi_{nmk} = \Psi_{nmk}(r, z)\, e^{ik\theta} = P_n^{|k|}(\eta/c_{nmk})\, P_n^{|k|}(\mu)\, e^{ik\theta}, \tag{2.12.7}$$

where

$$c_{nmk} = (1-\tfrac{1}{4}\lambda_{nmk}^2)^{-\frac{1}{2}}$$

and λ_{nmk} is the mth eigenvalue solution of the transcendental equation

$$kP_n^{|k|}(\tfrac{1}{2}\lambda) = 2\left(1-\frac{\lambda^2}{4}\right)\frac{d}{d\lambda}P_n^{|k|}(\tfrac{1}{2}\lambda). \tag{2.12.8}$$

The eigenfunction is specified by an index triple (n, m, k); n and k vary over all integers but m has a finite range,

$$m = 1, \ldots, \mathcal{N}_{nk}.$$

(The symbol, $|k|$, is used to achieve a minor algebraic simplification.)

It is important to note that on the surface of the sphere,

$$\mu = \cos\Theta, \quad \eta = \tfrac{1}{2}\lambda_{nmk}\, c_{nmk},$$

all the functions Ψ_{nmk} (m varying, n, k fixed) reduce to the same zonal harmonic

$$\Psi_{nmk} = P_n^{|k|}(\tfrac{1}{2}\lambda_{nmk})\, P_n^{|k|}(\cos\Theta), \tag{2.12.9}$$

where Θ is the polar angle.

Let the eigenvalue λ correspond to index k then (2.12.8) implies that $-\lambda$ corresponds to index $-k$. The meaning is clear: inertial modes are travelling waves, each propagates in a definite direction, clockwise or counterclockwise about the rotation axis. There are no standing oscillations. In general, the following relationships hold:

$$(\Phi_{nmk}, \Psi_{nmk}, \lambda_{nmk}, s_{nmk,1}) = (\Phi^{\dagger}_{nm-k}, \Psi_{nm-k}, -\lambda_{nm-k}, s^{\dagger}_{nm-k,1}). \tag{2.12.10}$$

The last factor is the modal decay number introduced in (2.5.8).

The product of Legendre functions above, is really an elementary polynomial in r and z. Toomre (see [1]) and Kudlick [100] have given general polynomial formulas for spherical and spheroidal modes; the particular form for the sphere is

$$\mathrm{P}_n^{|k|}(\eta/\mathrm{c}_{nmk})\,\mathrm{P}_n^{|k|}(\mu) = \left(\frac{(2n)!}{2^n n!\,(n-|k|)!}\right)^2 \left(\frac{\lambda_{nmk}\,\mathrm{z}}{2}\right)^{\mathfrak{e}_{nk}} \left(\frac{\mathrm{r}}{\mathrm{c}_{nmk}}\right)^{|k|}$$

$$\times \prod_{j=1}^{\mathcal{N}_{nk}} \left\{\mathfrak{y}_j^2\left(1-\frac{\lambda_{nmk}^2}{4}\right)\mathrm{r}^2 + \frac{\lambda_{nmk}^2}{4}(1-\mathfrak{y}_j^2)\,\mathrm{z}^2 + \mathfrak{y}_j^2(\mathfrak{y}_j^2-1)\right\}. \tag{2.12.11}$$

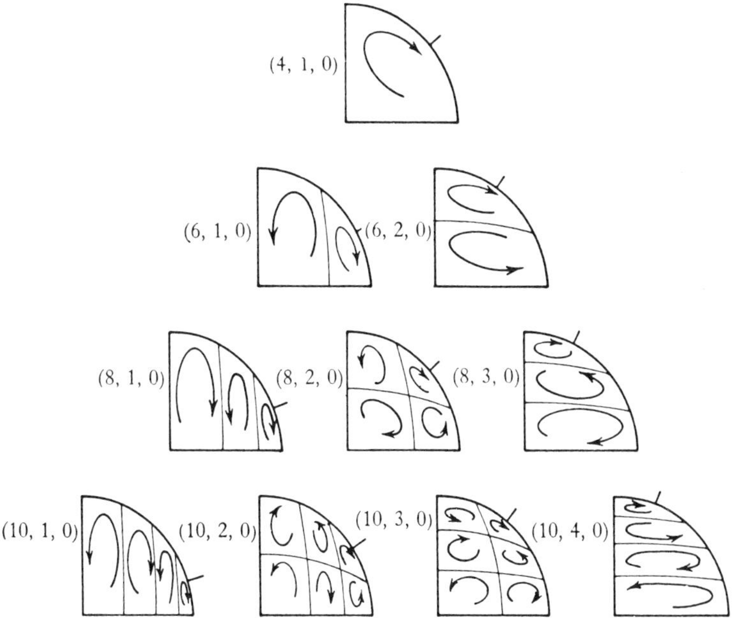

Fig. 2.7. Circulation patterns for the low order spherical modes.

Here
$$\mathfrak{e}_{nk} = \begin{cases} 0 & \text{if} \quad n-|k| \text{ is even,} \\ 1 & \text{if} \quad n-|k| \text{ is odd;} \end{cases}$$

$$\mathcal{N}_{nk} = n-|k|-\mathfrak{e}_{nk},$$

and $\mathfrak{y}_j$, $j = 1, \ldots, \mathcal{N}_{nk}$, are the $n-|k|-\mathfrak{e}_{nk}$ real distinct zeros of $\mathrm{P}_n^{|k|}(\mathfrak{y})$, in the interval $(0, 1)$, excluding $\mathfrak{y} = 0$ and $\mathfrak{y} = 1$. Some of the axially symmetric modes are illustrated in fig. 2.7.

It is advantageous to express the general orthogonality integral

(2.7.10), in terms of the newly defined functions. A particularly useful form is

$$\iint r\,dr\,dz\left\{\frac{\partial}{\partial r}\Psi_{nmk}\frac{\partial}{\partial r}\Psi_{n'm'k}+\frac{k^2}{r^2}\Psi_{nmk}\Psi_{n'm'k}\right.$$

$$\left.+\left(1+\frac{4}{\lambda_{nmk}\lambda_{n'm'k}}\right)\frac{\partial}{\partial z}\Psi_{nmk}\frac{\partial}{\partial z}\Psi_{n'm'k}\right\}=0, \quad (2.12.12)$$

for $(n, m) \neq (n', m')$. Modes associated with different absolute values of the index k are orthogonal by virtue of the relationship

$$\int_0^{2\pi} \exp i(k+k')\,\theta\,d\theta = 0, \quad k \neq -k'.$$

General results for the ellipsoidal container, comparable to the foregoing, have been obtained by Kudlick[100], and in specific instances by Hough[88], and Cartan[29].

The computation of viscous effects, specifically the modal decay factor according to (2.9.12) and (2.9.13), is laborious but straightforward. Table 2.1 lists some of the numerical results and the experimental findings to date.

Table 2.1

Mode identification (n, m, k)	Eigenvalue λ_{nmk}	Decay Factor $s_{nmk,1}$	Experimental values λ_{nmk}	$\mathscr{R}e\, s_{nmk,1}$
(2, 1, 1)	1·0	−2·62+0·259i	1·0	−2·82
(4, 1, 0)	1·309	−3·38+0·434i	1·302	−3·88
(4, 1, 1)	0·820	−3·87+0·315i	—	—
(4, 2, 1)	1·708	−2·64+0·504i	—	—
(4, 3, 1)	0·612	−3·95+0·180i	—	—
(6, 1, 0)	0·938	−4.64+0·329i	0·935	−5·17
(6, 2, 0)	1·660	−3·50+0·568i	—	—
(8, 1, 0)	0·726	−5·62+0·266i	0·724	−6·32
(8, 2, 0)	1·354	−4·87+0·487i	—	—
(8, 3, 0)	1·800	−3·53+0·632i	—	—

The first of these modes, (2, 1, 1), is especially interesting since it represents rigid rotation about an axis other than the rotation axis of the sphere itself. This is, of course, a non-trivial inviscid mode when viewed in the co-ordinate system of the rotating sphere and it has, at times, been called the spin-over mode. Practically speaking it is the easiest mode to excite, for all that is required is a slight impulsive change in the direction of the rotation axis of a rigidly rotating sphere. The experimental value of the decay factor quoted above was determined in this manner by W. Malkus (private communication) and W. G. Wing (an internal report of the Sperry Rand Corp.). Precession may be viewed as a sequence of infinitesimal

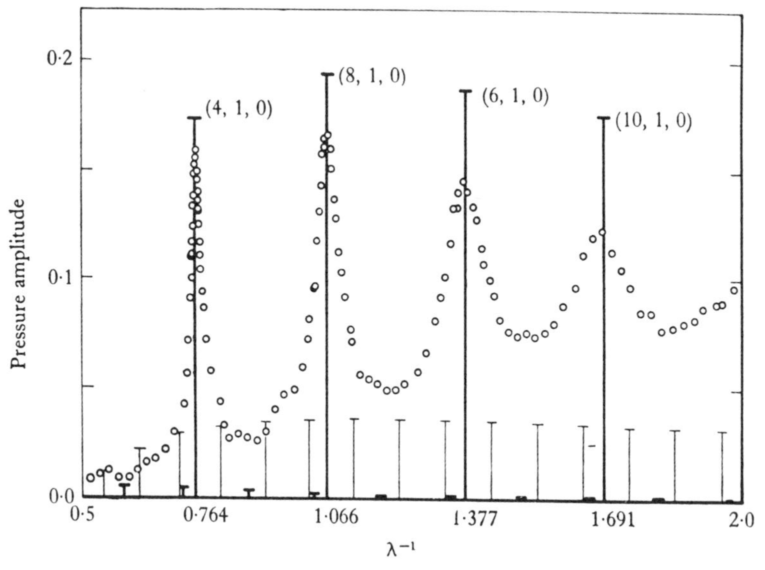

Fig. 2.8. Peak to peak amplitude versus $1/\lambda$ showing the resonance of the symmetric spherical modes. Oscillation amplitude is 8°; measurements were made at the centre of the sphere, [1].

changes of this sort and it is anticipated that this mode will play a key role in that problem. For this reason, the specific form of the eigenfunction is recorded now:

$$\Psi_{211} = rz. \tag{2.12.13}$$

The velocity and pressure are

$$\mathbf{Q}_{211}\, e^{it} = e^{i(\theta+t)}(z\hat{\mathbf{r}} + iz\hat{\boldsymbol{\theta}} - r\hat{\mathbf{k}}), \quad \Phi_{211}\, e^{it} = irz\, e^{i(\theta+t)}. \tag{2.12.14}$$

Aldridge and Toomre[1] used a somewhat novel procedure to resonate the inertial modes in a sphere. A sphere was subjected to a small oscillation as it (and the oscillator too) rotated with constant speed. The pressure response was then measured as the forcing frequency was varied over a wide range. Fig. 2.8, exhibiting the reduced pressure at the pole versus excitation frequency, shows definite resonance peaks at the eigenvalue positions predicted by theory. The mechanics of the resonance are simple. The oscillating sphere produces an oscillating Ekman layer which in turn establishes a small normal flux at the boundary forcing internal motion. At certain frequencies, this time-dependent flux is tuned with the interior modes and resonance follows. The eigenfrequencies are located rather precisely and the agreement with theory is excellent. Once the system is resonating, the excitation is stopped and a record is made of the time decay of the oscillation. This affords an accurate means of determining the exponential decay factor. The width of the response curves provides another method (but a less accurate one) of calculating the same number. The values obtained are in good agreement with theory but do show slight discrepancies, always in the direction of a faster decay time. Some instability or turbulence in the boundary layer could account for this error by raising the effective value of the kinematic viscosity in this region. The most recent observations do indicate some striations in the boundary layer but the flow appears to be essentially laminar. However, agreement is close enough to make this an issue of low priority for present purposes and it will not be pursued any further.

It should also be noted that there is no evidence from these experiments, as yet, of any eruptions in the viscous layer at critical latitudes.

The theory of forced oscillations in a sphere is developed in § 2.14 and formulas appear there for the pressure response and peak amplitude. In particular, these results will provide a complete analytical basis for the experiments just described.

2.13. Precession and the problem of forced motions

Consider the motion of a fluid completely filling a closed, rapidly rotating container which is moved about its mass centre non-uniformly. (The flow within a rotating, precessing spheroid is a typical problem of interest.) Departures from a state of rigid rotation are assumed small and the entire container surface is to be an envelope of closed contours of constant total height.

The choice of a co-ordinate system to describe the fluid motion determines the manner in which the applied forces and torques appear as inhomogeneities in the equations of motion. If the container is non-symmetric and of rather arbitrary shape, then a system of axes fixed in the body is really the only feasible one to employ. However, for certain symmetrical shapes, other co-ordinate frames can be used, but in any case the analytical description of the container surface, its position in space, should not involve the time.

Let (x, y, z) be a co-ordinate system *fixed* in a container which rotates with a prescribed *time-dependent* angular velocity $\boldsymbol{\Omega}(t)$. Since the rotation vector is assumed to be nearly constant in inertial space, it may also be represented in the body frame as

$$\boldsymbol{\Omega}(t) = \Omega(\hat{\mathbf{k}} + \varepsilon \boldsymbol{\delta}(t)) \tag{2.13.1}$$

where the unit vector $\hat{\mathbf{k}}$ defines the z direction, $\boldsymbol{\delta}(t)$ is the departure from rigid rotation and ε is small.

The boundary value problem can be made dimensionless using $[\![L, \Omega^{-1}, \varepsilon\Omega L]\!]$ to characterize the length of the container, the time, and the excess fluid velocity. Any externally applied conservative forces are combined with the centrifugal acceleration to form the reduced pressure. Therefore, the dimensionless boundary value problem[1] is

$$\left.\begin{aligned} \nabla\cdot\mathbf{q} &= 0, \\ \frac{\partial}{\partial t}\mathbf{q} + \varepsilon\mathbf{q}\cdot\nabla\mathbf{q} + 2(\hat{\mathbf{k}} + \varepsilon\boldsymbol{\delta}(t))\times\mathbf{q} &= -\nabla p + \mathbf{r}\times\frac{d}{dt}\boldsymbol{\delta} + E\nabla^2\mathbf{q}, \end{aligned}\right\} \tag{2.13.2}$$

with $\mathbf{q} = 0$ on Σ; $\mathbf{q} = \mathbf{q}_*$ at $t = 0$ and $E = \nu/\Omega L^2$. The linear theory is obtained by setting the Rossby number, ε, equal to zero.

The unforced initial value problem, $\boldsymbol{\delta}(t) \equiv 0$ in (2.13.2), has already been solved in § 2.10, but the contribution to the final motion from this source is unimportant because the original state is dissipated in the time scale $E^{-\frac{1}{2}}$. After the spin-up time has elapsed, only the direct response to the inertial body force remains. It is sufficient to determine a particular solution of the forced motion problem for this can then be added to the general solution of the homogeneous initial value problem to obtain the general solution of the complete boundary value problem.

The effects of viscosity are often of secondary importance in

[1] See the footnote on p. 6.

problems involving a non-conservative force field. Consider then, the response of an inviscid fluid to an oscillatory body force

$$\mathbf{N}(\mathbf{r})\,e^{iat}.$$

(The response to a general time-dependent body force can be found by the principle of superposition—a Fourier integral over all frequencies.) Hence, a particular solution of the following boundary value problem is required:

$$\nabla\cdot\mathbf{q} = 0,$$

$$\frac{\partial}{\partial t}\mathbf{q} + 2\hat{\mathbf{k}}\times\mathbf{q} = -\nabla p + \mathscr{R}e\,\mathbf{N}\,e^{iat}, \tag{2.13.3}$$

with $\mathbf{q}\cdot\hat{\mathbf{n}} = 0$ on Σ. This can be determined by setting

$$\left.\begin{aligned} \mathbf{q} &= \mathscr{R}e\,\mathbf{Q}\,e^{iat}, \\ p &= \mathscr{R}e\,\Phi\,e^{iat}, \end{aligned}\right\} \tag{2.13.4}$$

so that

$$\left.\begin{aligned} \nabla\cdot\mathbf{Q} &= 0, \\ ia\mathbf{Q} + 2\hat{\mathbf{k}}\times\mathbf{Q} &= -\nabla\Phi + \mathbf{N}, \end{aligned}\right\} \tag{2.13.5}$$

with

$$\mathbf{Q}\cdot\hat{\mathbf{n}} = 0 \quad \text{on } \Sigma.$$

The velocity is assumed to be a superposition of all the modal amplitude functions, inertial and geostrophic:

$$\mathbf{Q} = \mathbf{q}_0 + \sum_m B_m\,\mathbf{Q}_m; \tag{2.13.6}$$

the boundary condition is then automatically satisfied, as is the continuity equation. The substitution of this expression in (2.13.5) and the determination of the coefficients B_m using the known orthogonality properties of the modes yields

$$B_m = \frac{\displaystyle\int \mathbf{N}\cdot\mathbf{Q}_m^{\dagger}\,dV}{(ia - \lambda_m)\displaystyle\int \mathbf{Q}_m\cdot\mathbf{Q}_m^{\dagger}\,dV}. \tag{2.13.7}$$

The function $\mathbf{q}_0$ is obtained from the mean circulation theorem and the result is the same as replacing $\mathbf{q}_*$ and τ in (2.10.3) by $\mathbf{N}/ia$ and zero:

$$\mathbf{q}_0 = \frac{\mathbf{n}_T\times\mathbf{n}_B}{iaJ(h)}\oint_{\mathfrak{C}}\langle\mathbf{N}\rangle\cdot d\mathbf{s}. \tag{2.13.8}$$

Unless the frequency a is close to an eigenvalue λ_m, we may take

$\mathfrak{s}_m \simeq \mathrm{i}\lambda_m$ with no loss of accuracy. However, at a resonance setting, the inclusion of the viscous decay factor is necessary to maintain a finite, viscosity limited, resonant amplitude.

The fluid motion inside an arbitrarily shaped precessing container can now be found, but for definiteness, we shall focus our attention on the following practical problem. The container is firmly fixed on turntable A in fig. 2.9, which is itself mounted upon another, B. Both tables revolve at constant speeds; A with frequency Ω and B with frequency ω. Initially, A rests squarely upon B so that the rotation vectors $\mathbf{\Omega}$ and $\boldsymbol{\omega}$ are in the same direction. The rigidly

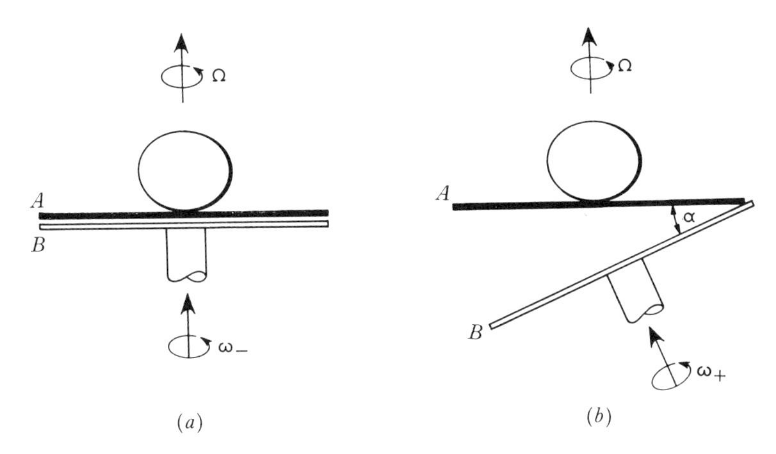

Fig. 2.9. Laboratory arrangement for precession. The rotation rates of turntables A and B are Ω and ω. Precession is initiated by tilting table B through angle α.

rotating fluid is disturbed by impulsively turning or tilting B through a small angle α, as shown. The container then begins to precess in the required manner.

Denote by $\boldsymbol{\omega}_-$ and $\boldsymbol{\omega}_+$ the vector $\boldsymbol{\omega}$ just before and after the impulsive change in inclination angle. The original state of rigid fluid rotation must be reconciled with the new orientation of the shell and a potential flow develops instantaneously to adjust the velocity to the new boundary conditions. As viewed from the *laboratory* frame, the velocity of the shell changes from $(\boldsymbol{\omega}_- + \mathbf{\Omega}) \times \mathbf{r}$ to $(\boldsymbol{\omega}_+ + \mathbf{\Omega}) \times \mathbf{r}$. Thus, the particle velocity immediately after the switch is representable *in the inertial frame* as

$$\mathbf{q}(\mathbf{r}, 0_+) = \nabla\Lambda + (\boldsymbol{\omega}_- + \boldsymbol{\Omega}) \times \mathbf{r}.$$

The boundary condition requires

$$\hat{\mathbf{n}}\cdot\mathbf{q}(\mathbf{r}, 0_+) = (\boldsymbol{\omega}_+ + \boldsymbol{\Omega}) \times \mathbf{r}\cdot\hat{\mathbf{n}} \quad \text{on } \Sigma$$

so that

$$\hat{\mathbf{n}}\cdot\nabla\Lambda = (\boldsymbol{\omega}_+ - \boldsymbol{\omega}_-) \times \mathbf{r}\cdot\hat{\mathbf{n}} \quad \text{on } \Sigma$$

is the appropriate constraint for the potential flow:

$$\nabla^2\Lambda = 0.$$

Once Λ is known, the initial condition in any reference frame is readily deduced.[1]

In a co-ordinate system *fixed in the container*,

$$\boldsymbol{\Omega} = \Omega\hat{\mathbf{k}}, \ \boldsymbol{\omega}_- = \omega\hat{\mathbf{k}},$$

and $\boldsymbol{\omega} = \boldsymbol{\omega}_+ = -\omega\sin\alpha(\sin\Omega t\,\hat{\mathbf{i}} + \cos\Omega t\,\hat{\mathbf{j}}) + \omega\cos\alpha\hat{\mathbf{k}}$.

Hence

$$\mathit{\Omega}(t) = \boldsymbol{\Omega} + \boldsymbol{\omega} = (\Omega + \omega\cos\alpha)\,\hat{\mathbf{k}} - \omega\sin\alpha\,(\sin\Omega t\,\hat{\mathbf{i}} + \cos\Omega t\,\hat{\mathbf{j}}),$$

and the following identifications with (2.13.1) can be made if α is small and $\sigma = \Omega/(\omega + \Omega)$: $\mathit{\Omega} = \Omega + \omega$; $\varepsilon = \sigma\alpha$;

$$\boldsymbol{\delta}(t) = -\frac{1-\sigma}{\sigma}(\sin\sigma t\,\hat{\mathbf{i}} + \cos\sigma t\,\hat{\mathbf{j}}). \tag{2.13.9}$$

The linear form of the inviscid momentum equation (2.13.2) is then

$$\frac{\partial}{\partial t}\mathbf{q} + 2\hat{\mathbf{k}} \times \mathbf{q} = -\nabla p - (1-\sigma)(-\cos\sigma t\,\hat{\mathbf{i}} + \sin\sigma t\,\hat{\mathbf{j}}) \times \mathbf{r}. \tag{2.13.10}$$

This compares exactly with (2.13.3) by setting

$$\left.\begin{aligned} a &= \sigma, \\ \mathbf{N} &= (1-\sigma)\,ie^{i\theta}(z\hat{\mathbf{r}} + iz\hat{\boldsymbol{\theta}} - r\hat{\mathbf{k}}); \end{aligned}\right\} \tag{2.13.11}$$

here a change to cylindrical co-ordinates has been introduced.

[1] Actually, the explicit determination of Λ is not required because the solution of the initial value problem depends primarily on the structure of the original vorticity field. Since there is no vorticity in a potential motion, this component of the initial flow has a zero effect on the calculation of the velocity as a superposition of normal modes, a procedure detailed in § 2.10.

If the container is a sphere, then $\mathbf{N}$ is actually proportional to the eigenfunction $\mathbf{Q}_{211}$ identified in (2.12.14) of the last section. This mode is designated simply as $\mathbf{Q}_1$ for present purposes and the corresponding eigenvalue by

$$\lambda_1 = \lambda_{211} = \mathrm{i} - E^{\frac{1}{2}}(2{\cdot}62 - 0{\cdot}259\mathrm{i}).$$

It follows directly from (2.13.7) that all the amplitude coefficients B_m are zero for $m \neq 1$ and

$$B_1 = -\frac{1-\sigma}{1-\sigma+\mathrm{i}E^{\frac{1}{2}}(2{\cdot}62-0{\cdot}259\mathrm{i})}.$$

This equation is most useful when the sphere precesses very slowly, i.e., $1-\sigma \cong \frac{\omega}{\Omega} = O(E^{\frac{1}{2}})$, and the excitation and modal frequencies are almost identical. However, if σ is not unity or nearly so, then we may set $E = 0$ with no loss of accuracy. In this case the velocity is

$$\mathbf{q} = -\mathscr{Re}\, e^{\mathrm{i}(\theta+\sigma t)}(z\hat{\mathbf{r}} + \mathrm{i}z\,\hat{\boldsymbol{\theta}} - r\hat{\mathbf{k}}),$$

which is a representation of rigid rotation about the original rotation axis $\boldsymbol{\omega}_- + \boldsymbol{\Omega}$, viewed from axes *fixed* in the body. Of course, in the absence of viscosity, a spherical shell cannot communicate any of its rotational motion to the interior fluid which remains unaffected. Consideration of boundary layer effects is required to determine the nature of the forced flow. The internal motion driven by the small Ekman layer flux can still be of unit magnitude in conditions that produce resonance and this will be discussed shortly.

Consider next the precession of a spheroidal casing

$$\frac{r^2}{r_0^2}+\frac{z^2}{z_0^2} = 1$$

and take r_0 to be the characteristic length L so that in dimensionless notation Σ is

$$r^2+\frac{z^2}{l_0^2} = 1, \quad (l_0 = z_0/r_0).$$

Kudlick[100] found all the inertial modes for this configuration, following the procedure developed in [73]. The mode most affected by the motion of the shell, when both the eccentricity, e_*, and the precession frequency are small, is that one which reduces to $\mathbf{Q}_{211}$ or $\mathbf{N}$ when $e_* = 0$. This is the mode

$$\mathbf{Q} = \left(\frac{z}{2-\lambda}\hat{\mathbf{r}} + \frac{\mathrm{i}z}{2-\lambda}\hat{\boldsymbol{\theta}} - \frac{r}{\lambda}\hat{\mathbf{k}}\right)e^{\mathrm{i}\theta}, \qquad (2.13.12)$$

whose inviscid eigenvalue is

$$\lambda = 2\frac{1+e_*}{2+e_*}; \quad e_* = \frac{1}{l_0^2} - 1.$$

The decay factor is also a function of eccentricity, i.e.,

$$s = 2i\frac{1+e_*}{2+e_*} + E^{\frac{1}{2}} s_{,1}(e_*),$$

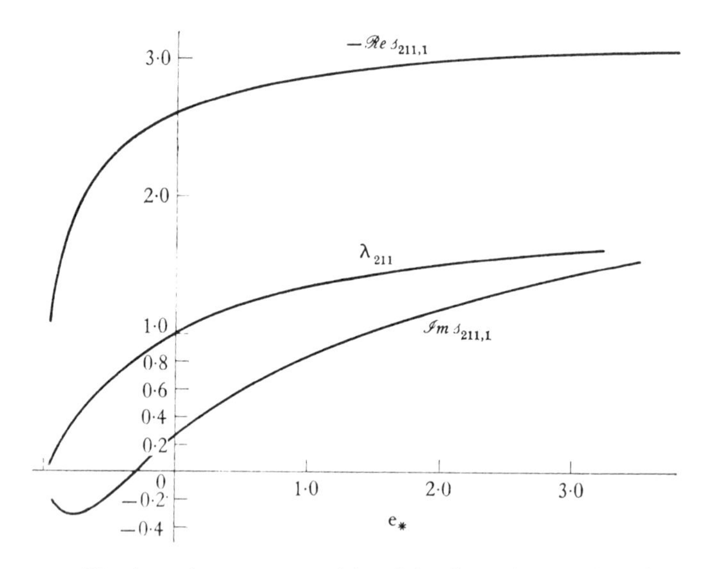

Fig. 2.10. The dependence on eccentricity of the eigenvalue and decay factor of spheroidal mode (2, 1, 1).

and $s_{,1}(e_*)$ is shown in fig. 2.10. It follows that the amplitude is

$$B = (1-\sigma)\left[\sigma - 2\frac{1+e_*}{2+e_*} + iE^{\frac{1}{2}} s_{,1}(e_*)\right]^{-1} \frac{\lambda\left(1+\dfrac{l_0^2\lambda}{2-\lambda}\right)}{\left(1+\dfrac{l_0^2\lambda^2}{(2-\lambda)^2}\right)}.$$

For small values of e_*, $O(e_*)$ terms can be neglected when compared directly with terms of $O(1)$. The decay factor for the sphere may then replace $s_{,1}$ so that

$$B \cong (1-\sigma)\left[\sigma - 2\left(\frac{1+e_*}{2+e_*}\right) - E^{\frac{1}{2}}i(2{\cdot}62 - 0{\cdot}259i)\right]^{-1}. \tag{2.13.13}$$

A 'resonance' is produced when

$$\sigma = 2\left(\frac{1+e_*}{2+e_*}\right)$$

and the amplitude is then limited only by viscous processes. Since e_* is small, this condition is approximately

$$\frac{\omega}{\Omega} = -\frac{e_*}{2}$$

showing that large amplitudes can result from a slow retrograde precession. This result was obtained by Stewartson and Roberts [166, 192] using a procedure based on the Laplace transform. Resonance of the higher modes occurs at larger precession frequencies, but we will deal with this in § 3.9.

As a final example, consider an axially symmetric container rotating uniformly but also oscillating slightly about its symmetry axis. In the laboratory arrangement shown in fig. 2.9 (*a*), table *A* rotates uniformly but *B* oscillates so that

$$\boldsymbol{\omega} = \varepsilon\Omega \sin \omega t\, \hat{\mathbf{k}}.$$

Hence
$$\boldsymbol{\Omega}(t) = \Omega(1+\varepsilon \sin \omega t)\, \hat{\mathbf{k}}$$

and in the notation of (2.13.1) (suitably scaled)

$$\boldsymbol{\delta}(t) = \sin \frac{\omega}{\Omega} t\, \hat{\mathbf{k}}.$$

The forcing function appearing in (2.13.3) is now

$$\mathbf{N} = -\frac{\omega}{\Omega} \hat{\mathbf{k}} \times \mathbf{r} \tag{2.13.14}$$

and the frequency is $\alpha = \omega/\Omega$. For *symmetrical container shapes*, $\mathbf{N}$ is a geostrophic amplitude function and accordingly must be orthogonal to all the inertial modes. The solution of the problem in this case is determined from (2.13.8):

$$\mathbf{q} = \mathcal{Re}\, i\hat{\mathbf{k}} \times \mathbf{r}\, e^{i\alpha t}. \tag{2.13.15}$$

Obviously, very little information has been obtained because in the last equation $\mathbf{q}$ is just the scaled form of $-\boldsymbol{\omega} \times \mathbf{r}$. The oscillation has been removed from the problem leaving a primary velocity that is simply one of rigid rotation in *inertial* space at frequency Ω, i.e. $\mathbf{\Omega}\times\mathbf{r}$. This is reasonable because the interaction of the symmetric casing with the fluid is here a purely viscous process; the shell does not push the liquid about. It is then essential to determine the secondary circulation established by the Ekman layer by correcting (2.13.15) for viscosity at the surface Σ. A large interior flow can still be produced if the boundary layer suction is in tune with or resonates an inertial mode. Equation (2.13.15) really implies that the co-ordinate system rotating with angular speed Ω is just as appropriate as the body frame for the analysis of this particular type of oscillation.

It becomes imperative to study the problem wherein circulations are produced by the boundary layers alone. When motion is forced by suction at the bounding surface, the general inviscid boundary value problem consists of the homogeneous form of (2.13.3) with

$$\mathbf{q}\cdot\hat{\mathbf{n}} = \mathscr{R}e\,\mathrm{F}\,\mathrm{e}^{\mathrm{i}\alpha t} \quad \text{on } \Sigma.$$

As before, only excitation at a single frequency α need be considered. The production of resonant modes by oscillating a rotating container is formulated as a problem of this type, as is any calculation of a secondary circulation.

The solution of (2.13.15) with $\mathbf{N} = 0$ is required subject to the condition

$$\mathbf{Q}\cdot\hat{\mathbf{n}} = \mathrm{F} \quad \text{on } \Sigma. \tag{2.13.16}$$

One method is to convert the inhomogeneous boundary condition into a body force and to apply the results developed earlier. A general approach would be to determine a particular function $\mathbf{Q}_P$ having the properties

$$\nabla\cdot\mathbf{Q}_P = 0 \quad \text{and} \quad \mathbf{Q}_P\cdot\hat{\mathbf{n}} = \mathrm{F} \quad \text{on } \Sigma.$$

For definiteness, $\mathbf{Q}_P$ is assumed to be a potential flow,

$$\nabla\times\mathbf{Q}_P = 0.$$

By setting

$$\mathbf{Q} = \mathbf{Q}_P + \mathbf{Q}',$$

it follows that $\mathbf{Q}'$ satisfies the boundary value problem (2.13.5) with

$$\mathbf{N} = -(\mathrm{i}\alpha\mathbf{Q}_P + 2\hat{\mathbf{k}}\times\mathbf{Q}_P).$$

The desired conversion is obtained. The solution is then

$$\mathbf{Q}' = \mathbf{q}_0' + \sum_m \mathrm{B}_m'\,\mathbf{Q}_m,$$

where the coefficients are determined from (2.13.7) and (2.13.8). It follows that

$$\left.\begin{aligned} \mathbf{q}_0' &= \frac{2}{i\alpha J(h)} \mathbf{n}_T \times \mathbf{n}_B \int (F_T |\mathbf{n}_T| + F_B |\mathbf{n}_B|)\, d\Sigma_P, \\ B_m' &= -\frac{\displaystyle\int \Phi_m^\dagger F \, d\Sigma}{(i\alpha - \mathcal{S}_m)\displaystyle\int \mathbf{Q}_m \cdot \mathbf{Q}_m^\dagger \, dV}, \end{aligned}\right\} \tag{2.13.17}$$

where Σ_P is the projection onto the x, y plane of the area bounded by contour $\mathfrak{C}$. This is a particularly useful form to compute the resonant response that occurs when $\alpha = \lambda_m$, an inviscid eigenvalue. The simplest way to calculate the viscosity limited modal amplitude is to include the $O(E^{\frac{1}{2}})$ viscous correction to $\mathcal{S}_m$ in the above formula.

It may not be an easy matter to find $\mathbf{Q}_P$, and a direct approach would be useful. For certain symmetrical containers rotating about their symmetry axes, another method of solving the problem, based on the natural modes, was devised in [73] and extended in [100].

This approach employs the pressure as the primary variable and utilizes the expansion

$$\Phi = \varphi_0(x, y) + \sum_m A_m \Phi_m(\mathbf{r}). \tag{2.13.18}$$

Here φ_0 is the geostrophic pressure and Φ_m is the pressure amplitude function of the *m*th inertial mode. The fluid velocity cannot be represented so simply as a superposition of modes $\mathbf{Q}_m$ because the boundary condition is inhomogeneous.

If the rotation axis is also a symmetry axis of the container, then, with considerable effort, it is possible to determine the response of the geostrophic component, φ_0, separately from that of the inertial waves. The following relationship can be established:

$$\begin{aligned} \oint_{\mathfrak{C}} \langle \mathbf{Q} \rangle \cdot \mathbf{ds} &= \frac{2h(r)}{(4-\alpha^2)} \frac{\partial \varphi_0}{\partial h} \oint_{\mathfrak{C}} |\mathbf{n}_T \times \mathbf{n}_B| \, ds \\ &= \frac{2}{i\alpha} \int (|\mathbf{n}_T|\, F_T + |\mathbf{n}_B|\, F_B)\, d\Sigma_P. \end{aligned}$$

To obtain the amplitudes of the inertial modes in a symmetric container, an infinite system of linear equations must be solved:

$$A_j d_j + \frac{1}{2i}\sum_{m \neq j} A_m \frac{(\alpha-\lambda_m)(\alpha-\lambda_j)}{\lambda_m+\lambda_j}\int \Phi_j^\dagger \hat{\mathbf{n}}\cdot\hat{\mathbf{k}}\times\nabla\Phi_m \, d\Sigma$$

$$= i\alpha\left(1-\frac{\alpha^2}{4}\right)\int \Phi_j^\dagger F \, d\Sigma, \quad (2.13.19)$$

for $j = 1, 2, \ldots$, where

$$d_j = \int dV\left[\frac{\alpha^2-\lambda_j^2}{4}|\nabla\Phi_j|^2 + \frac{i}{2}(\alpha-\lambda_j)\nabla\Phi_j\cdot\nabla\Phi_j^\dagger\times\hat{\mathbf{k}}\right].$$

By utilizing the representation

$$\Phi_j = \Psi_j(r, z)\, e^{ij\theta},$$

the equations reduce to relationships only among those modes having the same index j. This can mean that the system reduces to finite order as it indeed does when the container is a spheroid. The particular calculation for the sphere is given in the next section and the details for the spheroid are found in [100]. Stewartson and Roberts [166, 192] analysed the same problem by transform methods.

A solution completely in terms of the natural modes may not be the best or most suitable representation for every rotating fluid motion. Since the inviscid equations are, after all, hyperbolic, contained flows can and do exhibit weak spatial 'discontinuities' in the fluid interior. (Witness the flow pattern of fig. 1.3.) It is well known that the Fourier synthesis of a function with discontinuities is very inefficient. Wood [226] studied the inviscid limit of the viscous problem for the *non-resonant* precession of a cylinder. The interior flow is shown to be weakly discontinuous across characteristics which emanate from the corner regions and the pattern formed is extremely sensitive to the dimensions of the container. Convergence of the modal series must be equally sensitive.

2.14. Resonance in a sphere

The theory of the last section can be illustrated by considering the resonant oscillations in a sphere, Greenspan [73].

Suppose the sphere moves about its fixed centre in a definite manner but that this motion is viewed from a *co-ordinate system which rotates with uniform angular velocity* $\mathbf{\Omega}$ as described on p. 75. The spherical shell appears then to be slipping on the surface of the contained fluid mass. The boundary layers on the container surface excite the internal modes by means of a small induced mass flux and the boundary condition on Σ for the inviscid interior velocity has the general form

$$\mathbf{q}\cdot\hat{\mathbf{n}} = \mathrm{ru}+\mathrm{zw} = \mathrm{F}(\theta, \Theta, \mathrm{t}), \tag{2.14.1}$$

where Θ is the polar angle. The function $\mathrm{F}(\theta, \Theta, \mathrm{t})$ is made known by a boundary layer analysis that relates the induced flux to the prescribed container motion. It is sufficient to consider the response to an excitation of the form

$$\mathrm{F}_k(\Theta)\, \mathrm{e}^{\mathrm{i}(k\theta+\alpha t)};$$

the arbitrary function in (2.14.1) can be constructed from a complete superposition of these individual components.

Let the velocity of the spherical surface be

$$\mathscr{U} = \mathscr{U}_{\mathrm{w}}(\mathrm{r}, \mathrm{z}) \exp \mathrm{i}(k\theta + \alpha\mathrm{t}). \tag{2.14.2}$$

The corresponding boundary layer velocity is calculated from (2.9.4) to be

$$\begin{aligned}\tilde{\mathbf{q}} &= \tfrac{1}{2}(\mathscr{U} - \mathrm{i}\hat{\mathbf{r}} \times \mathscr{U}) \exp\{-(\mathrm{i}[\alpha + 2\cos\Theta])^{\frac{1}{2}}\zeta\} \\ &\quad + \tfrac{1}{2}(\mathscr{U} + \mathrm{i}\hat{\mathbf{r}} \times \mathscr{U}) \exp\{-(\mathrm{i}[\alpha - 2\cos\Theta])^{\frac{1}{2}}\zeta\}\end{aligned} \tag{2.14.3}$$

and the induced mass flux is

$$\mathrm{F}_k(\Theta) \exp \mathrm{i}(k\theta + \alpha\mathrm{t}) = -\mathrm{E}^{\frac{1}{2}} \int_0^\infty \hat{\mathbf{n}}\cdot\nabla\times(\hat{\mathbf{n}}\times\tilde{\mathbf{q}})\, d\zeta. \tag{2.14.4}$$

The flux is $O(\mathrm{E}^{\frac{1}{2}})$ because the interaction between the container and the fluid is purely viscous. With $\mathrm{F}_k(\Theta)$ identified, the particular form of (2.13.19) can be written explicitly using the formula for the modes, (2.12.7):

$$\Phi_{nmk} = \Psi_{nmk}\, \mathrm{e}^{\mathrm{i}k\theta} = \mathrm{P}_n^{|k|}(\eta/\mathrm{c}_{nmk})\, \mathrm{P}_n^{|k|}(\mu)\, \mathrm{e}^{\mathrm{i}k\theta}.$$

The complete triple index notation is used; for definite values of n and k, index m varies between one and an integer $\mathcal{N}_{nk}$, which represents the total number of acceptable eigenvalues. The eigenfunctions corresponding to these indices all reduce to the same zonal harmonic on the surface of the sphere. This enables the following surface integral to be evaluated simply:

$$\int_0^\pi \Psi_{nmk}\Psi_{n'm'k} \sin\Theta \, d\Theta$$
$$= \frac{2}{2n+1}\frac{(n+|k|)!}{(n-|k|)!} P_n^{|k|}(\tfrac{1}{2}\lambda_{nmk}) P_n^{|k|}(\tfrac{1}{2}\lambda_{n'm'k}) \delta_{nn'}.$$

With this information, (2.13.19) is expressible as (see [73])

$$A_{nmk}\, d_{nmk} - 4k \sum_{\substack{j=1 \\ j\neq m}}^{\mathcal{N}_{nk}} A_{njk} \frac{(\alpha-\lambda_{njk})(\alpha-\lambda_{nmk})}{\alpha^2(\lambda_{njk}+\lambda_{nmk})} \frac{(n+|k|)!}{(n-|k|)!}$$
$$\times \frac{P_n^{|k|}(\tfrac{1}{2}\lambda_{njk}) P_n^{|k|}(\tfrac{1}{2}\lambda_{nmk})}{(2n+1)} = -\frac{(4-\alpha^2)}{i\alpha}\int_0^\pi F_k(\Theta)\Psi_{nmk}\sin\Theta\, d\Theta,$$
$$m = 1, \ldots, \mathcal{N}_{nk}, \quad (2.14.5)$$

where $$d_{nmk} = 4\left(\frac{1}{\lambda_{nmk}^2} - \frac{1}{\alpha^2}\right)\iint\left(\frac{\partial}{\partial z}\Psi_{nmk}\right)^2 r\, dr\, dz$$
$$+ 2k\left(\frac{1}{\alpha} - \frac{1}{\lambda_{nmk}}\right)\int_0^\pi \Psi_{nmk}^2 \sin\Theta\, d\Theta. \quad (2.14.6)$$

There are then just $\mathcal{N}_{nk}$ equations in the $\mathcal{N}_{nk}$ unknowns, A_{nmk}, $m = 1, \ldots, \mathcal{N}_{nk}$ and the system is exactly soluble. The amplitude coefficients are determined in a straightforward manner.

Upon inspection of these formulas, we conclude that resonance occurs whenever the excitation frequency, α, is a characteristic value, λ_m. However, viscous processes maintain a finite amplitude response at all times. The appropriate viscous corrections may be included in all the preceding equations simply by replacing λ_m, the inviscid eigenvalue, by the more precise value, $-i\sigma_m = \lambda_m - iE^{\frac{1}{2}}\sigma_{m,1}$.

In the case of axially symmetric motion, $k = 0$, the solution is particularly simple because the equations (2.14.5) uncouple. The experiments of Aldridge and Toomre[1] utilize an axially symmetric oscillation to resonate the interior fluid and the amplitude responses in this situation can now be completely determined.

As a concrete illustration, let the motion of the boundary be

$$\mathcal{U} = \hat{\mathbf{k}} \times \mathbf{r}\, e^{i\alpha t} = \sin\Theta\; e^{i\alpha t}\,\hat{\boldsymbol{\theta}}, \quad (2.14.7)$$

which is, of course, entirely in the circumferential direction. The first step is to determine the normal flux into the Ekman layer established by boundary motion and this computation is similar to that in § 2.9. Since $k = 0$, the modal amplitudes can be calculated separately (see also (2.13.17)). The subscript notation may be dropped; ψ can then represent an arbitrary mode, A, its amplitude, and λ, the corresponding eigenvalue. Hence, (2.14.5) becomes

$$A = -2^{-\frac{3}{2}}\left(1-\frac{\alpha^2}{4}\right)\frac{\alpha\lambda^2 E^{\frac{1}{2}}}{(\alpha+\lambda)(\alpha-\lambda)}\frac{I_1}{I_2}, \tag{2.14.8}$$

where $$I_1 = \int_0^\pi d\Theta \frac{\partial \Psi}{\partial \Theta}\sin^2\Theta\left\{|\alpha+2\cos\Theta|^{-\frac{1}{2}} - |\alpha-2\cos\Theta|^{-\frac{1}{2}} - i\left[\frac{\alpha+2\cos\Theta}{|\alpha+2\cos\Theta|^{\frac{3}{2}}} - \frac{\alpha-2\cos\Theta}{|\alpha-2\cos\Theta|^{\frac{3}{2}}}\right]\right\}, \tag{2.14.9}$$

$$I_2 = \iint\left(\frac{\partial\Psi}{\partial z}\right)^2 r\,dr\,dz.$$

The last integration is over the volume of the unit sphere.

Consider the resonance of the mode

$$\Psi = \Psi_{410} = z^2 - \tfrac{1}{2}z^4 - 2r^2z^2 + \tfrac{2}{3}r^2 - \tfrac{1}{3}r^4$$

for which $\lambda_{410} = 1{\cdot}309$, $\mathfrak{s}_{410,1} = -3{\cdot}38 + 0{\cdot}434i$. To compute the resonant amplitude, the excitation frequency is taken to be

$$\alpha = \lambda_{410}$$

and λ in (2.14.8) is *replaced* by the complete eigenvalue $-i\mathfrak{s}_{410}$; i.e.,

$$-i\mathfrak{s}_{410} = \lambda_{410} - iE^{\frac{1}{2}}\mathfrak{s}_{410,1}.$$

The integrations are elementary and the result is

$$A = A_{410} = 0{\cdot}452 + 0{\cdot}055i.$$

The results of Aldridge and Toomre [1] show that the theoretical and experimental values for the pressure amplitude are in excellent agreement as the value of the Rossby number is reduced towards zero. Experimental values are always less than those predicted, but the discrepancies do not require further attention.

2.15. Motion in a cylinder

The cylindrical container provides another excellent illustration of the general theory in the special case (see § 2.11) in which the total height, h, is everywhere constant, i.e., $h = 1$. This problem was first studied by Kelvin[95], later by Bjerknes and Solberg[9] and by many others since then.

Geostrophic motion in this geometry is two-dimensional, as always, with

$$\mathbf{q}_0 = \tfrac{1}{2}\hat{\mathbf{k}} \times \nabla\varphi_0. \qquad (2.15.1)$$

The governing equation, (2.11.5), reduces to

$$\nabla^2\varphi_0 = 2\hat{\mathbf{k}}\cdot\nabla\times\langle\mathbf{q}_*\rangle \qquad (2.15.2)$$

and the boundary condition at the lateral surface, $r = a$, is $(\partial/\partial\theta)\,\varphi_0 = 0$ or simply

$$\varphi_0(a, \theta) = 0. \qquad (2.15.3)$$

Rotational constraints on motion of this type are considerably relaxed. Any surface contour is a geostrophic contour and the nature of the geostrophic flow depends entirely on the initial setting.

The boundary value problem for the inviscid inertial oscillations, $\Phi\, e^{i\lambda t}$, consists of equation (2.12.4) once again with the particular boundary conditions

$$\left.\begin{aligned} \frac{\partial\Phi}{\partial z} &= 0 \quad \text{on} \quad z = 0, z = 1; \\ i\lambda\frac{\partial\Phi}{\partial r} + \frac{2}{r}\frac{\partial\Phi}{\partial\theta} &= 0 \quad \text{on} \quad r = a. \end{aligned}\right\} \qquad (2.15.4)$$

Solutions of this eigenvalue problem, determined by the method of separation of variables, are

$$\Phi_{nmk}(r, \theta, z) = J_{|k|}(\xi_{nmk}\, r/a)\cos(n\pi z)\exp(ik\theta), \qquad (2.15.5)$$

for $k = 0$ and $n, k = \pm 1, \pm 2, \ldots;$

$$\lambda_{nmk} = 2(1 + \xi^2_{nmk}/n^2\pi^2a^2)^{-\frac{1}{2}} \qquad (2.15.6)$$

and ξ_{nmk} is the mth *positive* solution of the transcendental equation

$$\xi\frac{d}{d\xi}J_{|k|}(\xi) + k(1 + \xi^2/n^2\pi^2a^2)^{\frac{1}{2}}J_{|k|}(\xi) = 0. \qquad (2.15.7)$$

The negative eigenvalues are obtained using the fact that $-\lambda$ corresponds to $-k$ if λ corresponds to k so that the calculation of all positive eigenvalues completely determines the spectrum. The same value of λ does not in general correspond to both $\pm k$ and the modal solutions represent propagating waves.

Some of the remarks made in §2.11 can now be substantiated.

For example, from the nature of the eigenvalue equation, it is not difficult to show that the inviscid spectrum is denumerable but dense in the interval $|\lambda| \leqslant 2$. This conclusion does not pertain to the real viscous spectrum because the basic concept of the boundary layer theory, that each mode is separable into an inviscid component and a boundary layer perturbation, fails when the effective wave length of the oscillation is of the order of the boundary layer thickness, $E^{\frac{1}{2}}$. The error made by neglecting viscosity in the interior can be estimated roughly upon substituting

$$p = J_0(\xi r/a)\, e^{i(n\pi z+\lambda t)}$$

into the exact pressure equation (2.2.2) and by then determining λ in terms of ξ. It follows that

$$\lambda = 2(1+\xi^2/n^2\pi^2a^2)^{-\frac{1}{2}} + iE(n^2\pi^2+\xi^2/a^2).$$

The first term is essentially the inviscid eigenvalue; the correction is of a comparable magnitude when the modal wave number, κ, is large,

$$\kappa = (n^2\pi^2+\xi^2/a^2)^{\frac{1}{2}} \doteq E^{-\frac{1}{2}}.$$

Hence, viscous effects are very important when the modal wave *length* is small. The correction indicates that true eigenvalue positions are displaced from the real axis, approaching $i\infty$ as ξ or n becomes large. Therefore, a dense spectrum arises because

$$\lim_{\xi,\, n\to\infty}\ \lim_{E\to 0} \neq \lim_{E\to 0}\ \lim_{\xi,\, n\to\infty}.$$

The inviscid approximation, $E = 0$, locates only the projection of the exact eigenvalue position onto the real axis and it is the projected positions that constitute a dense set. Similar conclusions undoubtedly hold for all configurations. The situation in the neighbourhood of the origin, $\lambda = 0$, has not been fully explored.

An experiment by Fultz[60], designed to resonate the axially symmetric modes, $k = 0$, yielded measurements of frequency in remarkable agreement with theoretical predictions. Resonance was produced by varying the frequency of a small oscillating disk, placed strategically within the cylinder. Several criteria for resonance were employed, although the maximum response was quite sensitive to frequency changes and could be located very accurately. The findings are presented in the accompanying table. Figs. 2.11 and 2.12 show typical modal oscillations.

Table 2.2

Mode (n, m, k)	$a = \dfrac{\text{radius}}{\text{height}}$	Eigenvalue	
		Theory	Experiment
(1, 1, 0)	2·0	1·708	1·754
	0·667	0·959	0·966
	0·4	0·623	0·617
(1, 2, 0)	2·0	1·487	1·351
	0·667	0·572	0·581
	0·286	0·254	0·253
(1, 3, 0)	2·0	1·051	1·052
	1·0	0·590	0·595
	0·4	0·245	0·243
(1, 4, 0)	0·5	0·234	0·234
(2, 1, 0)	2·0	1·913	1·923
	0·4	1·096	1·099
(2, 2, 0)	0·667	1·025	1·020
	0·4	0·674	0·676
(3, 1, 0)	1·0	1·852	1·887
	0·4	1·402	1·408

Unfortunately, no measurements were made by Fultz of the viscous decay rate nor have any appeared since. The theoretical boundary layer corrections due to the top and bottom walls of the container can be determined as in previous sections. However, the boundary layer at the sidewall has a radically different structure from the Ekman layer because it evolves from a more subtle balance of forces than that of Coriolis force versus pressure gradient. Vertical shear layers are studied at great length in § 2.18; their role in the viscous dissipation process has only recently been examined by Kudlick [100]. The spin-up time still characterizes the dissipation rate of both geostrophic and inertial modes. However, as the radius of the cylinder is increased, the dissipation time for the *inertial* modes becomes progressively larger and ultimately is E^{-1} for the configuration of infinite concentric disks. This accounts for the three phases of spin-up in that idealized geometry, the last at $t \doteq E^{-1}$ being the viscous decay of residual modal oscillations. Fig. 2.13 shows the eigenvalue and decay factor versus a for the first mode $(1, 1, 0)$.

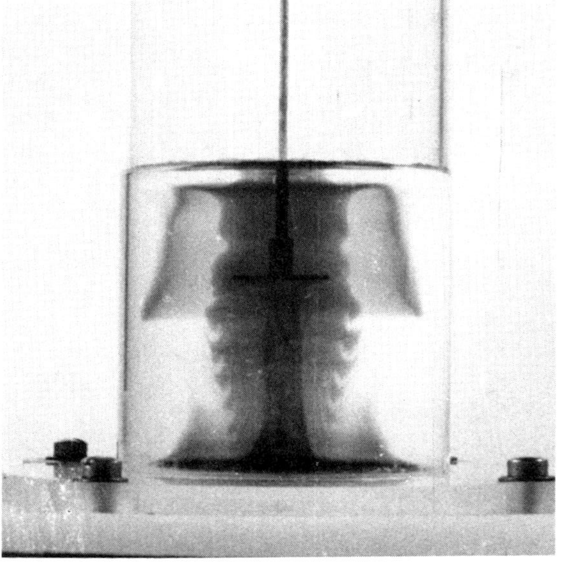

Fig. 2.11. Cylindrical axisymmetric inertial mode (2, 1, 0); photographed by Fultz [60]. Rotation period—4·04 sec.; period of disk oscillation—3·189 sec.; tank diameter and height—8·23 cm. and 8·25 cm; disk range—2 cm.

Fig. 2.12. Cylindrical axisymmetric inertial mode (3, 1, 0). Rotation period—4·055 sec.; period of disk oscillation—2·612 sec.; tank diameter and height—8·23 cm. and 8·25 cm; disk range—1·45 cm.

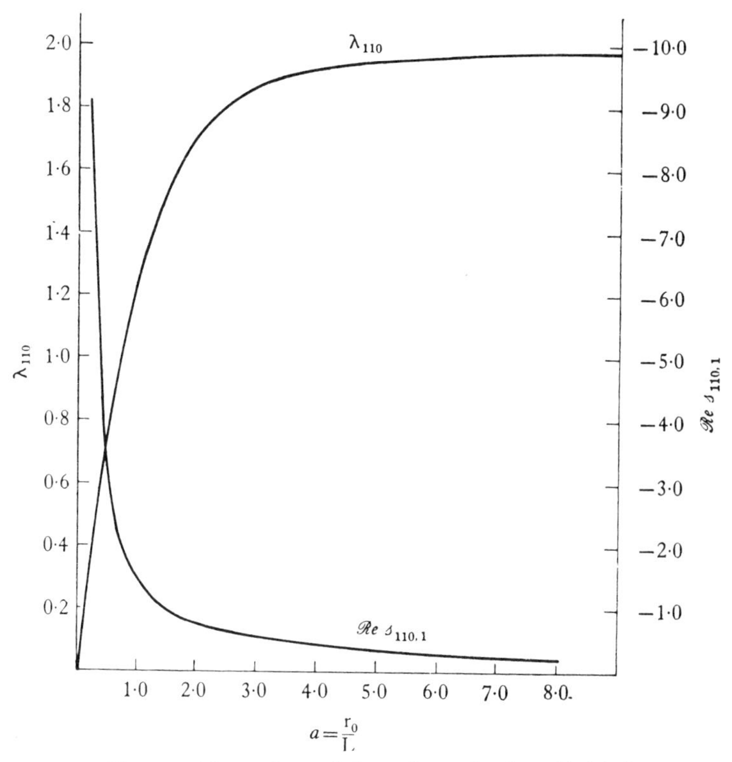

Fig. 2.13. Eigenvalue and decay factor for the cylindrical mode (1, 1, 0) versus the radius of the tank.

2.16. Rossby waves: part one

The general theory applies to any container that is a smooth envelope of closed geostrophic contours. However, it is easy to construct shapes for which there are no such closed curves of constant height and the question then immediately arises as to how the theory should be modified. Significant new effects appear and these concern us now.

Consider the container shown in fig. 2.14 (which shall be called a 'sliced cylinder') where the inclination angle, α, is small but $\alpha \gg E^{\frac{1}{4}}$. For definiteness, let the top surface be $z = 1$, and

$$z = y \tan \alpha$$

be the equation of the base plate. There are no closed contours of constant height in this configuration and no geostrophic mode is possible. However, geostrophic motion in the special case of the right circular cylinder, $\alpha = 0$, is quite arbitrary and we are faced with a problem in which a slight difference of geometry means a reduction of an infinity of possible flows to none.

It will be shown by means of a perturbation analysis in α that an infinite number of new low frequency depth-independent inertial oscillations arise when $\alpha \neq 0$. These waves assume the position

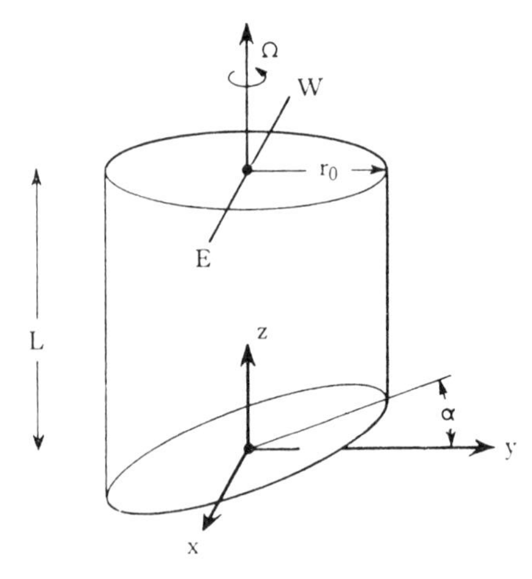

Fig. 2.14. The sliced cylinder. The shallowest region is north along the positive y axis; east and west are marked accordingly.

formerly held by the single geostrophic mode; they possess vorticity (the mean circulation theorem does not apply to this configuration) and together with the rest of the inertial modes are capable of synthesizing any initial velocity distribution.

Waves of this type were examined by Rossby[174] in a geophysical context and bear his name. The particular problem under discussion now was studied by Pedlosky and Greenspan[148].

Once again, solutions of the inviscid *dimensionless* equations,

$$\left.\begin{aligned} \frac{\partial}{\partial t}\mathbf{q} + 2\hat{\mathbf{k}} \times \mathbf{q} &= -\nabla p, \\ \nabla \cdot \mathbf{q} &= 0, \end{aligned}\right\} \tag{2.16.1}$$

satisfying the boundary condition $\mathbf{q} \cdot \hat{\mathbf{n}} = 0$ on the surface Σ are sought of the form

$$\mathbf{q} = \mathbf{Q}\, e^{i\lambda t}, \quad p = \Phi\, e^{i\lambda t}. \tag{2.16.2}$$

As might be expected, each of the inviscid inertial modes determined in the previous section has its counterpart in the altered container shape that is but slightly modified by the small inclination angle α. Discussion of these eigenfunctions is omitted.

Our concern is with the appearance of low frequency inertial waves, all of which degenerate into the general geostrophic flow when α is zero. To this end, set $E = 0$ and let

$$\left.\begin{aligned} \lambda &= \alpha\lambda^{(1)} + \alpha^2\lambda^{(2)} + \ldots, \\ \mathbf{Q} &= \mathbf{Q}^{(0)} + \alpha\mathbf{Q}^{(1)} + \ldots, \\ \Phi &= \Phi^{(0)} + \alpha\Phi^{(1)} + \ldots. \end{aligned}\right\} \tag{2.16.3}$$

Upon substituting these expansions into (2.16.1), it follows that

$$\left.\begin{aligned} 2\hat{\mathbf{k}} \times \mathbf{Q}^{(0)} &= -\nabla\Phi^{(0)}, \\ \nabla\cdot\mathbf{Q}^{(0)} &= 0, \end{aligned}\right\} \tag{2.16.4}$$

with $$\mathbf{Q}^{(0)}\cdot\hat{\mathbf{k}} = 0 \quad \text{on } z = 0,\ z = 1$$

and $$\mathbf{Q}^{(0)}\cdot\hat{\mathbf{r}} = 0 \quad \text{on } r = a.$$

The next order problem is

$$\left.\begin{aligned} i\lambda^{(1)}\mathbf{Q}^{(0)} + 2\hat{\mathbf{k}} \times \mathbf{Q}^{(1)} &= -\nabla\Phi^{(1)}, \\ \nabla\cdot\mathbf{Q}^{(1)} &= 0, \end{aligned}\right\} \tag{2.16.5}$$

with conditions $$\mathbf{Q}^{(1)}\cdot\hat{\mathbf{k}} = 0 \quad \text{on } z = 1,$$

$$-\left(y\frac{\partial}{\partial z}\mathbf{Q}^{(0)} + \mathbf{Q}^{(1)}\right)\cdot\hat{\mathbf{k}} + \mathbf{Q}^{(0)}\cdot\hat{\mathbf{j}} = 0 \quad \text{on } z = 0 \tag{2.16.6}$$

and $$\mathbf{Q}^{(1)}\cdot\hat{\mathbf{r}} = 0 \quad \text{on } r = a.$$

It follows from (2.16.4) that

$$\mathbf{Q}^{(0)} = \tfrac{1}{2}\hat{\mathbf{k}} \times \nabla\Phi^{(0)}, \quad \frac{\partial\Phi^{(0)}}{\partial z} = 0 \tag{2.16.7}$$

and $$\Phi^{(0)} = 0 \quad \text{on } r = a.$$

The primary velocity is independent of height and geostrophic. In fact, to this order, the solution is *any one* of the possible geostrophic motions in the right circular cylinder. Consideration of the next problem of the sequence, (2.16.5) and (2.16.6), is necessary to specify it exactly.

The order α velocity can be expressed in terms of the pressure function $\Phi^{(0)}$ by taking the curl of the momentum equation in (2.16.5) and integrating it with respect to z. The result is

$$\mathbf{Q}^{(1)} = (\tfrac{1}{4}i\lambda^{(1)}(z-1)\,\nabla^2\Phi^{(0)})\,\hat{\mathbf{k}} + \mathbf{A}(x,y), \tag{2.16.8}$$

where $\mathbf{A}$ is arbitrary. However, since $\mathbf{Q}^{(1)}\cdot\hat{\mathbf{k}} = 0$ on $z = 1$,

$$\mathbf{A}\cdot\hat{\mathbf{k}} = 0.$$

A single equation for $\Phi^{(0)}$ is obtained by substituting (2.16.8) for $\mathbf{Q}^{(1)}$ in (2.16.6), the boundary condition at the inclined surface:

$$\nabla^2\Phi^{(0)} + \frac{2}{i\lambda^{(1)}}\frac{\partial}{\partial x}\Phi^{(0)} = 0, \tag{2.16.9}$$

with $\Phi^{(0)} = 0$ on $r = a$. This is the equation studied by Rossby[174].

The solutions of this eigenvalue problem are

$$\left.\begin{aligned} \Phi^{(0)}_{mn} &= J_m\!\left(\xi_{mn}\frac{r}{a}\right)\exp i\left(m\theta + \xi_{mn}\frac{r}{a}\cos\theta\right),\\ \lambda^{(1)}_{mn} &= a/\xi_{mn}, \end{aligned}\right\} \tag{2.16.10}$$

where $J_m(\xi_{mn}) = 0$ and m ranges over all integer values, positive and negative. These waves have no vertical structure, they possess mean circulation and completely replace the geostrophic mode. In addition, they satisfy all properties (such as orthogonality) that the other inertial waves do.

The situation is entirely similar for any shaped lower surface for which closed geostrophic contours do not exist as long as the deviation from a level plane is small. If

$$z = -\alpha\,\mathfrak{d}(x,y)$$

is the bottom boundary, then the equation for the pressure amplitude corresponding to (2.16.9) is

$$\nabla^2\Phi^{(0)} - \frac{2}{i\lambda^{(1)}}\nabla\mathfrak{d}\cdot(\hat{\mathbf{k}}\times\nabla\Phi^{(0)}) = 0. \tag{2.16.11}$$

Although the boundary value problem becomes more difficult, the general conclusions remain the same. If the container has no closed geostrophic contours, then Rossby waves arise to compensate for the loss of a steady geostrophic flow. If the container has closed geostrophic contours, Rossby waves may still be possible, but they do not then possess mean circulation. The residual geostrophic mode contains all the mean circulation and the Rossby modes have the same character as the regular inertial modes. For example, if $\mathfrak{d} = \sqrt{(x^2+y^2)}$, then circles of constant r are geostrophic contours. A geostrophic mode exists but Rossby wave modes are also present:

$$\Phi^{(0)}_{mn}\, e^{i\alpha\lambda^{(1)}_{mn}t} = J_{2m}\left(\eta_{mn}\sqrt{\frac{r}{a}}\right) \exp i(m\theta + \alpha\lambda^{(1)}_{mn}\, t),$$

where $$J_{2m}(\eta_{mn}) = 0 \quad \text{and} \quad \lambda^{(1)}_{mn} = \frac{8ma}{\eta^2_{mn}}.$$

In this case, Rossby waves propagate along geostrophic contours but have no net circulation around these closed curves unless $m = 0$. However, when $m = 0$, $\lambda^{(1)}_{mn} = 0$ and the solution degenerates into one of the infinite number of purely geostrophic modes.

Spin-up in the sliced cylinder is an interesting example [148], in which fluid rotates rigidly at time zero, and the rotation rate of the container is *increased* impulsively. The initial state, as viewed from the *faster rotating* co-ordinate system, excites all the Rossby waves and for that matter only these depth independent inertial modes (if $O(\alpha)$ effects are neglected).

Let $\mathbf{q}_*$ represent the initial state; it follows (see p. 72) that to $O(\alpha)$,

$$\left.\begin{aligned} \mathbf{q}_* &= -r\,\hat{\boldsymbol{\theta}} = \tfrac{1}{2}\hat{\mathbf{k}}\times\nabla p_* \\ \text{or} \quad \nabla p_* &= -2r\,\hat{\mathbf{r}}. \end{aligned}\right\} \tag{2.16.12}$$

The general solution of the problem in terms of the Rossby waves (velocity and pressure to order one, frequency to order α) is

$$\mathbf{q} = \sum A_{mn}\, \mathbf{Q}_{mn} \exp \frac{ia\alpha}{\xi_{mn}}\, t, \tag{2.16.13}$$

$$p = \sum A_{mn}\, \Phi_{mn} \exp \frac{ia\alpha}{\xi_{mn}}\, t, \tag{2.16.14}$$

with $$\mathbf{Q}_{mn} = \tfrac{1}{2}\hat{\mathbf{k}}\times\nabla\Phi_{mn}; \tag{2.16.15}$$

here the superscript notation is dropped. The orthogonality relationship is used to determine the Fourier coefficients from the initial velocity distribution. The calculation is straightforward and the result is

$$A_{mn} = \frac{2i^{-m}a^2}{\xi^2_{mn}}. \tag{2.16.16}$$

The basic pressure is computed with a little algebraic manipulation and it follows that

$$\frac{\mathrm{p}}{a^2} = 8 \sum_{m=0}^{\infty} \sum_{n=1}^{\infty} \frac{e_m}{\xi_{mn}^2} \mathrm{J}_m \left(\xi_{mn} \frac{\mathrm{r}}{a}\right)$$

$$\times \cos\left(\xi_{mn} \frac{\mathrm{r}}{a} \cos\theta + \frac{\alpha a t}{\xi_{mn}} - \frac{m\pi}{2}\right) \cos m\theta, \quad (2.16.17)$$

where $e_m = 1, m \neq 0$, and $e_m = \frac{1}{2}, m = 0$. Another form convenient for computation is

$$-\frac{\mathrm{p}}{a^2} = \left(\frac{\mathrm{r}}{a}\right)^2 - 1 + 16 \sum_{m=0}^{\infty} \sum_{n=1}^{\infty} \frac{e_m}{\xi_{mn}^2} \mathrm{J}_m \left(\xi_{mn} \frac{\mathrm{r}}{a}\right) \cos m\theta$$

$$\times \sin\left(\xi_{mn} \frac{\mathrm{r}}{a} \cos\theta - \frac{m\pi}{2} + \frac{\alpha a t}{2\xi_{mn}}\right) \sin \frac{\alpha a t}{2\xi_{mn}}. \quad (2.16.18)$$

The development of the flow is shown in fig. 2.15 which displays the summation of the preceding series at three times. Fig. 2.16 is a sequence of photos which correspond to the spin-down of a sliced cylinder, as illuminated by a horizontal slit beam and photographed from above. It is seen, from either of these, that a wave in the form of a vortex appears at the right and proceeds to the left side of the container, from position E to W in fig. 2.14. As time increases, viscosity dissipates the motion. In this connection, all Rossby waves decay with the same decay factor appropriate for flow in the right circular cylinder. Viscous corrections may be made in (2.16.17) or (2.16.18) by multiplying the right-hand sides by $\exp(-2\mathrm{E}^{\frac{1}{2}}\mathrm{t})$. Although the phase speed is unidirectional, the group velocity is not since the system is dispersive. There is no net energy flux from east to west.

The corresponding spin-up motion in the cylinder of unit height is, to lowest order, just a slow decay of rigid rotation,

$$\left.\begin{aligned} \mathbf{q} &= -\mathrm{r}\hat{\boldsymbol{\theta}} \exp(-2\mathrm{E}^{\frac{1}{2}}\mathrm{t}), \\ \frac{\mathrm{p}}{a^2} &= \left(1 - \frac{\mathrm{r}^2}{a^2}\right) \exp(-2\mathrm{E}^{\frac{1}{2}}\mathrm{t}). \end{aligned}\right\} \quad (2.16.19)$$

Thus, a marked change in the flow pattern occurs when the top surface is inclined slightly. The purely azimuthal flow, in the case $\alpha = 0$, is replaced by a set of vorticity-carrying waves that propagate in the east to west direction across the cylinder

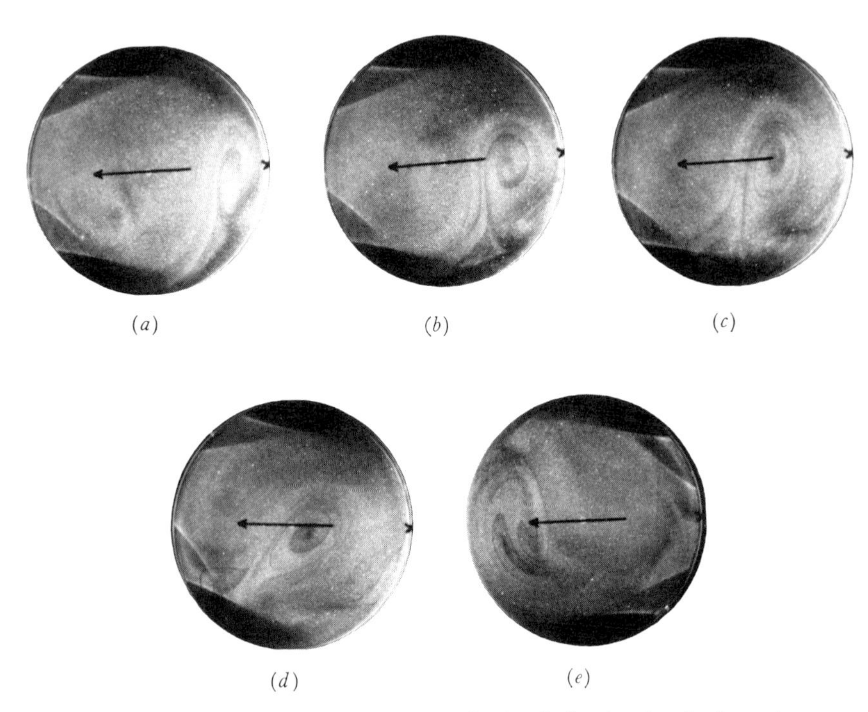

Fig. 2.16. A spin-down experiment with a sliced cylinder showing the formation and propagation of a Rossby wave in the predicted manner, see fig. 2.15.

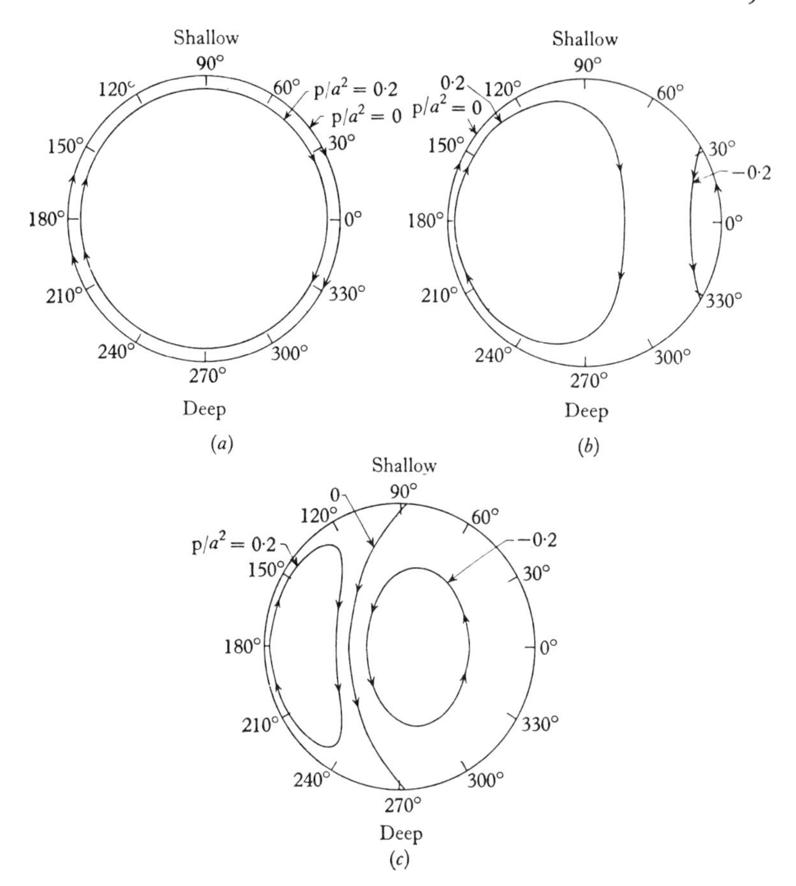

Fig. 2.15. Spin-up in the sliced cylinder. Constant pressure contours are shown at three times $\alpha at = 0, \pi, 2\pi$. The counter vortex forms at the eastern edge of the basin and propagates westward.

2.17. Steady motions and Ekman layers

Small steady motions in a rotating fluid can be produced in many ways. A simple method is to make various sections of the bounding surface rotate at slightly different angular velocities. For example, the two hemispheres of a spherical container can be made to rotate separately, likewise for the top, bottom and lateral surfaces of a cylinder. Obviously, other means of producing motion are available. Fluid may be injected through a permeable wall (§ 2.19) and it is not difficult to arrange and control temperature and salinity distributions with the same net effect. In this section, the motions discussed are produced primarily by movement of the boundaries.

The mathematical apparatus to solve these problems has already been developed in § 2.6 and need only be recapitulated at this time. If, for the present, we consider only a container whose surface area is an envelope of closed constant-height contours, then the basic steady geostrophic flow is given by (2.6.7):

$$\mathbf{q}_0 = -\frac{1}{2}\left[\frac{d}{dh}\varphi_0(h)\right]\mathbf{n}_T \times \mathbf{n}_B. \tag{2.17.1}$$

Note that the same equations govern $\mathbf{q}_1$ (i.e., $\mathbf{q} = \mathbf{q}_0 + E^{\frac{1}{2}}\mathbf{q}_1 + \ldots$) and both primary and secondary flows are independent of height, z.

The Ekman layer separates the inviscid flow from the container wall and the mass flux induced thereby drives the secondary interior motion. The structure of the boundary layer need not be explicitly calculated because only the effect on the inviscid interior is of importance. It was established in (2.6.13) that the normal flux at the surface Σ is

$$\hat{\mathbf{n}}\cdot\mathbf{q}_1 = -\tfrac{1}{2}\hat{\mathbf{n}}\cdot\nabla\times\left\{\left[\hat{\mathbf{n}}\times(\mathbf{q}_0-\mathscr{U})+\frac{\hat{\mathbf{n}}\cdot\hat{\mathbf{k}}}{|\hat{\mathbf{n}}\cdot\hat{\mathbf{k}}|}(\mathbf{q}_0-\mathscr{U})\right]|\hat{\mathbf{n}}\cdot\hat{\mathbf{k}}|^{-\frac{1}{2}}\right\}_\Sigma \tag{2.17.2}$$

and this formula relates the inviscid velocity components on Σ with the tangential wall velocity $\mathscr{U}$. The bracket notation $\{\ \}_\Sigma$ indicates that the enclosed expression is evaluated on the boundary. Another version of this relationship, convenient for certain purposes which do not utilize expansions in powers of $E^{\frac{1}{2}}$, is

$$\hat{\mathbf{n}}\cdot\mathbf{q} = -\frac{E^{\frac{1}{2}}}{2}\hat{\mathbf{n}}\cdot\nabla\times\left\{\left[\hat{\mathbf{n}}\times(\mathbf{q}-\mathscr{U})+\frac{\hat{\mathbf{n}}\cdot\hat{\mathbf{k}}}{|\hat{\mathbf{n}}\cdot\hat{\mathbf{k}}|}(\mathbf{q}-\mathscr{U})\right]|\hat{\mathbf{n}}\cdot\hat{\mathbf{k}}|^{-\frac{1}{2}}\right\}_\Sigma. \tag{2.17.3}$$

This equation is valid as long as the boundary layer is an Ekman layer and variations of $\mathbf{q}$ occur on a distance scale that is large compared to $E^{\frac{1}{2}}$. In fact, to be consistent, terms smaller than $E^{\frac{1}{2}}$ *must* all be neglected because the formula is *not* accurate beyond this degree of approximation.

The arbitrary function, $(d/dh)\varphi_0$, is determined from the boundary conditions in essentially the manner used previously, except that now $\mathbf{q}_0$ must be replaced everywhere by $\mathbf{q}_0 - \mathscr{U}$, and the motion is steady. Therefore, with $\mathscr{U} = \mathscr{U}_T$ on $z = f$, $\mathscr{U} = \mathscr{U}_B$ on $z = -g$, it follows from (2.6.20) that

$$\oint_{\mathfrak{C}} \mathbf{ds} \cdot \{(\mathbf{q}_0 - \mathscr{U}_T + \hat{\mathbf{n}}_T \times (\mathbf{q}_0 - \mathscr{U}_T))\, |\hat{\mathbf{n}}_T \cdot \hat{\mathbf{k}}|^{-\frac{1}{2}} + (\mathbf{q}_0 - \mathscr{U}_B - \hat{\mathbf{n}}_B \times (\mathbf{q}_0 - \mathscr{U}_B))\, |\hat{\mathbf{n}}_B \cdot \hat{\mathbf{k}}|^{-\frac{1}{2}}\} = 0, \quad (2.17.4)$$

whence

$$-\frac{1}{2}\frac{d\varphi_0}{dh} = \frac{1}{I(h)} \oint_{\mathfrak{C}} \mathbf{ds} \cdot \{(\mathscr{U}_T + \hat{\mathbf{n}}_T \times \mathscr{U}_T)\, |\hat{\mathbf{n}}_T \cdot \hat{\mathbf{k}}|^{-\frac{1}{2}} + (\mathscr{U}_B - \hat{\mathbf{n}}_B \times \mathscr{U}_B)\, |\hat{\mathbf{n}}_B \cdot \hat{\mathbf{k}}|^{-\frac{1}{2}}\}, \quad (2.17.5)$$

where $I(h)$ is defined in (2.6.24). The velocity $\mathbf{q}_0$ is then given by (2.17.1).

The flow component normal to constant-depth contours, $\mathbf{v}_\perp$, is determined from (2.17.2). Let

$$\mathbf{q}_1 = \mathbf{v}_\parallel + \mathbf{v}_\perp$$

and note that $(\partial/\partial z)\mathbf{q}_1 = 0$. Therefore, at $z = f$

$$\hat{\mathbf{n}}_T \cdot \mathbf{v}_\perp = \hat{\mathbf{n}}_T \cdot \mathbf{q}_1 = \tfrac{1}{2}\hat{\mathbf{n}}_T \cdot \nabla \times \{(\hat{\mathbf{n}}_T \times (\mathscr{U}_T - \mathbf{q}_0) + (\mathscr{U}_T - \mathbf{q}_0))\, |\hat{\mathbf{n}}_T \cdot \hat{\mathbf{k}}|^{-\frac{1}{2}}\}_\Sigma, \quad (2.17.6)$$

and at $z = -g$

$$\hat{\mathbf{n}}_B \cdot \mathbf{v}_\perp = \hat{\mathbf{n}}_B \cdot \mathbf{q}_1 = \tfrac{1}{2}\hat{\mathbf{n}}_B \cdot \nabla \times \{(\hat{\mathbf{n}}_B \times (\mathscr{U}_B - \mathbf{q}_0) - (\mathscr{U}_B - \mathbf{q}_0))\, |\hat{\mathbf{n}}_B \cdot \hat{\mathbf{k}}|^{-\frac{1}{2}}\}_\Sigma; \quad (2.17.7)$$

both scalar components of $\mathbf{v}_\perp$ are found from this pair of equations.

A simple problem will illustrate the procedure. Let the upper surface of an axially symmetric container, $z = f(r)$, rotate with angular velocity $(1+\varepsilon)\,\Omega$ while the lower surface, $z = -g(r)$, rotates with frequency Ω. The relative, scaled wall speed in the frame rotating with angular velocity Ω is then

$$\left.\begin{aligned} \mathscr{U}_T &= r\,\hat{\boldsymbol{\theta}}, \\ \mathscr{U}_B &= 0. \end{aligned}\right\} \quad (2.17.8)$$

The geostrophic velocity is $\mathbf{q}_0 = v_0(r)\,\hat{\boldsymbol{\theta}}$ and

$$\frac{v_0(r)}{r} = \frac{\left(1 + \left(\frac{df}{dr}\right)^2\right)^{\frac{1}{4}}}{\left(1 + \left(\frac{df}{dr}\right)^2\right)^{\frac{1}{4}} + \left(1 + \left(\frac{dg}{dr}\right)^2\right)^{\frac{1}{4}}}; \quad (2.17.9)$$

the secondary circulation is strictly in the vertical direction, $\mathbf{q}_1 = w_1(r)\hat{\mathbf{k}}$, with

$$w_1(r) = \frac{1}{2r}\frac{d}{dr}\left[r^2 \frac{\left(1+\left(\frac{df}{dr}\right)^2\right)^{\frac{1}{4}}\left(1+\left(\frac{dg}{dr}\right)^2\right)^{\frac{1}{4}}}{\left(1+\left(\frac{df}{dr}\right)^2\right)^{\frac{1}{4}}+\left(1+\left(\frac{dg}{dr}\right)^2\right)^{\frac{1}{4}}}\right]. \quad (2.17.10)$$

It is apparent that the top and bottom Ekman layers are in direct communication with each other. The local efflux from one is exactly matched at the same radial position by the mass influx to the other. This point by point balance of the two boundary layers is a consequence of the Taylor–Proudman theorem.

The formulas for the unit sphere,

$$v_0 = \frac{r}{2}, \quad (2.17.11)$$

$$w_1 = \frac{4-3r^2}{8(1-r^2)^{\frac{5}{4}}}, \quad (2.17.12)$$

show that the main body of fluid rotates at the average rate of the two hemispheres. A fluid particle emerging from the bottom Ekman layer spirals vertically upward about a cylinder of fixed radius. There is no interior radial motion and the fluid element is returned to its original position entirely within the boundary layer.

The description for the cylindrical container is virtually the same, with one important exception. There is no Ekman layer along a vertical wall and the boundary layer there has a quite different structure. Its principal function, however, is still to return the fluid from top to bottom and the net mass transport there must be the same as that carried in the horizontal boundary layers. This merely reflects the fact that the Ekman layers in this example are responsible for all the motion. (Another function of the sidewall boundary layer is to adjust the fluid velocity to that of the vertical boundary. Vertical shear layers also arise whenever the boundaries or the boundary conditions are permitted to be discontinuous.)

It is a false impression that motion within vertical layers is always of secondary importance compared with that in the Ekman layers. The nature of the flow there is of vital concern in many significant physical problems. For this reason and to complete our understanding of rotating flows, the next two sections are devoted to their analysis.

Containers having no closed geostrophic contour lines present many difficulties and the flow patterns therein are radically changed as a consequence. Consider the sliced cylinder configuration, fig. 2.14, with fluid motion produced by rotating the top disk at a slightly different speed from that of the rest of the container. Our analysis will show that the flow along constant-height contours is only $O(E^{\frac{1}{2}})$ in contrast to the unit velocity differential of the different surfaces of the container. Very little motion is communicated to the fluid by the driving mechanism.

The reason for this is that a velocity of unit magnitude would have to be directed along geostrophic contours which in this case intersect the container wall. An estimate of the applied and induced stresses indicates that there is no viscous boundary layer which can accept so large a mass transport under the circumstances assumed. In order to satisfy the condition $\mathbf{q}\cdot\hat{\mathbf{n}} = 0$ on Σ, it follows that

$$\mathbf{q}_0 \equiv 0.$$

However, the normal component of the interior velocity can be $O(E^{\frac{1}{2}})$ at impermeable walls because this is consistent with the mass flux established by horizontal and vertical layers.

The velocity component of this order that is perpendicular to the geostrophic contours can be determined directly from (2.17.6) and (2.17.7) by setting $\mathscr{U}_B = 0$, $\mathscr{U}_T = r\,\hat{\boldsymbol{\theta}}$ and $\mathbf{q}_0 = 0$. The component parallel to these constant-depth lines is obtained from the continuity equation as an arbitrary function of distance along the contour. Using the co-ordinate system introduced earlier for this particular container (p. 85), it follows that

$$\mathbf{q} = E^{\frac{1}{2}}\mathbf{q}_1 = E^{\frac{1}{2}}\left(\frac{\mathscr{X}(y)}{\sin\alpha}\hat{\mathbf{i}} + \operatorname{ctn}\alpha\,\hat{\mathbf{j}} + \hat{\mathbf{k}}\right). \qquad (2.17.13)$$

The inclination angle, α, must be large compared to $E^{\frac{1}{2}}$ for the asymptotic perturbation analysis to hold but no other restriction is required. Evidently, the changeover from the axially symmetric motion of unit magnitude produced when $\alpha = 0$, to the effective blocking that occurs for non-zero inclination is accomplished at very small values of the tilt angle, say

$$\alpha E^{-\frac{1}{2}} \doteq 1.$$

The transition regime has not been completely analysed.

The explicit determination of the function $\mathscr{X}(y)$ requires a careful analysis of the sidewall boundary layers and this appears in § 2.20. By proving that $\mathscr{X}(y) = O(1)$, we will have verified all the remarks made here.

Only minor modifications are required to solve the problem in which there is a weak mass flux through the container surface. Let

$$\mathbf{q}\cdot\hat{\mathbf{n}} = -\mathscr{S},$$

(the magnitude of the prescribed source distribution $\mathscr{S}$ is used to scale the velocity field); of course, the net transport into the container is always zero:

$$\int \mathscr{S}\, d\Sigma = 0.$$

Moreover, since the resultant interior motion is generally larger by a multiplicative factor $E^{-\frac{1}{2}}$ than the mass flux producing it, we assume that

$$\mathscr{S} = E^{\frac{1}{2}}\mathscr{M}$$

in order to assure the validity of a linear theory in all circumstances.

With these conditions, the appropriate form of (2.17.2) is

$$\mathbf{q}_1\cdot\hat{\mathbf{n}} = -\mathscr{M} - \tfrac{1}{2}\hat{\mathbf{n}}\cdot\nabla\times\left\{\left(\hat{\mathbf{n}}\times\mathbf{q}_0 + \frac{\hat{\mathbf{n}}\cdot\hat{\mathbf{k}}}{|\hat{\mathbf{n}}\cdot\hat{\mathbf{k}}|}\mathbf{q}_0\right)|\hat{\mathbf{n}}\cdot\hat{\mathbf{k}}|^{-\frac{1}{2}}\right\}. \tag{2.17.14}$$

The evaluation of this formula at the top and bottom surfaces of the container yields two equations for the velocity components normal to geostrophic contours because, to O(E), the velocity is independent of height.

The geostrophic velocity is given by

$$\mathbf{q}_0 = -\frac{2}{I(h)}\left[\int \mathscr{M}_T\, d\Sigma_T' + \int \mathscr{M}_B\, d\Sigma_B'\right]\mathbf{n}_T\times\mathbf{n}_B, \tag{2.17.15}$$

where I(h) is defined in (2.6.24) and Σ_T', Σ_B' are the sections of the top and bottom surfaces enclosed by the geostrophic contour $\mathfrak{C}$.

The last equation is readily interpreted once it is established that the mass transport across geostrophic contours in the interior can only be O(E). Since $\mathbf{q}_0$ is directed along geostrophic contours, and

$$\mathbf{q}_1 = \tfrac{1}{2}\hat{\mathbf{k}}\times\nabla\varphi_1 + w_1\hat{\mathbf{k}},$$

simple vector manipulations and integral theorems yield the desired conclusion:

$$\int (\mathbf{q}_0 + E^{\frac{1}{2}}\mathbf{q}_1)\cdot\hat{\mathbf{n}}_L \, d\Sigma_L = 0.$$

Here, Σ_L is the lateral surface of the vertical cylinder through contour $\mathfrak{C}$, $\hat{\mathbf{n}}_L$ is its unit normal ($\hat{\mathbf{n}}_L \cdot \hat{\mathbf{k}} = 0$) and Σ_L, Σ'_T, Σ'_B form a closed volume V_L.

If the surface integrals appearing in (2.17.15) do not add identically to zero, then mass conservation within V_L requires that fluid either be supplied to, or withdrawn from, the rest of the container volume. In this situation, fluid must be transported across geostrophic contours, the lines of constant total height h. However, this cannot be done by means of an inviscid, interior circulation according to the arguments set forth in the preceding paragraph. Hence, we conclude that the necessary flux takes place entirely within the Ekman boundary layer, whose existence, in turn, requires the interior velocity field to be of unit magnitude.

If the surface integrals in (2.17.15) always cancel each other, the primary geostrophic velocity, $\mathbf{q}_0$, is zero since there is no need of an Ekman layer to redistribute mass across constant-height contours. Under these conditions, the normal velocity at the wall may even be of unit magnitude because the interior velocity is of comparable size.

Source-sink flows are reconsidered in §2.19.

2.18. Vertical boundary layers

The steady sidewall boundary layer (or free shear layer) evolves from a different and more delicate balance of forces than that responsible for the Ekman layer. The result is a highly complex structure which consists of two intense transition regions, one inside the other. Abrupt changes and discontinuities in the azimuthal velocity component are made smooth and continuous across the outer viscous layer whose scaled thickness is approximately $E^{\frac{1}{4}}$. An inner layer, of thickness $E^{\frac{1}{3}}$, permits an $O(E^{\frac{1}{2}})$ vertical mass flux and is often the means by which mass is transported from one Ekman layer to another.

A general inspection of the basic equations and a comparison of the magnitudes of various terms can reveal a great deal about vertical shear layers. Usually, such a scaling analysis furnishes the motivation for a more formal approach, but in the present case, it must be said that the reverse is nearer to the truth. The basic physical processes are by no means obvious and the insight supplied by a precise analysis seems to be an essential ingredient in an 'order of magnitude' discussion. Be that as it may, the dimensional analysis of the equations of motion is discussed first.

Consider a rather general state of steady fluid flow in a cylindrical container whose different surfaces can rotate independently. The fundamental equation for the reduced pressure, (2.2.2), is

$$E^2(\nabla^2)^3\,p + 4\frac{\partial^2 p}{\partial z^2} = 0. \tag{2.18.1}$$

Obviously, the two terms of this equation are comparable only in regions where rapid changes of pressure occur. Near a horizontal boundary, say $z = 0$, a stretched variable may be introduced, $\zeta_1 = E^{-a}z$, so that $\nabla^2 \simeq E^{-2a}\,\partial^2/\partial\zeta_1^2$; (2.18.1) is then approximately

$$E^{2-6a}\frac{\partial^6 p}{\partial\zeta_1^6} + 4E^{-2a}\frac{\partial^2 p}{\partial\zeta_1^2} = 0.$$

A balance of terms requires $a = \frac{1}{2}$, which is, of course, the value appropriate for the Ekman layer at such surfaces. Other possibilities exist. At the cylindrical wall, $r = 1$, set $\zeta_2 = E^{-a}(1-r)$, whence, to lowest order, (2.18.1) becomes

$$E^{2-6a}\frac{\partial^6 p}{\partial\zeta_2^6} + 4\frac{\partial^2 p}{\partial z^2} = 0. \tag{2.18.2}$$

The equality of terms now requires $a = \frac{1}{3}$, which indicates that a vertical boundary layer of thickness $E^{\frac{1}{3}}$ is a distinct possibility. Additional information can be obtained by scaling the original equations of motion themselves with the purpose of making each new dependent variable of unit magnitude. However, since complete generality really presents unnecessary complications at this stage of the investigation, an additional restriction is added. Discussion is limited to axially symmetric configurations and all functions are assumed to be independent of the azimuthal angle θ.

If the dependent variables are scaled according to the rule

$$(u(r,z), v, w, p) \Rightarrow (E^b\tilde{U}(\zeta_2, z), E^c\tilde{V}, E^d\tilde{W}, E^e\tilde{P}),$$

then upon substituting these expressions into the basic equations (2.2.1) and retaining only the dominant terms, it follows that

$$\left.\begin{aligned} -\underline{E^{b-a}\frac{\partial \tilde{U}}{\partial \zeta_2}} + \underline{E^{d}\frac{\partial \tilde{W}}{\partial z}} &= 0, \\ \underline{2E^{c}\tilde{V}} + \underline{E^{e-a}\frac{\partial \tilde{P}}{\partial \zeta_2}} + E^{1-2a+b}\frac{\partial^2 \tilde{U}}{\partial \zeta_2^2} &= 0, \\ \underline{2E^{b}\tilde{U}} - \underline{E^{1-2a+c}\frac{\partial^2 \tilde{V}}{\partial \zeta_2^2}} &= 0, \\ \underline{E^{e}\frac{\partial \tilde{P}}{\partial z}} - \underline{E^{1-2a+d}\frac{\partial^2 \tilde{W}}{\partial \zeta_2^2}} &= 0. \end{aligned}\right\} \qquad (2.18.3)$$

A boundary layer, of thickness $E^{\frac{1}{3}}$, carrying an $O(E^{\frac{1}{2}})$ vertical mass transport is one possibility with the index setting

$$(a, b, c, d, e) = (\tfrac{1}{3}, \tfrac{1}{2}, \tfrac{1}{6}, \tfrac{1}{6}, \tfrac{1}{2}).$$

In this case, the underlined terms above, when equated to zero, constitute the appropriate boundary layer equations. Note that the radial momentum equation expresses a balance of Coriolis force with the radial pressure gradient only.

Since $v = O(E^{\frac{1}{6}})$, a boundary layer of this type cannot be fitted to an arbitrary outside flow for which this component is of unit magnitude. For this purpose, another overall balance of terms must be sought, one which is entirely consistent with the $E^{\frac{1}{3}}$-layer. In particular, consistency requires the magnitude of the radial mass transport to remain unchanged and $O(E^{\frac{1}{2}})$ in the new layer.

Equation (2.18.2) implies that in any vertical shear layer not of thickness $E^{\frac{1}{3}}$

$$\frac{\partial^2 \tilde{P}}{\partial z^2} = 0.$$

An examination of the exponents in (2.18.3) leads to the conclusion that

$$(a, b, c, d, e) = (\tfrac{1}{4}, \tfrac{1}{2}, 0, \tfrac{1}{4}, \tfrac{1}{4})$$

is a parameter setting of a flow meeting all the requirements. The new boundary layer thickness is $E^{\frac{1}{4}}$ and the boundary layer equations in terms of the stretched co-ordinate

$$\zeta_3 = E^{-\frac{1}{4}}(1 - r)$$

are the same as those identified in (2.18.3) for the $E^{\frac{1}{3}}$-layer excepting the last equation. This is replaced by

$$\frac{\partial \tilde{P}}{\partial z} = 0.$$

(The radial shear term, $(\partial^2/\partial r^2)\,u$, is negligible in *both* boundary layers and can be safely omitted.) Although this boundary layer also has a vertical mass transport of order $E^{\frac{1}{2}}$, it *cannot* alone satisfy all the sidewall boundary conditions and thereby obviate the need of a narrower transition regime. The inherent vertical uniformity of the flow in the thicker layer implied by

$$\frac{\partial \tilde{V}}{\partial z} = 0 = \frac{\partial \tilde{U}}{\partial z},$$

in general necessitates a non-zero radial flux for small values of ζ_3 (but large ζ_2). Therefore, the $E^{\frac{1}{4}}$-layer serves to match v with the interior flow but an $E^{\frac{1}{3}}$-layer is required to satisfy the boundary conditions at the wall.

There are other possibilities. Mass injection problems in the next section give rise to boundary layers with index settings $(\frac{1}{3}, \frac{7}{12}, \frac{1}{4}, \frac{1}{4}, \frac{7}{12})$ and $(\frac{1}{3}, 0, -\frac{1}{3}, -\frac{1}{3}, 0)$. In fact, any setting of the form $(\frac{1}{3}, b, b-\frac{1}{3}, b-\frac{1}{3}, b)$ is legitimate; the value of b depends on the particular problem. In view of this, there is little more to be gained from a general scaling analysis and it is abandoned now in favour of direct analytical procedures.

Morrison and Morgan[132] used transform methods to study the slow motion of a disk in a viscous, rotating fluid. At about the same time, Proudman[155] dealt with the free shear layers formed when two concentric spheres rotate at slightly different speeds. However, the nature of vertical boundary layers is perhaps best illustrated by examining the concentric disk configuration discussed by Stewartson[188]. (The methods used here are somewhat different because both Fourier transforms and boundary layer techniques are employed simultaneously.)

Two identical parallel plates rotate uniformly about the same axis, as shown in fig. 2.17. An inner disk of each plate can spin independently although the differential between rotation rates is always kept small. Two problems are considered. The first, or symmetrical problem, corresponds to the situation in which both inner disks have an excess rotational speed, i.e., $(1+\varepsilon)\,\Omega$. In the anti-symmetrical problem, the speed of one disk is $(1+\varepsilon)\,\Omega$ and that of the other is $(1-\varepsilon)\,\Omega$.

The velocity outside the shear layers is known from earlier work in § 2.17. In the inviscid interior, the fluid rotates rigidly with the average rotation rate of the top and bottom disks.

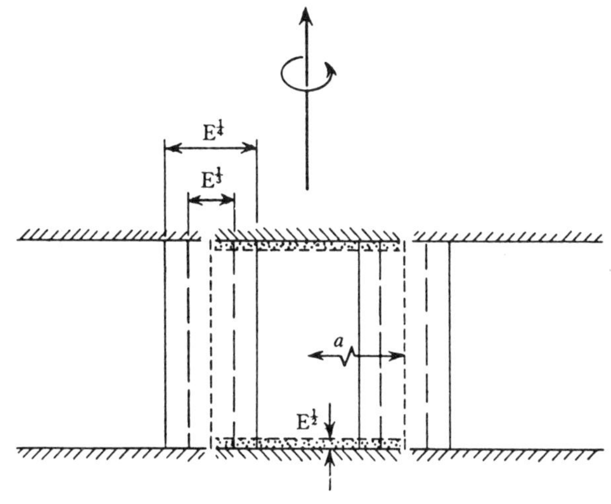

Fig. 2.17. A drawing of shear layers (in exaggerated proportion) that arise in the concentric disk configuration.

If separation distance L between plates is used as the characteristic length, $[\![L, \Omega^{-1}, \varepsilon\Omega L]\!]$, and

$$\mathbf{q} = -\nabla \times (\chi(r, z)\,\hat{\boldsymbol{\theta}}) + v(r, z)\,\hat{\boldsymbol{\theta}},\text{[1]} \tag{2.18.4}$$

then the *dimensionless* equations of motion, in the co-ordinate frame *rotating with angular speed* Ω, are

$$\left.\begin{aligned} E\mathscr{G}v - 2\frac{\partial \chi}{\partial z} &= 0, \\ E\mathscr{G}^2\chi + 2\frac{\partial v}{\partial z} &= 0, \end{aligned}\right\} \tag{2.18.5}$$

[1] The conventional definition of a stream function,

$$u = \frac{1}{r}\frac{\partial \psi}{\partial z}, \quad w = -\frac{1}{r}\frac{\partial \psi}{\partial r}, \quad \psi = r\chi,$$

is often employed. The two functions are interchangeable and the choice of one over the other is made on the basis of overall convenience in algebraic manipulation or to facilitate reference with source material.

where
$$\mathscr{G} = \frac{\partial}{\partial r}\frac{1}{r}\frac{\partial}{\partial r}r + \frac{\partial^2}{\partial z^2}.$$

The boundary conditions are

$$u = w = 0, \quad v = V_T = r\mathbb{S}(1-r/a) \qquad \text{on } z = 1,$$

$$u = w = 0, \quad v = V_B = \pm r\mathbb{S}(1-r/a) \quad \text{on } z = 0;$$

here $a = r_0/L$, and the choice of sign corresponds to the symmetrical or anti-symmetrical problems. $\mathbb{S}(x)$ is the unit step function:

$$\mathbb{S}(x) = 1, \quad x > 0,$$
$$= 0, \quad x < 0.$$

Although a direct transform attack leads to the solution easily enough, certain simplifications and definite advantages are attained by using what is already known about the Ekman layer on each plate. The structure of the Ekman layer need not be calculated again; its effect on the interior regime, in the form of an induced flux, is already given in (2.17.3). For example, at the bottom plate, $\hat{\mathbf{n}} = -\hat{\mathbf{k}}$, and the wall velocity is $\mathscr{U} = V_B\hat{\boldsymbol{\theta}}$; the normal component of the interior velocity at the wall is then

$$w_B = \tfrac{1}{2}E^{\frac{1}{2}}\frac{1}{r}\frac{\partial}{\partial r}r(v_B - V_B).$$

(The subscript serves to identify the location.) At the top, $z = 1$, the unit normal is $\hat{\mathbf{n}} = \hat{\mathbf{k}}$; hence,

$$w_T = -\tfrac{1}{2}E^{\frac{1}{2}}\frac{1}{r}\frac{\partial}{\partial r}r(v_T - V_T).$$

However,
$$w = -\frac{1}{r}\frac{\partial}{\partial r}(r\chi),$$

and these conditions can be reduced to

$$\left.\begin{aligned} \chi_B + \tfrac{1}{2}E^{\frac{1}{2}}(v_B - V_B) &= 0 \quad \text{at } z = 0, \\ \chi_T - \tfrac{1}{2}E^{\frac{1}{2}}(v_T - V_T) &= 0 \quad \text{at } z = 1. \end{aligned}\right\} \tag{2.18.6}$$

These equations provide the appropriate boundary conditions on the entire flow *outside* the Ekman layers, as long as changes in this region occur over a length scale large compared to $E^{\frac{1}{2}}$, which is certainly the case for both vertical shear layers.

The interior equations can be approximated by

$$\left.\begin{aligned} E\frac{\partial^2 v}{\partial \mathfrak{x}^2} - 2\frac{\partial \chi}{\partial z} &= 0, \\ E\frac{\partial^4 \chi}{\partial \mathfrak{x}^4} + 2\frac{\partial v}{\partial z} &= 0, \end{aligned}\right\} \tag{2.18.7}$$

where

$$\mathfrak{x} = a - r, \tag{2.18.8}$$

because all rapid vertical variations have already been accounted for in the Ekman layer calculation. Rapid changes in the radial direction can still occur but then only the most highly differentiated terms are of any significance. The variable $\mathfrak{x}$, the unstretched boundary layer co-ordinate, is allowed to range between plus and minus infinity and this approximation introduces negligible errors.

Functions v and χ will not be resolved into interior and boundary layer components (e.g., $v = v_I + \tilde{v}$) because there is no special advantage to this decomposition in the problem under discussion for the methods employed.

The anti-symmetrical problem, (2.18.7) with conditions (2.18.6) written in a form appropriate for the boundary layer region near $r = a$,

$$\chi_B + \tfrac{1}{2}E^{\frac{1}{2}}[v_B + a\mathbf{S}(\mathfrak{x})] = 0 \quad \text{on } z = 0,$$

$$\chi_T - \tfrac{1}{2}E^{\frac{1}{2}}[v_T - a\mathbf{S}(\mathfrak{x})] = 0 \quad \text{on } z = 1,$$

is solved by transform methods. It is readily deduced that

$$\frac{v}{a} = +\frac{E^{\frac{1}{2}}\mathfrak{x}}{2\pi i\,|\mathfrak{x}|}\int_{-\infty}^{\infty} \frac{e^{i|\mathfrak{x}|\xi}\sinh\{\frac{1}{2}E\xi^3(z-\frac{1}{2})\}}{[2\cosh\frac{1}{4}E\xi^3 + \xi E^{\frac{1}{2}}\sinh\frac{1}{4}E\xi^3]}\,d\xi, \tag{2.18.9}$$

$$\frac{\chi}{a} = -\frac{E^{\frac{1}{2}}}{2\pi i}\int_{-\infty}^{\infty} \frac{e^{i\mathfrak{x}\xi}}{\xi}\,\frac{\cosh\{\frac{1}{2}E\xi^3(z-\frac{1}{2})\}}{[2\cosh\frac{1}{4}E\xi^3 + \xi E^{\frac{1}{2}}\sinh\frac{1}{4}E\xi^3]}\,d\xi, \tag{2.18.10}$$

where the integration contour passes beneath the origin.

The singularities of the integrand are located at the approximate positions given by

$$\cosh\tfrac{1}{4}E\xi^3 = 0;$$

hence

$$\xi_n = \left(\frac{(4n+2)\pi i}{E}\right)^{\frac{1}{3}}$$

The residue calculation yields

$$\frac{\mathrm{v}}{a} = -\frac{2^{\frac{1}{3}}\mathrm{E}^{\frac{1}{6}}}{3\pi^{\frac{2}{3}}}\left(\frac{a-\mathrm{r}}{|a-\mathrm{r}|}\right)\sum_{n=0}^{\infty}\frac{(-1)^n \sin[\pi(2n+1)(\mathrm{z}-\frac{1}{2})]}{(2n+1)^{\frac{2}{3}}} D_n, \quad (2.18.11)$$

$$\frac{\chi}{a} = -\frac{\mathrm{E}^{\frac{1}{2}}}{2}\left\{\mathcal{S}(1-\mathrm{r}/a) - \frac{2}{3\pi}\left(\frac{a-\mathrm{r}}{|a-\mathrm{r}|}\right)\sum_{n=0}^{\infty}\frac{(-1)^n \cos[\pi(2n+1)(\mathrm{z}-\frac{1}{2})]}{2n+1} F_n\right\}, \quad (2.18.12)$$

with
$$D_n = \mathrm{e}^{-\gamma_n|a-\mathrm{r}|} - 2\mathrm{e}^{-\frac{1}{2}\gamma_n|a-\mathrm{r}|}\cos\left(\frac{\sqrt{3}}{2}\gamma_n|a-\mathrm{r}| - \frac{\pi}{3}\right),$$

$$F_n = \mathrm{e}^{-\gamma_n|a-\mathrm{r}|} + 2\mathrm{e}^{-\frac{1}{2}\gamma_n|a-\mathrm{r}|}\cos\left(\frac{\sqrt{3}}{2}\gamma_n|a-\mathrm{r}|\right),$$

and
$$\gamma_n = \left(\frac{(4n+2)\pi}{\mathrm{E}}\right)^{\frac{1}{3}}. \quad (2.18.13)$$

We see that when the central disks are counter-rotated, only an $\mathrm{E}^{\frac{1}{3}}$-thickness vertical boundary layer is produced. Ekman layers form at the top and bottom plates, but the interior azimuthal velocity is zero almost everywhere except in the neighbourhood of the discontinuity at $\mathrm{r} = a$. A single vertical layer allows for the mass transport from one plate to the other.

The situation for the symmetrical problem, in which the velocity excess of each disk is the same, is somewhat different. Here, the interior azimuthal velocity is $\mathrm{v} = \mathcal{S}(1-\mathrm{r}/a)$ and the change of unit magnitude experienced by this component across $\mathrm{r} = a$, requires an $\mathrm{E}^{\frac{1}{4}}$-layer. The solution of the transform problem, this time with the negative sign in the boundary condition, is

$$\frac{\mathrm{v}}{a} = -\frac{\mathrm{i}\mathrm{E}^{\frac{1}{2}}}{4\pi}\oint_{-\infty}^{\infty}\frac{\mathrm{e}^{\mathrm{i}\mathrm{r}\xi}\cosh[\frac{1}{2}\mathrm{E}\xi^3(\mathrm{z}-\frac{1}{2})]}{[\sinh\frac{1}{4}\mathrm{E}\xi^3 + \frac{1}{2}\mathrm{E}^{\frac{1}{2}}\xi\cosh\frac{1}{4}\mathrm{E}\xi^3]}\,d\xi.$$

The singularities of the integrand are located at positions given by

$$\sinh\tfrac{1}{4}\mathrm{E}\xi^3 \simeq 0,$$

so that in this case the roots are

$$\xi_n = \left(\frac{4n\pi\mathrm{i}}{\mathrm{E}}\right)^{\frac{1}{3}}.$$

In addition, there are distinct zeros of the denominator near the points

$$\xi_0 = \left(-\frac{2}{E^{\frac{1}{2}}}\right)^{\frac{1}{2}}.$$

Hence, the residue computation yields the following results for the velocity field beyond the Ekman layers:

$$\frac{v}{a} = \frac{r}{a}\mathcal{S}(1 - r/a) - \frac{1}{2}\frac{a-r}{|a-r|}\exp(-2^{\frac{1}{2}}|a-r|\,E^{-\frac{1}{4}})$$

$$-\frac{2^{\frac{1}{3}}E^{\frac{1}{6}}}{3}\frac{a-r}{|a-r|}\sum_{n=1}^{\infty}\frac{(-1)^n}{(2\pi n)^{\frac{2}{3}}}\cos[2n\pi(z-\tfrac{1}{2})]\,G_n; \qquad (2.18.14)$$

$$\frac{\chi}{a} = -\tfrac{1}{2}E^{\frac{1}{2}}\frac{a-r}{|a-r|}\Big\{(z-\tfrac{1}{2})\exp(-2^{\frac{1}{2}}E^{-\frac{1}{4}}\,|a-r|)$$

$$+\frac{1}{3\pi}\sum_{n=1}^{\infty}\frac{(-1)^n}{n}\sin[2n\pi(z-\tfrac{1}{2})]\,H_n\Big\}; \qquad (2.18.15)$$

with $$G_n = e^{-\beta_n|a-r|} - 2e^{-\frac{1}{2}\beta_n|a-r|}\cos\left(\frac{\sqrt{3}}{2}\beta_n|a-r| - \frac{\pi}{3}\right),$$

$$H_n = e^{-\beta_n|a-r|} + 2e^{-\frac{1}{2}\beta_n|a-r|}\cos\left(\frac{\sqrt{3}}{2}\beta_n\,|a-r|\right),$$

and $$\beta_n = \left(\frac{4n\pi}{E}\right)^{\frac{1}{3}}. \qquad (2.18.16)$$

An inspection of these formulas shows that the symmetrical flow involves *two* vertical shear layers, one to adjust the discontinuity in azimuthal velocity, the other to transport mass. (Some care is required in differentiating χ to obtain u and w.) Thus, the main features of free shear layers discussed previously are confirmed. An $E^{\frac{1}{4}}$-layer arises whenever an adjustment in azimuthal velocity is necessary and a shear layer of thickness $E^{\frac{1}{3}}$ is created to satisfy transport requirements.

The concentric sphere configuration, analysed by Proudman[155], exhibits all of these phenomena because the basic physical processes are almost unaffected by a change in geometry. However, some difficulty is encountered in the neighbourhood of a vertical tangency, the equator of the inner sphere. The exact structure of the linear boundary layer there is a very difficult question which has only recently been examined by Stewartson[191]. He finds that there is a new boundary layer of thickness $E^{\frac{1}{7}}$ in addition to layers of thicknesses $E^{\frac{1}{3}}$ and $E^{\frac{1}{4}}$. The function of this new vertical layer is to remove a singularity in the gradient of the azimuthal velocity component. The Ekman boundary layer is modified within a latitudinal belt surrounding the equator of width $E^{\frac{1}{5}}$; elsewhere, the theory is unchanged. Obviously, the situation is becoming incredibly complicated. The multiplicity of scales puts a formal expansion theory in the realm of the improbable, for such an attempt would seemingly involve powers of $E^{\frac{1}{28}}$ (and even $E^{\frac{1}{3}\cdot\frac{1}{4}\cdot\frac{1}{5}\cdot\frac{1}{7}}$).

At finite values of the Rossby number, the flow from the boundary layer on the sphere may penetrate the interior as a jet in the equatorial plane instead of turning into a free vertical shear layer. The non-linear problem for the sphere is considered in Chapters 3 and 5.

2.19. Steady motions and vertical shear layers

The theme of this section is fluid motions in which vertical shear layers are of primary importance. To begin, a few general results are deduced about steady, inviscid, rotational flows. This is followed by the detailed analysis of the boundary layers on a right circular cylinder with a surface distribution of fluid sources (Barcilon [5], Hide [83]) leading to the solution of the general problem of this type. Finally, several other problems, each of which touches upon an interesting aspect of vertical shear layers, are discussed briefly.

Consider the problem of *steady* motion within a uniformly rotating cylinder, of unit dimensionless height, whose horizontal cross-section is shown in fig. 2.18. The Taylor–Proudman theorem implies immediately that, under these conditions, the interior velocity $\mathbf{q}_I$ is independent of z to O(E) terms. Furthermore, the boundary conditions on $\mathbf{q}_I$ at the end plates are, from (2.17.3),

$$w_I = -\tfrac{1}{2}E^{\frac{1}{2}}\hat{\mathbf{k}}\cdot\nabla\times\mathbf{q}_I \quad \text{at } z = 1,$$

$$w_I = \tfrac{1}{2}E^{\frac{1}{2}}\hat{\mathbf{k}}\cdot\nabla\times\mathbf{q}_I \quad \text{at } z = 0.$$

By adding and subtracting these formulas, it becomes obvious that, to this order, the Ekman layer is non-divergent, i.e., $w_I = 0$, and

$$\hat{\mathbf{k}}\cdot\nabla\times\mathbf{q}_I = 0.$$

The last expression is equivalent to

$$\nabla\times\mathbf{q}_I = 0, \tag{2.19.1}$$

and since $\nabla\cdot\mathbf{q}_I = 0$ as well, the interior motion must be a potential flow *in the rotating co-ordinate frame*. The pressure acts like a stream

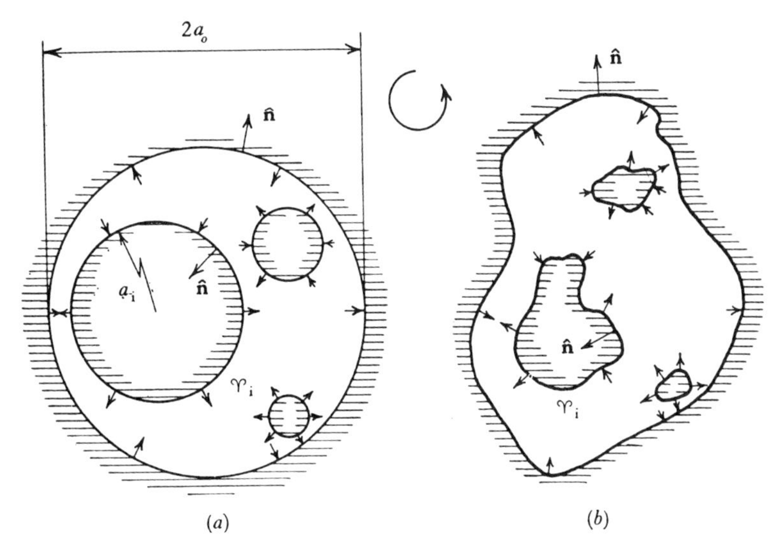

Fig. 2.18. Horizontal cross-sections of multiply-connected containers with mass effusion from the vertical walls: (*a*) Container sidewalls are all right circular cylinders. (*b*) Container sidewalls are arbitrary vertical cylinders.

function for the completely horizontal interior flow since

$$\mathbf{q}_I = \tfrac{1}{2}\hat{\mathbf{k}}\times\nabla p_I, \tag{2.19.2}$$

and it satisfies

$$\nabla^2 p_I = 0. \tag{2.19.3}$$

But p_I, unlike a stream function, is a physical variable and must be a single-valued function of position. This is equivalent to the assertion that no *net* mass transport can take place by means of an interior motion. The reasoning is as follows: the 'mass' flux to the interior from an arbitrarily shaped cylinder, fig. 2.18(*b*), is

$$\mathscr{M} = -\oint_{\Upsilon}\mathbf{q}_I\cdot\hat{\mathbf{n}}\,ds,$$

but by virtue of (2.19.2), this can be written as

$$\cdot\mathscr{M} = \frac{1}{2}\oint_{\Upsilon} \nabla p_I \cdot \mathbf{ds} = \frac{1}{2}\oint_{\Upsilon} \frac{\partial p_I}{\partial s}\, ds.$$

The last integral is zero because the integrand is an exact differential and the pressure is a single-valued function of position.

Therefore, any net transport of mass in a multiply-connected configuration must take place entirely within the vertical and horizontal boundary layers. In a simply-connected geometry, there is no

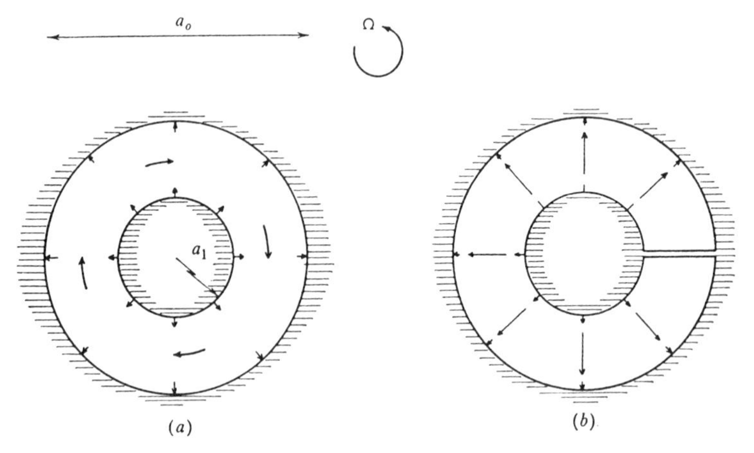

Fig. 2.19. (*a*) Horizontal cross-section of a multiply-connected container of concentric cylinders. (*b*) The same configuration made simply-connected by a radial wall.

question of a *net* flux from the whole bounding surface. The same conclusions were drawn in § 2.17 (p. 96).

The two configurations shown in fig. 2.19 have radically different flow patterns when the inner cylinder of each is a uniform source of fluid. In case (*a*), a multiply-connected cross-section, the transport is confined to the viscous layers; the interior flow is a potential vortex. Configuration (*b*) is simply-connected; the thin radial barrier supports a pressure jump and a purely radial interior flow occurs.

The nature of the interior potential motion is dictated by the boundary conditions on the walls of the container. Just as the investigation of Ekman layers resulted in equivalent boundary conditions on $\mathbf{q}_I$, the analysis of the sidewall boundary layer has the same ultimate goal. The effective two-dimensional source strength at the cylinder Σ_i in fig. 2.18(*b*) and the *circulation* about it on contour Υ_i are the important data to be obtained from a boundary layer analysis. Barcilon[5] found that if

$$\mathbf{q} = -\mathscr{S}_i\,\hat{\mathbf{n}} \quad \text{is the boundary condition on } \Sigma_i,$$

then the equivalent constraint on the interior flow at this surface is

$$\frac{\partial p_I}{\partial s} = -2\langle\mathscr{S}_i\rangle = -2\int_0^1 \mathscr{S}_i\,dz. \tag{2.19.4}$$

Moreover, it will also be shown that the interior circulation about Υ_i is prescribed by

$$\Gamma_i = \oint_{\Upsilon_i} \mathbf{q}_I\cdot\mathbf{ds} = -\frac{1}{2}\oint_{\Upsilon_i}\frac{\partial p_I}{\partial n}\,ds = -E^{-\frac{1}{2}}\oint_{\Upsilon_i}\langle\mathscr{S}_i\rangle\,ds. \tag{2.19.5}$$

The solution of (2.19.3) subject solely to condition (2.19.4) at each surface Σ_i, with

$$\sum_i \int \mathscr{S}_i\,d\Sigma_i = 0,$$

is not unique. Any harmonic function that assumes the constant value c_i on Σ_i meets the requirement of (2.19.4). (Note that a *single* constant can always be added to p_I without affecting the velocity solution.) However, the problem has a unique solution when the value of the circulation about each internal cylinder is also specified as in (2.19.5). (All but one of the constants are then determinable.)

Conformal mappings can be used to obtain the interior solution but this aspect of the problem receives little attention here.

The particular configuration shown in fig. 2.18(*a*) is examined first, since the results derived on this basis can be generalized quite easily to apply the geometry of fig. 2.18(*b*). Our methods differ from those employed in [5] and several new results are uncovered as a consequence.

Let the typical injection velocity be characterized by $\varepsilon\Omega L$ and take $[\![L, \Omega^{-1}, \varepsilon\Omega L]\!]$ as the scaling rule. The linear *dimensionless* problem is then

$$\left.\begin{aligned} \nabla\cdot\mathbf{q} &= 0,\\ 2\hat{\mathbf{k}}\times\mathbf{q} &= -\nabla p + E\nabla^2\mathbf{q}, \end{aligned}\right\} \tag{2.19.6}$$

subject to the conditions that on the surface Σ_i

$$\mathbf{q} = -\mathscr{S}_i\hat{\mathbf{n}} \quad (i=0, 1, \ldots, N); \tag{2.19.7}$$

on the end plates, $z = 0$ and $z = 1$

$$\mathbf{q} \equiv 0. \tag{2.19.8}$$

The distributed source strength $\mathscr{S}_i$ is a function of position and it is positive for local mass efflux and negative for an influx. Furthermore, overall mass conservation requires that

$$\sum_{i=0}^{N} \int \mathscr{S}_i \, d\Sigma_i = 0. \tag{2.19.9}$$

Another restriction will be necessary in order to keep the Rossby number of the interior flow truly small: the *net* mass transport from any surface is assumed to be $O(E^{\frac{1}{2}})$,

$$\int \mathscr{S}_i \, d\Sigma_i = O(E^{\frac{1}{2}}), \tag{2.19.10}$$

but the source function $\mathscr{S}_i$ itself may be of unit magnitude.

The exact viscous conditions on the end plates are replaced by equivalent constraints on the flow outside the Ekman layers in the usual manner. These are, by (2.17.3),

$$\left.\begin{aligned} w &= -\tfrac{1}{2}E^{\frac{1}{2}}\hat{\mathbf{k}}\cdot\nabla\times\mathbf{q} \quad \text{at } z = 1, \\ \text{and} \qquad w &= \tfrac{1}{2}E^{\frac{1}{2}}\hat{\mathbf{k}}\cdot\nabla\times\mathbf{q} \quad \text{at } z = 0. \end{aligned}\right\} \tag{2.19.11}$$

Consider any one of the inner cylinders and take its centre to be the origin of a cylindrical co-ordinate system. Let

$$u = u_I + \tilde{u}, \quad v = v_I + \tilde{v}, \quad w = \tilde{w}, \quad p = p_I + \tilde{p}, \tag{2.19.12}$$

where $\tilde{u}$, etc. describe the boundary layer at the vertical surface of the cylinder, $r = a$, (subscripts are unnecessary).

The equations governing the boundary layer functions in the vicinity of $r = a$ are, with $x = r - a$, approximately

$$2\tilde{v} = \frac{\partial \tilde{p}}{\partial x}, \tag{2.19.13}$$

$$2\tilde{u} = -\frac{1}{a}\frac{\partial \tilde{p}}{\partial \theta} + E\frac{\partial^2 \tilde{v}}{\partial x^2}, \tag{2.19.14}$$

$$0 = -\frac{\partial \tilde{p}}{\partial z} + E\frac{\partial^2 \tilde{w}}{\partial x^2}, \tag{2.19.15}$$

$$\frac{\partial \tilde{u}}{\partial x} + \frac{1}{a}\frac{\partial \tilde{v}}{\partial \theta} + \frac{\partial \tilde{w}}{\partial z} = 0. \tag{2.19.16}$$

The radial shear term is unimportant in the vertical boundary layer and is neglected for this reason; the range of x is considered to be from zero at the wall, to infinity in the interior. (Actually, x is the unstretched boundary layer co-ordinate.) It follows that

$$\left.\begin{aligned} \tilde{u} &= -\frac{1}{2a}\frac{\partial \tilde{p}}{\partial \theta} + \frac{E}{4}\frac{\partial^3 \tilde{p}}{\partial x^3}, \\ \tilde{v} &= \frac{1}{2}\frac{\partial \tilde{p}}{\partial x}; \end{aligned}\right\} \quad (2.19.17)$$

$$\left.\begin{aligned} E\frac{\partial^4 \tilde{p}}{\partial x^4} + 4\frac{\partial \tilde{w}}{\partial z} &= 0, \\ E\frac{\partial^2 \tilde{w}}{\partial x^2} - \frac{\partial \tilde{p}}{\partial z} &= 0. \end{aligned}\right\} \quad (2.19.18)$$

Note that equations (2.19.18) are essentially the same as those studied in the last section, (2.18.7). The asymmetry inherent in the general problem enters the equations solely as an azimuthal derivative of pressure in the expression for $\tilde{u}$. The fact that (2.19.18) does not involve the tangential co-ordinate means that the boundary layer calculation for the cylinder applies as well to any vertical boundary of arbitrary cross-section if x and $a\theta$ are replaced by the co-ordinates in the normal and circumferential directions.

The boundary conditions at $r = a$, $x = 0$, which are

$$\tilde{u} + u_I = \mathscr{S}(\theta, z), \quad \tilde{v} + v_I = 0, \quad \text{upon substitution}$$

become

$$\frac{1}{a}\frac{\partial \tilde{p}}{\partial \theta} - \frac{E}{2}\frac{\partial^3 \tilde{p}}{\partial x^3} = -\frac{1}{a}\frac{\partial p_I}{\partial \theta} - 2\mathscr{S}(\theta, z), \quad (2.19.19)$$

$$\frac{\partial \tilde{p}}{\partial x} = -\frac{\partial p_I}{\partial r}; \quad (2.19.20)$$

in addition, $\tilde{w} = 0$ at this surface.

The remaining conditions are

$$\left.\begin{aligned} \tilde{w} &\cong \tfrac{1}{4}E^{\frac{1}{2}}\frac{\partial^2 \tilde{p}}{\partial x^2} \quad \text{at } z = 0, \\ \tilde{w} &\cong -\tfrac{1}{4}E^{\frac{1}{2}}\frac{\partial^2 \tilde{p}}{\partial x^2} \quad \text{at } z = 1. \end{aligned}\right\} \quad (2.19.21)$$

and

The solution of this boundary value problem is obtained using the methods introduced by Stewartson[188]. First, all the separable solutions of the form

$$\left.\begin{aligned}\tilde{w} &= A\,e^{-\gamma x}\sin(az+b),\\ \tilde{p} &= B\,e^{-\gamma x}\cos(az+b),\end{aligned}\right\} \qquad (2.19.22)$$

are determined. It follows from the equations of motion that

$$a = \frac{E\gamma^3}{2}, \qquad (2.19.23)$$

whereas, the boundary conditions imply that

$$\left.\begin{aligned}\tan\left(\frac{E\gamma^3}{2}+b\right) &\cong \frac{E^{\frac{1}{2}}\gamma}{2},\\ \tan b &\cong -\frac{E^{\frac{1}{2}}\gamma}{2}.\end{aligned}\right\} \qquad (2.19.24)$$

The values of γ satisfying the preceding transcendental relationship are proportional to $E^{-\frac{1}{3}}$ in all cases except one, when $\gamma \doteq E^{-\frac{1}{4}}$. The former correspond to the $E^{\frac{1}{3}}$ boundary layer and are

$$\gamma_{n1} = \gamma_n = \left(\frac{2n\pi}{E}\right)^{\frac{1}{3}}, \quad \gamma_{n2} = \gamma_n e^{i\pi/3},$$

$$\gamma_{n3} = \gamma_n e^{-i\pi/3}, \quad (n = 1, 2, \ldots); \qquad (2.19.25)$$

with $\qquad a_{n1} = n\pi, \quad a_{n2} = a_{n3} = -n\pi.$

The phase is negligible for these values. The other root is associated with the $E^{\frac{1}{4}}$-layer and is

$$\gamma_0 = 2^{\frac{1}{2}}E^{-\frac{1}{4}} \qquad (2.19.26)$$

with $\qquad a_0 = 2^{\frac{1}{2}}E^{\frac{1}{4}} \quad \text{and} \quad b_0 = -2^{-\frac{1}{2}}E^{\frac{1}{4}}.$

Therefore, the general solution is

$$\tilde{w} = 2^{\frac{1}{2}}E^{\frac{1}{4}}A_0(\theta)\,(z-\tfrac{1}{2})\exp(-2^{\frac{1}{2}}E^{-\frac{1}{4}}x)$$
$$+\sum_{n=1}^{\infty}[A_{n1}(\theta)\,e^{-\gamma_{n1}x}+A_{n2}(\theta)\,e^{-\gamma_{n2}x}+A_{n3}(\theta)\,e^{-\gamma_{n3}x}]\sin(n\pi z), \qquad (2.19.27)$$

$$\tilde{p} = -2^{\frac{1}{2}}E^{\frac{1}{4}}A_0(\theta)\exp(-2^{\frac{1}{2}}E^{-\frac{1}{4}}x)$$
$$+2\sum_{n=1}^{\infty}\left[-\frac{A_{n1}(\theta)}{\gamma_{n1}}e^{-\gamma_{n1}x}+\frac{A_{n2}(\theta)}{\gamma_{n2}}e^{-\gamma_{n2}x}+\frac{A_{n3}(\theta)}{\gamma_{n3}}e^{-\gamma_{n3}x}\right]\cos(n\pi z). \qquad (2.19.28)$$

The unknown coefficients, functions of θ, are determined from the boundary conditions and it follows from (2.19.20), for example, that

$$A_0(\theta) = -\frac{1}{2}\frac{\partial p_I}{\partial r}, \tag{2.19.29}$$

$$A_{n1}(\theta) - A_{n2}(\theta) - A_{n3}(\theta) = 0. \tag{2.19.30}$$

Thus, $-A_0(\theta)$ is the interior azimuthal velocity component at the wall.

Application of condition $\tilde{w} = 0$ at $x = 0$, leads to

$$A_{n1}(\theta) + A_{n2}(\theta) + A_{n3}(\theta) = 2^{\frac{1}{2}}E^{\frac{1}{4}}\left(\frac{1+\cos n\pi}{n\pi}\right)A_0(\theta), \tag{2.19.31}$$

and the remaining boundary condition expressed in (2.19.19) implies that

$$\left.\begin{aligned}
&\frac{2^{\frac{1}{2}}E^{\frac{1}{4}}}{a}\frac{dA_0}{d\theta} + 2E^{\frac{1}{2}}A_0 = 2\langle\mathscr{S}(\theta, z)\rangle + \frac{1}{a}\frac{\partial p_I}{\partial\theta},\\
&-\frac{1}{\gamma_{n1}}\left(\frac{1}{a}\frac{dA_{n1}}{d\theta} + n\pi A_{n1}\right) + \frac{1}{\gamma_{n2}}\left(\frac{1}{a}\frac{dA_{n2}}{d\theta} - n\pi A_{n2}\right)\\
&+\frac{1}{\gamma_{n3}}\left(\frac{1}{a}\frac{dA_{n3}}{d\theta} - n\pi A_{n3}\right) = -2\int_0^1 \mathscr{S}(\theta, z)\cos n\pi z\,dz.
\end{aligned}\right\} \tag{2.19.32}$$

Explicit formulas for the coefficients, the solutions of the first order ordinary differential equations which are periodic in θ, are readily determined. Of greater immediate importance, is to obtain *the boundary condition on the interior flow* by eliminating $A_0(\theta)$ between (2.19.29) and the first equation in (2.19.32). This yields

$$\frac{2^{-\frac{1}{2}}E^{\frac{1}{4}}}{a}\frac{\partial^2 p_I}{\partial r\,\partial\theta} + E^{\frac{1}{2}}\frac{\partial p_I}{\partial r} + \frac{1}{a}\frac{\partial p_I}{\partial\theta} = -2\langle\mathscr{S}\rangle, \tag{2.19.33}$$

on $r = a$. Upon integrating this expression about the perimeter of the cylinder, the first and third terms are eliminated leaving

$$E^{\frac{1}{2}}\int_0^{2\pi}\frac{\partial p_I}{\partial r}d\theta = -2\int_0^{2\pi}\langle\mathscr{S}\rangle\,d\theta. \tag{2.19.34}$$

The right-hand side of this formula is proportional to the *net* mass flux emanating from this surface and if the interior flow is to be of unit magnitude, then the condition expressed in (2.19.10) is necessary,

$$\int_0^{2\pi} \langle \mathscr{S} \rangle \, d\theta = O(E^{\frac{1}{2}}).$$

The left-hand side of (2.19.34) is proportional to the circulation about the cylinder and, in fact,

$$\Gamma = -E^{-\frac{1}{2}} a \int_0^{2\pi} \langle \mathscr{S} \rangle \, d\theta. \tag{2.19.35}$$

The generalization of this formula to (2.19.5) for an arbitrary cross-section is immediate.

As long as restriction (2.19.10) is in force, the principal balance of terms in (2.19.33) is

$$\frac{1}{a} \frac{\partial p_I}{\partial \theta} = -2 \langle \mathscr{S} \rangle. \tag{2.19.36}$$

This is the boundary condition on the normal component of the interior velocity at $r = a$ and it obviously depends only on the nature of the depth-averaged source distribution at the wall. The generalization of this formula is (2.19.4).

Equation (2.19.36) is accurate to $O(E^{\frac{1}{4}})$; a better approximation would be

$$\frac{1}{a} \frac{\partial p_I}{\partial \theta} = -2 \left[\langle \mathscr{S} \rangle - \frac{1}{2\pi} \int_0^{2\pi} \langle \mathscr{S} \rangle \, d\theta \right]. \tag{2.19.37}$$

The vertical boundary layers act to adjust the motion produced by the arbitrary mass efflux from the walls to a depth-averaged state that is acceptable to the interior potential field. This implies that any mean transport of mass takes place only within the boundary layers. The vertical layers shift half of the average efflux from any vertical cylinder to each of the Ekman layers at the horizontal end plates. As far as the interior motion is concerned, the effect of the boundaries is the same as that of a distribution of two-dimensional fluid dipoles and higher multipoles.

The monopole character of the applied source distribution on a vertical cylinder is completely transformed by the adjacent shear layers into an interior circulation which is a potential vortex. A positive source induces a relative negative vortex about itself through the action of the Coriolis force; a sink gives rise to a positive circulation. (In fact, similar effects on the interior flow could be obtained by substituting a suitable source-sink flow at the outer wall of the container for that on a rigid internal cylinder. See the discussion at the end of § 2.17.)

These results apply to any of the interior cylinders and to the outer boundary, too, if $\mathfrak{x}(=a_0-\mathrm{r})$ replaces $\mathfrak{x}$ and the appropriate changes in sign are made. The modifications are of a rather trivial nature.

An inspection of the boundary layer functions corresponding to a general source distribution, $\mathscr{S}=O(1)$, shows that in the layer of thickness $E^{\frac{1}{3}}$

$$\tilde{u}=O(1),\quad \tilde{v}=O(E^{-\frac{1}{3}}),\quad \tilde{w}=O(E^{-\frac{1}{3}}),\quad \tilde{p}=O(1),$$

whereas in the $E^{\frac{1}{4}}$-layer

$$\tilde{u}=O(E^{\frac{1}{4}}),\quad \tilde{v}=O(1),\quad \tilde{w}=O(E^{\frac{1}{4}}),\quad \tilde{p}=O(E^{\frac{1}{4}}).$$

The inner layer can be marked by considerable recirculation of fluid, the magnitude of which indicates that non-linear interactions are probably important in this region.

As an illustrative example of some importance, consider the concentric cylinder configuration shown in fig. 2.19 (*a*), where the inner (outer) boundary is a *uniform* source (sink) of fluid. In accordance with the restriction on the net mass transport, let

$$\mathscr{S}_1=\frac{\mathscr{M}E^{\frac{1}{2}}}{2\pi a_1}$$

at the inner cylinder $\mathrm{r}=a_1$; it follows from (2.19.9) that

$$\mathscr{S}_0=-\frac{\mathscr{M}E^{\frac{1}{2}}}{2\pi a_0}\quad \text{on } \mathrm{r}=a_0.$$

The lowest order interior pressure field is the solution of (2.19.3), which satisfies the conditions (2.19.34) and (2.19.36):

$$\nabla^2 p_I=0,$$

with
$$\frac{\partial p_I}{\partial\theta}=0\quad \text{on } \mathrm{r}=a_0 \quad\text{and}\quad \mathrm{r}=a_1;$$

$$\int_0^{2\pi}\frac{\partial p_I}{\partial \mathrm{r}}\,d\theta=-\frac{2\mathscr{M}}{a_1},\quad \text{at } \mathrm{r}=a_1.$$

In this symmetrical geometry, the pressure can only be a function of the radial co-ordinate, so that

$$p_I=c\ln(\mathrm{r}). \qquad (2.19.38)$$

The constant is evaluated from the prescribed circulation about the inner cylinder and the result is

$$\mathrm{c} = -\frac{\mathscr{M}}{\pi}.$$

The primary interior velocity is entirely circumferential:

$$\mathrm{u_I} = 0, \quad \mathrm{v_I} = -\frac{\mathscr{M}}{2\pi \mathrm{r}}, \quad \mathrm{w_I} = 0. \tag{2.19.39}$$

The potential flow is that due to a *negative* line vortex of strength $-\mathscr{M}/2\pi$, located at the centre of the inner cylinder.

The calculation of the boundary layer structure at $\mathrm{r} = a_1$ is routine and the results are

$$\begin{aligned}\tilde{\mathrm{w}} = {} & \frac{\mathrm{E}^{\frac{1}{4}}\mathscr{M}}{2^{\frac{1}{2}}\pi a_1}(\mathrm{z}-\tfrac{1}{2})\exp\left[-2^{\frac{1}{2}}\mathrm{E}^{-\frac{1}{4}}(\mathrm{r}-a_1)\right] \\ & + \frac{\mathrm{E}^{\frac{1}{4}}\mathscr{M}}{2^{\frac{3}{2}}\pi^2 a_1}\sum_{n=1}^{\infty}\frac{\sin 2n\pi \mathrm{z}}{n}\left\{\mathrm{e}^{-\gamma_{2n}(\mathrm{r}-a_1)} - \frac{2}{\sqrt{3}}\mathrm{e}^{-\frac{1}{2}\gamma_{2n}(\mathrm{r}-a_1)}F_n\right\},\end{aligned} \tag{2.19.40}$$

with $\quad F_n = \sin\left(\dfrac{\sqrt{3}}{2}\gamma_{2n}(\mathrm{r}-a_1)\right) + \sin\left(\dfrac{\sqrt{3}}{2}\gamma_{2n}(\mathrm{r}-a_1) - \dfrac{\pi}{3}\right);$

$$\begin{aligned}\tilde{\mathrm{p}} = {} & -\frac{\mathrm{E}^{\frac{1}{4}}\mathscr{M}}{2^{\frac{1}{2}}\pi a_1}\exp\left[-2^{\frac{1}{2}}\mathrm{E}^{-\frac{1}{4}}(\mathrm{r}-a_1)\right] \\ & - \frac{\mathrm{E}^{\frac{1}{4}}\mathscr{M}}{2^{\frac{1}{2}}\pi^2 a_1}\sum_{n=1}^{\infty}\frac{\cos 2n\pi \mathrm{z}}{n\gamma_{2n}}\left\{\mathrm{e}^{-\gamma_{2n}(\mathrm{r}-a_1)} + \frac{2}{\sqrt{3}}\mathrm{e}^{-\frac{1}{2}\gamma_{2n}(\mathrm{r}-a_1)}G_n\right\},\end{aligned} \tag{2.19.41}$$

with $\quad G_n = \sin\left(\dfrac{\sqrt{3}}{2}\gamma_{2n}(\mathrm{r}-a_1)\right) + \sin\left(\dfrac{\sqrt{3}}{2}\gamma_{2n}(\mathrm{r}-a_1) + \dfrac{\pi}{3}\right);$

where $$\gamma_{2n} = \left(\frac{4n\pi}{\mathrm{E}}\right)^{\frac{1}{3}}.$$

The structure of the layers at the outer cylinder may be obtained from these formulas by replacing $\mathscr{M}$ by $-\mathscr{M}$ and $\mathrm{r}-a_1$ by $a_0-\mathrm{r}$.

Magnitudes associated with the $\mathrm{E}^{\frac{1}{3}}$-layer are

$$\tilde{\mathrm{u}} = O(\mathrm{E}^{\frac{7}{12}}), \quad \tilde{\mathrm{v}} = O(\mathrm{E}^{\frac{1}{4}}), \quad \tilde{\mathrm{w}} = O(\mathrm{E}^{\frac{1}{4}}), \quad \tilde{\mathrm{p}} = O(\mathrm{E}^{\frac{7}{12}});$$

the appearance of the factor $\mathrm{E}^{\frac{1}{12}}$ indicates that a consistent formal expansion procedure might have to be based on this parameter, (see p. 106, Barcilon[5], Greenspan and Howard[75], and Kudlick [100]).

Fig. 2.20. Source-sink flow in a pie-shaped vessel with a paraboloidal base, [196]. There are no closed geostrophic contours in this configuration and mass transport is via a western boundary layer.

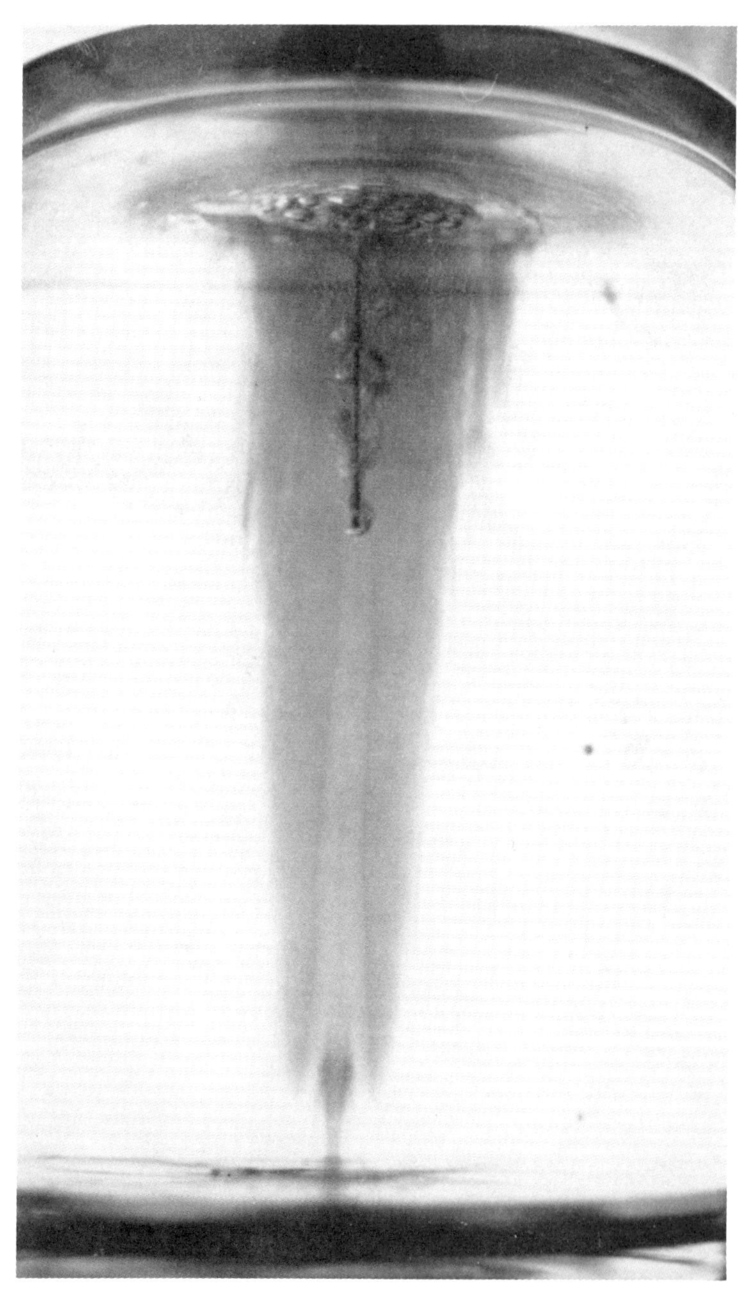

Fig. 2.21. The rotational effects of an upwards convection, produced by rising bubbles, extend to the base plate where a 'tornado' is formed, Turner [212].

Experimental confirmation of the main theoretical results has recently been obtained by Hide[83]. Annular flows of the type described have also been examined by Lewellen[105] for distinctly non-linear conditions, and by Carrier[27] for a compressible medium.

Stommel, Arons and Faller[196], and Faller[45] investigated source-sink flows in a pie-shaped container whose bottom surface is a section of a paraboloid. The total depth is

$$\mathrm{h} = \left(1 + b\left(\frac{\mathrm{r}}{a}\right)^2\right),$$

so that no closed geostrophic contours exist. The topography severely restricts the inviscid motion because the fluid is forced across constant-depth contours. Consequently, a basic geostrophic motion cannot occur and the flow pattern differs markedly from those just studied. A 'western' boundary layer, seen in fig. 2.20, is then an intrinsic part of the circulation pattern and acts as the main carrier of mass from the source to the sink.

A similar but simpler problem is analysed in detail in the next section. The strong resemblance that these laboratory flows bear to oceanic currents and circulations is the subject of Chapter 5.

Turner[212] simulated convection in clouds by the buoyant rise of bubbles introduced into a rotating container as shown in fig. 2.21. The effect of the disturbance is comparable to that produced by a small disk rotating with a slight excess speed at the bubble entry point. Both devices induce a secondary circulation of fluid, one through the bubble rise, the other by a convergence into the Ekman layer. In either case, the small flow establishes a columnar motion that is shielded from the main fluid by vertical shear layers. Turner's purpose was to show that convection high above the ground can produce a strong surface effect (a tornado) by means of the Taylor column mechanism.

A motion similar to that described in § 1.1 and shown in fig. 1.2 (*b*) was analyzed by Jacobs[92] who determined the fine structure of the multiple shear layers that separate the Taylor–Proudman column from the rest of the fluid. This was apparently the first work to consider a layer whose structure depended on azimuthal angle.

Free shear layers are often a prominent feature of non-linear flows and their stability characteristics may be of some importance in geophysical problems. Fig. 3.19, p. 175, clearly shows internal shear layers in the case of a sphere which is rotating and slowly precessing. These seem to arise from a non-linear wave interaction, but since this is properly part of the next chapter, this discussion will be continued there.

2.20. A 'wind-driven' circulation

In § 2.17, we began a discussion of forced motion in the rotating sliced cylinder (Pedlosky and Greenspan[148]) that could only be completed after the essentials of sidewall boundary layers were mastered. It may be recalled that motion is produced by rotating the level top disk at a rate slightly different from that of the rest of the container. Since the bottom lid is tilted at an angle α, constant-depth contours are not closed curves and the interior flow is very much restricted, being only $O(E^{\frac{1}{2}})$ compared to the unit magnitude velocity differential between elements of the boundary surface. The interior velocity was shown in (2.17.13) to be

$$\mathbf{q} = E^{\frac{1}{2}}\left[\frac{\mathscr{X}(y)}{\sin\alpha}\,\hat{\mathbf{i}} + \operatorname{ctn}\alpha\,\hat{\mathbf{j}} + \hat{\mathbf{k}}\right], \tag{2.20.1}$$

and the determination of the unknown function is the objective of this section.

Spin-up in this geometry, § 2.16, involves the production of Rossby waves and the entire process bears a remarkable similarity to phenomena revealed in studies of oceanic circulation based on the β-plane approximation, (see § 5.5). The resemblance is not fortuitous and the comparison of results obtained from this simple model with those derived from various theories in oceanography will be stressed at a later time. However, for the present, it should be noted that the variation of vertical depth in the sliced cylinder plays the same role as the latitudinal variation of the Coriolis force does in oceanography (that component tangent to the surface of the earth). This is the very foundation of the analogy. In the problem under discussion, the disk-driven fluid effectively models oceanic circulation caused by the wind-stress. The resultant flow pattern will be seen to have a close connection with geophysical motions.

In mathematical terms, the problem requires the determination of the sidewall boundary layer that matches correctly to the interior flow given above. Once again, the important restrictions,

$$1 \gg \alpha \gg E^{\frac{1}{2}},$$

are set and the analysis proceeds along the lines laid out in §2.19.

Within the sidewall layer, the cylindrical components of velocity, (u, v, w), satisfy, to a high degree of precision, equations (2.19.13) to (2.19.16), (with $\mathfrak{x} = a - r$ replacing the variable x).

The boundary conditions at $\mathfrak{x} = 0$ are

$$u = v = w = 0,$$

and as $\mathfrak{x} \to \infty$ (the interior regime),

$$\left.\begin{aligned} u &\to \frac{E^{\frac{1}{2}}}{\alpha}\{\mathscr{X}(a\sin\theta)\cos\theta + \sin\theta\}, \\ v &\to \frac{E^{\frac{1}{2}}}{\alpha}\{-\mathscr{X}(a\sin\theta)\sin\theta + \cos\theta\}, \\ w &\to E^{\frac{1}{2}}. \end{aligned}\right\} \quad (2.20.2)$$

The appropriate conditions just outside the Ekman layers at the top and bottom plates, $z = 1$, $z = a\alpha\sin\theta$, follow from the general formula (2.17.3), with $\mathscr{U}_T = r\hat{\boldsymbol{\theta}}$ and $\mathscr{U}_B = 0$. They are, respectively,

$$\left.\begin{aligned} w &= E^{\frac{1}{2}} + \tfrac{1}{2}E^{\frac{1}{2}}\left[\frac{\partial v}{\partial \mathfrak{x}} + \frac{1}{a}\frac{\partial u}{\partial \theta} + \frac{\partial w}{\partial z}\right], \\ -w + \alpha v\cos\theta + \alpha u\sin\theta &= \frac{E^{\frac{1}{2}}}{2}\left[\alpha\sin\theta\left(\frac{1}{a}\frac{\partial w}{\partial \theta} - \frac{\partial}{\partial z}(u+v)\right)\right. \\ &\quad \left. - \alpha\cos\theta\left(\frac{\partial w}{\partial \mathfrak{x}} + \frac{\partial}{\partial z}(v-u)\right) - \frac{\partial}{\partial \mathfrak{x}}(u + v + \alpha w\sin\theta)\right]. \end{aligned}\right\} \quad (2.20.3)$$

Here, as before, only the dominant derivatives are retained.

We anticipate that the interior velocity is $O(E^{\frac{1}{2}})$, in which case there is no need of a prominent boundary layer of thickness $E^{\frac{1}{4}}$ whose raison d'être is to eliminate a primary shear discontinuity. An $E^{\frac{1}{3}}$ boundary layer alone can fulfil the requirements of mass transport and the equations are scaled at once with

$$(u, v, w, p, \mathfrak{x})$$
$$\Rightarrow (E^{\frac{1}{2}}u_I + E^{\frac{1}{2}}\tilde{U}, E^{\frac{1}{2}}v_I + E^{\frac{1}{6}}\tilde{V}, E^{\frac{1}{2}}w_I + E^{\frac{1}{6}}\tilde{W}, p_I + E^{\frac{1}{2}}\tilde{P}, E^{\frac{1}{3}}\zeta),$$

where subscript I denotes an interior function evaluated at the wall, $r = a$. The complete boundary value problem for the perturbation boundary layer functions $\tilde{U}$, $\tilde{V}$, etc. is

$$\left.\begin{aligned} -2\tilde{V} &= \frac{\partial \tilde{P}}{\partial \zeta}, \\ 2\tilde{U} &= -\frac{1}{a}\frac{\partial \tilde{P}}{\partial \theta} + \frac{\partial^2 \tilde{V}}{\partial \zeta^2}, \\ 0 &= -\frac{\partial \tilde{P}}{\partial z} + \frac{\partial^2 \tilde{W}}{\partial \zeta^2}, \\ -\frac{\partial \tilde{U}}{\partial \zeta} + \frac{1}{a}\frac{\partial \tilde{V}}{\partial \theta} + \frac{\partial \tilde{W}}{\partial z} &= 0, \end{aligned}\right\} \tag{2.20.4}$$

with

$$\tilde{W} = 0, \quad \text{on } z = 1;$$

$$-\tilde{W} + \alpha \tilde{V}\cos\theta = 0, \quad \text{on } z = a\alpha\sin\theta; \tag{2.20.5}$$

$$\tilde{W} = \tilde{V} = 0, \quad \text{on } \zeta = 0 \quad \text{and} \quad \zeta = \infty;$$

$$\tilde{U} = -\frac{1}{\alpha}(\mathscr{X}(a\sin\theta)\cos\theta + \sin\theta), \quad \text{on } \zeta = 0; \tag{2.20.6}$$

$$\tilde{U} = 0, \quad \text{on } \zeta = \infty.$$

If the horizontal velocity components are eliminated from the formulation, and $\mathfrak{z} = z - 1$, then

$$\left.\begin{aligned} \frac{1}{4}\frac{\partial^4 \tilde{P}}{\partial \zeta^4} + \frac{\partial \tilde{W}}{\partial \mathfrak{z}} &= 0, \\ \frac{\partial \tilde{P}}{\partial \mathfrak{z}} - \frac{\partial^2 \tilde{W}}{\partial \zeta^2} &= 0. \end{aligned}\right\} \tag{2.20.7}$$

Separable solutions exist of the form

$$\left.\begin{aligned} \tilde{W} &= C\,e^{-m\zeta}\sin\frac{m^3}{2}\mathfrak{z}, \\ \tilde{P} &= -\frac{2}{m}C\,e^{-m\zeta}\cos\frac{m^3}{2}\mathfrak{z}, \end{aligned}\right\} \tag{2.20.8}$$

and the eigenvalues are the solutions of

$$\tan\frac{m^3}{2} = \alpha\cos\theta. \tag{2.20.9}$$

The roots of this expression with $\mathscr{Re}\,m > 0$ are:

$$m \cong |2\alpha\cos\theta|^{\frac{1}{3}} \quad \text{for } \cos\theta > 0;$$

$$m \cong |2\alpha\cos\theta|^{\frac{1}{3}}\left(\frac{1}{2} \pm \frac{\sqrt{3}}{2}i\right) \quad \text{for } \cos\theta < 0;$$

$$m \cong (2n\pi)^{\frac{1}{3}}\left(1, \frac{1}{2} \pm \frac{\sqrt{3}}{2}i\right)$$

when $$m^3 \cong \pm 2n\pi.$$

It is very important to note that of the roots near zero, two occur when $\cos\theta < 0$, but only one exists if $\cos\theta > 0$. As a consequence, it will be shown that a boundary layer develops only on that portion of the sidewall for which $\cos\theta < 0$, that is

$$\frac{\pi}{2} \leqslant \theta \leqslant \frac{3\pi}{2}.$$

The general solution for α small is

$$\begin{aligned}\tilde{W} = {} & \alpha C_{01}\mathfrak{z}\cos\theta\,\mathbb{S}(\cos\theta)\exp[-|2\alpha\cos\theta|^{\frac{1}{3}}\zeta] \\ & + \{\alpha C_{02}\mathfrak{z}\cos\theta\exp[-|2\alpha\cos\theta|^{\frac{1}{3}}e^{i\pi/3}\zeta] \\ & + \alpha C_{03}\mathfrak{z}\cos\theta\exp[-|2\alpha\cos\theta|^{\frac{1}{3}}e^{-i\pi/3}\zeta]\}\,\mathbb{S}(-\cos\theta) \\ & + \sum_{n=1}^{\infty}\{C_{n1}\exp[-(2n\pi)^{\frac{1}{3}}\zeta] - C_{n2}\exp[-(2n\pi)^{\frac{1}{3}}e^{i\pi/3}\zeta] \\ & \qquad - C_{n3}\exp[-(2n\pi)^{\frac{1}{3}}e^{-i\pi/3}\zeta]\}\sin n\pi\mathfrak{z},\end{aligned} \tag{2.20.10}$$

$$\begin{aligned}\tilde{P} = {} & -\frac{2C_{01}}{|2\alpha\cos\theta|^{\frac{1}{3}}}\mathbb{S}(\cos\theta)\exp[-|2\alpha\cos\theta|^{\frac{1}{3}}\zeta] \\ & - \frac{2}{|2\alpha\cos\theta|^{\frac{1}{3}}}\{C_{02}e^{-i\pi/3}\exp[-|2\alpha\cos\theta|^{\frac{1}{3}}e^{i\pi/3}\zeta] \\ & + C_{03}e^{i\pi/3}\exp[-|2\alpha\cos\theta|^{\frac{1}{3}}e^{-i\pi/3}\zeta]\}\,\mathbb{S}(-\cos\theta) \\ & - 2\sum_{n=1}^{\infty}\left\{\frac{C_{n1}}{(2n\pi)^{\frac{1}{3}}}\exp[-(2n\pi)^{\frac{1}{3}}\zeta] + \frac{C_{n2}}{(2n\pi)^{\frac{1}{3}}}\right. \\ & \qquad \times e^{-i\pi/3}\exp[-(2n\pi)^{\frac{1}{3}}e^{i\pi/3}\zeta] \\ & \left. + \frac{C_{n3}}{(2n\pi)^{\frac{1}{3}}}e^{i\pi/3}\exp[-(2n\pi)^{\frac{1}{3}}e^{-i\pi/3}\zeta]\right\}\cos n\pi\mathfrak{z}.\end{aligned} \tag{2.20.11}$$

The coefficients, which are functions of θ, are determined from the boundary conditions. For example, $\tilde{V} = 0$ at $\zeta = 0$ requires that

$$C_{01} = 0 \quad \text{for } \cos\theta > 0, \quad C_{02}+C_{03} = 0 \quad \text{for } \cos\theta < 0,$$

and
$$C_{n1}+C_{n2}+C_{n3} = 0.$$

Since $\tilde{W} = 0$ at $\zeta = 0$, the relationship

$$C_{n1}-C_{n2}-C_{n3} = 0$$

can be added to the preceding. Hence,

$$C_{01} = 0, \quad C_{02}+C_{03} = 0,$$
$$C_{n1} = 0, \quad C_{n2}+C_{n3} = 0,$$

and these formulas can be taken to hold for all θ in view of the presence of multiplicative step functions in (2.20.10) and (2.20.11). The final boundary condition at $\zeta = 0$,

$$\tilde{U} = -\frac{1}{\alpha}(\mathscr{X}(a\sin\theta)\cos\theta+\sin\theta)$$
$$= -\frac{1}{2a}\frac{\partial\tilde{P}}{\partial\theta}-\frac{1}{4}\frac{\partial^3\tilde{P}}{\partial\zeta^3},$$

leads to the differential equations

$$\frac{dC_{n2}}{d\theta}+n\pi a\, C_{n2} = 0, \tag{2.20.12}$$

$$\left\{\frac{\sqrt{3}}{a}\frac{d}{d\theta}\left(\frac{C_{02}}{|2\alpha\cos\theta|^{\frac{1}{3}}}\right)+\frac{\sqrt{3}}{2}|2\alpha\cos\theta|^{\frac{2}{3}}C_{02}\right\}\mathcal{S}(-\cos\theta) = i\tilde{U}. \tag{2.20.13}$$

The first of these implies that

$$C_{n2}(\theta) \equiv 0 \quad (n \geqslant 1), \tag{2.20.14}$$

because all functions must be periodic in θ. In the range $\cos\theta > 0$, the left-hand side of (2.20.13) is identically zero. Therefore, $\tilde{U}$ must vanish in this region too, and it follows that

$$\mathscr{X}(a\sin\theta) = -\frac{\sin\theta}{(1-\sin^2\theta)^{\frac{1}{2}}}$$

or
$$\mathscr{X}(y) = -\frac{y}{(a^2-y^2)^{\frac{1}{2}}}. \tag{2.20.15}$$

Thus, the unknown function $\mathscr{X}(y)$ is determined from the fact that *there is no boundary layer on the lateral surface for* $\cos\theta > 0$. Furthermore, for $\cos\theta < 0$,

$$\tilde{U} = -\frac{2}{\alpha}\sin\theta \tag{2.20.16}$$

and (2.20.13) can be integrated directly. If terms of $O(\alpha)$ are neglected, then

$$\frac{C_{02}}{|2\alpha\cos\theta|^{\frac{1}{3}}} = \frac{2ai\cos\theta}{\alpha\sqrt{3}}. \tag{2.20.17}$$

Here, an arbitrary constant has been eliminated by the condition that the meridional boundary layer flux, $\int_0^\infty \tilde{V}\,d\zeta$, must balance the interior transport across any diameter.

All the coefficients are now determined to $O(\alpha)$; the boundary layer solution for the pressure is

$$\tilde{P} = \mathscr{I}m\left\{\frac{8a}{\alpha\sqrt{3}}\cos\theta\left(\frac{1}{2}-\frac{\sqrt{3}}{2}i\right)\exp\left[-|2\alpha\cos\theta|^{\frac{1}{3}}\left(\frac{1}{2}+\frac{\sqrt{3}}{2}i\right)\zeta\right]\right\} \times \mathbb{S}(-\cos\theta). \tag{2.20.18}$$

The inviscid interior motion, for which the pressure is a stream function, is given by

$$p = \frac{2}{\alpha}E^{\frac{1}{2}}[x-(a^2-y^2)^{\frac{1}{2}}] \tag{2.20.19}$$

and the vertical velocity is $w = E^{\frac{1}{2}}$. The primary flow is a completely horizontal circulation of magnitude $E^{\frac{1}{2}}/\alpha$ which closes via a western boundary layer. The small vertical velocity is part of a secondary circulation feeding the Ekman layer whose structure may be determined by considering terms of higher order in α. The theoretical solution is shown in fig. 2.22(*a*).

Experiments carried out at M.I.T. by R. Beardsley confirm the main theoretical predictions concerning the location of the boundary layer and the general scale and appearance of the interior flow. However, a very interesting non-linear effect is observed, fig. 2.22(*b*), and this is the appearance of an instability in the form of a meandering 'jet' emanating from the boundary layer in the general location of the terminal point. Fig. 2.23 shows this phenomenon at different times; a detailed description is given on p. 243. It is hard not to compare this with the separation of the Gulf Stream at Cape Hatteras which may also be due to a changing topography. Indeed,

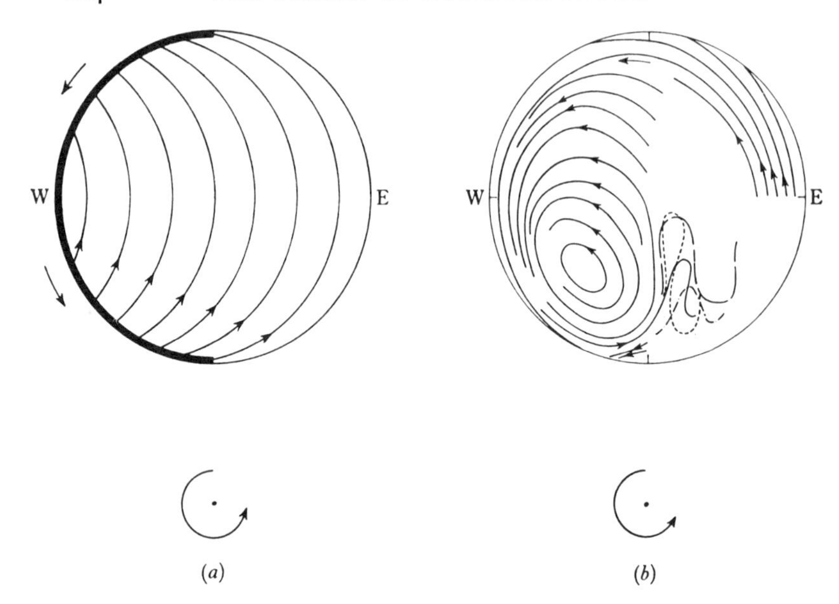

Fig. 2.22. (*a*) Theoretical circulation for the disk-driven motion in the sliced cylinder. (*b*) A drawing of observed circulation pattern.

a criterion has been advanced, [71] (see p. 242), governing the separation of ocean currents that is relevant to the laboratory model as well. This analogy with certain planetary motions is pursued further in chapter 5.

2.21. Some effects of stratification

Stable stratification introduces two new processes both of which tend to destroy the phenomena peculiar to rotating homogeneous fluids. In fact, with a density gradient that is large enough, all the familiar features can be eliminated: Ekman layers, two-dimensional motions, vertical shear layers and secondary circulations.

The buoyancy force acts to inhibit vertical motions and this implies a rather drastic diminution of the control exercised by the Ekman layers. The tendency of the viscous layers to produce a secondary circulation is counteracted by the bias towards horizontal motion in the gravitational field. As a consequence, the mechanism of vortex line stretching is rendered partially or totally ineffective and viscous diffusion becomes a dominant process once again. The

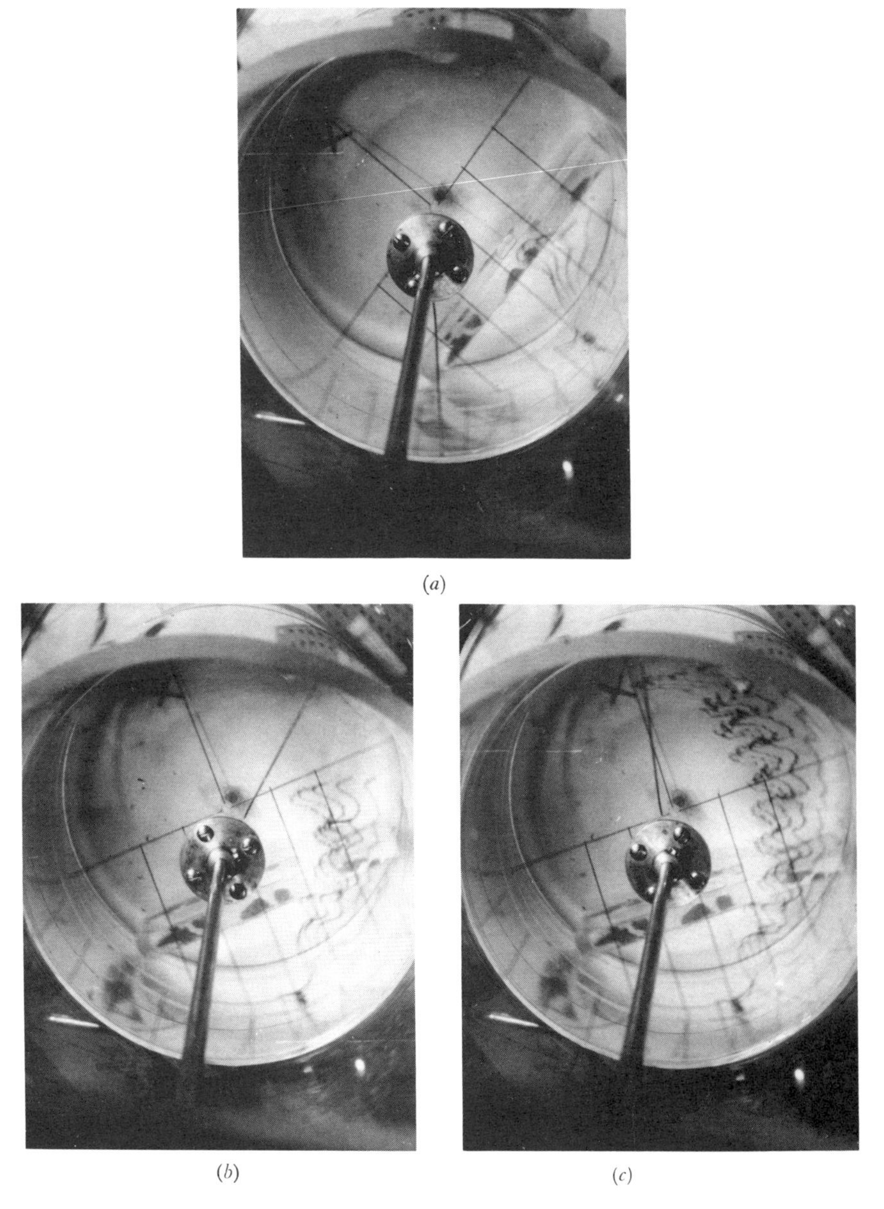

Fig. 2.23. The dye streaks, each at a different depth, show the development of 'meanders' at three times (*a*), (*b*), (*c*).

diffusion time scale, $E^{-1}\Omega^{-1}$, can then be as significant in characterizing transient motions as $E^{-\frac{1}{2}}\Omega^{-1}$, the spin-up time for homogeneous fluids. Finally, columnar motion is made impossible by the new capability for disturbances to propagate horizontally along the surfaces of constant density.

Of course, the class of motions which combines rotation and stratification has its own distinctive features and this section is devoted to a brief exploratory study of some of the rather complicated interactions that develop.

Our objectives are limited and it will be sufficient to consider the consequences of the simple linear theory for steady motions formulated in § 1.4. Assuming that the stratification is stable, and $\varepsilon = 0$, $F_R = 0$, $\varkappa = 0$, $T_e = z$, and $\mathbf{\Omega}$ is a vertical vector, the *dimensionless* equations, ((1.4.16) to (1.4.19)), are

$$\nabla \cdot \mathbf{q} = 0, \tag{2.21.1}$$

$$2\hat{\mathbf{k}} \times \mathbf{q} = -\nabla p + T\hat{\mathbf{k}} + E\nabla^2 \mathbf{q}, \tag{2.21.2}$$

$$\mathbf{q} \cdot \hat{\mathbf{k}} = \frac{E f_R}{\sigma_P} \nabla^2 T, \tag{2.21.3}$$

with $\mathbf{q}$ and T prescribed on the surface of the container.

If the Prandtl number is very small, $\sigma_P \ll E$ (or f_R is large), in which case the diffusion of heat is much more important than thermal convection, then the last equation is approximately

$$\nabla^2 T = 0. \tag{2.21.4}$$

In this special circumstance, the temperature is determined independently of the velocity field. Furthermore, if the thermal boundary condition is $T = 0$ on Σ, then the temperature is identically zero throughout the medium and the fluid motion is entirely unaffected by the equilibrium density field. On the other hand, when thermal processes, in this limit, do indeed modify the velocity distribution, the changes are neither radical nor profound. The problem can still be resolved in terms of viscous boundary layers and an inviscid interior. For example, the curl of (2.21.2) yields the vorticity equation,

$$\frac{\partial}{\partial z}\mathbf{q} = \tfrac{1}{2}\hat{\mathbf{k}} \times \nabla T,$$

(the 'thermal wind' equation in meteorology) and in particular

$$\frac{\partial w}{\partial z} = 0.$$

Thus, the vertical velocity established by Ekman layer suction is unchanged by the density field and the stretching of vortex lines remains a major action in the inviscid domain. However, the horizontal velocity components no longer conform to the Taylor–Proudman theorem and a spatial variation is dictated by the imposed temperature field.

The solution of any problem in this parameter range proceeds along the same lines as for homogeneous fluids. As an example, consider the motion in a rotating cylindrical tank with axially symmetric boundary conditions on the velocity and temperature: $\mathbf{q} = \mathscr{U} = r\hat{\boldsymbol{\theta}}$ on $z = 0$; $\mathbf{q} = 0$ elsewhere; $T = \mathscr{T}_T$ on $z = 1$, $\mathscr{T}_B$ on $z = 0$, and $\mathscr{T}_L$ on $r = a$. $T(r, z)$ is determined first and can therefore be presumed known. The viscous boundary layers on the horizontal plates $z = 0, 1$ are ordinary Ekman layers and the $O(E^{\frac{1}{2}})$ normal flux is calculated from (2.17.2) in the usual manner. It follows that the basic interior velocity is circumferential with

$$v_0(r, z) = \frac{r}{2} + \frac{1}{2}\int_0^z \frac{\partial T}{\partial r}\,dz - \frac{1}{4}\int_0^1 \frac{\partial T}{\partial r}\,dz. \tag{2.21.5}$$

The detailed description of the flow field is changed somewhat from the homogeneous fluid solution (the first term on the right-hand side), but the underlying mechanics are essentially the same.

We may conclude, as Carrier [25] did, that when diffusion overwhelms convection, the resultant flow is comparable to that of a homogeneous fluid.

In the other extreme, $\sigma_P \gg 1$ (or $f_R \ll 1$), (2.21.3) reduces to

$$w = \hat{\mathbf{k}}\cdot\mathbf{q} \equiv 0. \tag{2.21.6}$$

No vertical motion is then permitted in the stratified medium and all movement is confined to horizontal sheets. Steady flows in these conditions do not require viscous layers at either horizontal or vertical surfaces since viscosity is important throughout the fluid. In addition, the boundary condition on T must be surrendered because there is no diffusion of heat. For these special values, the problem of axially symmetric motion in the cylinder, posed above, becomes the following:

$$\left(\frac{\partial}{\partial r}\frac{1}{r}\frac{\partial}{\partial r}r + \frac{\partial^2}{\partial z^2}\right)v = 0, \tag{2.21.7}$$

with $v = 0$ on $r = a$, and $z = 1$; $v = r$ on $z = 0$. The other variables are given by

$$u = 0, \quad w = 0, \quad \frac{\partial p}{\partial r} = 2v, \quad T = \frac{\partial p}{\partial z}.$$

The complete solution is easily determined:

$$v = \sum_{n=1}^{\infty} \frac{2a}{\xi_{n1}} \frac{J_1\left(\xi_{n1}\frac{r}{a}\right)}{J_2(\xi_{n1})} \frac{\sinh\left((1-z)\frac{\xi_{n1}}{a}\right)}{\sinh\left(\frac{\xi_{n1}}{a}\right)}. \tag{2.21.8}$$

The particular solution for a configuration of two concentric infinite disks is

$$v = r(1-z), \tag{2.21.9}$$

which shows simply that the azimuthal velocity varies continuously with height. The effect of the spinning plate is totally absorbed by a redistribution of the density and velocity fields from one state of equilibrium to another. The stress is transmitted in diminishing proportion by the horizontal fluid laminas and the velocity remains in geostrophic balance. The motion bears little resemblance to that of a homogeneous fluid in the same conditions and this is generally true when convection overwhelms diffusion, [26].

Steady linear motions for which $\sigma_P \doteq 1$, $f_R \doteq 1$ have been shown by Barcilon and Pedlosky[6] to be very closely related with the extreme case $\sigma_P \gg 1$. In this case, equation (2.21.3) implies immediately that $w = O(E)$ and this relegates the function of the Ekman layers to a distinctly secondary and passive position. Without a vertical velocity of magnitude $E^{\frac{1}{2}}$, there can be no effective stretching of vortex lines and the fluid loses one of its most important means of exercising control.

A single equation for the pressure is

$$\nabla^2\left(\frac{E^2 f_R}{\sigma_P}\nabla^2\nabla^2\nabla^2 p + \nabla^2 p + \left(\frac{4f_R}{\sigma_P} - 1\right)\frac{\partial^2 p}{\partial z^2}\right) = 0, \tag{2.21.10}$$

which reduces properly to all the earlier versions. The structure of the boundary layers on horizontal or vertical surfaces can be determined from this, by scaling the variables appropriately at each wall. If a stretched coordinate is introduced near $z = 0$ so that

$$\frac{\partial}{\partial \zeta} = E^{-a}\frac{\partial}{\partial z},$$

it follows that $a = \frac{1}{2}$; a boundary layer of thickness $E^{\frac{1}{2}}$ can arise on this plane. More important is the conclusion that this is the principal boundary layer thickness—even at a vertical wall. Vertical shear layers of thicknesses $E^{\frac{1}{3}}$ and $E^{\frac{1}{4}}$ cannot exist in a fluid medium with the stated properties. The thinner $E^{\frac{1}{2}}$-layer is required in order for diffusion to change the density of a particle as it rises or sinks along the lateral boundary.

Consider next the problem of motion within a cylinder, the configuration studied by Barcilon and Pedlosky [6]. For definiteness, the boundary conditions are taken to be $\mathbf{q} = \mathscr{U}_B = r\hat{\boldsymbol{\theta}}$ on $z = 0$, $\mathbf{q} = 0$ elsewhere, and the temperature over the entire surface is prescribed arbitrarily. It is assumed, subject to confirmation, that the motion is composed of boundary layers and an inviscid core. The first objective then is to analyse the viscous layers in order to determine the equivalent boundary conditions for the interior flow, which is of major interest.

Let

$$\mathbf{q} = \mathbf{q}_0 + E^{\frac{1}{2}}\mathbf{q}_1 + \ldots + \tilde{\mathbf{q}}_0 + \ldots$$

with similar expansions for the pressure (i.e., $p = \varphi_0 + \ldots$) and temperature. The substitution of these series in (2.21.1) to (2.21.3) leads to a sequence of problems for the flow in the different regions.

The primary interior pressure function, φ_0, satisfies a reduced form of (2.21.10)

$$\nabla^2\left(\nabla^2\varphi_0 + \left(4\frac{f_R}{\sigma_P} - 1\right)\frac{\partial^2\varphi_0}{\partial z^2}\right) = 0. \tag{2.21.11}$$

Once φ_0 is calculated, the velocity and temperature are determined from

$$\mathbf{q}_0 = \tfrac{1}{2}\hat{\mathbf{k}} \times \nabla\varphi_0, \tag{2.21.12}$$

$$T_0 = \frac{\partial\varphi_0}{\partial z}. \tag{2.21.13}$$

According to (2.21.3), $$\frac{\partial^2\tilde{T}_0}{\partial\zeta^2} = 0$$

within a horizontal boundary layer. (Here $\zeta = E^{-\frac{1}{2}}z$ near the plate $z = 0$.) Hence, $\tilde{T}_0 \equiv 0$; there is no zeroth order temperature change across the boundary layer and the interior temperature must itself satisfy the prescribed boundary condition, $T_0 = \mathscr{T}_B$.

To this order of approximation, the horizontal viscous layer is an ordinary Ekman boundary layer and all results previously obtained apply. Therefore, the normal flux at z = 0 is, from (2.17.2),

$$w_1 = \tfrac{1}{2}\hat{\mathbf{k}}\cdot\nabla\times[\hat{\mathbf{k}}\times(\mathbf{q}_0-\mathscr{U}_B)+\mathbf{q}_0-\mathscr{U}_B],$$

whereas, at z = 1, $w_1 = -\tfrac{1}{2}\hat{\mathbf{k}}\cdot\nabla\times(\hat{\mathbf{k}}\times\mathbf{q}_0+\mathbf{q}_0)$.

But the interior vertical velocity is O(E) by (2.21.3), so that

$$w_1 = 0, \tag{2.21.14}$$

which shows that both the top and bottom Ekman layers are non-divergent. These conditions, written in terms of the interior pressure, are

$$\left.\begin{aligned}(\nabla^2-(\hat{\mathbf{k}}\cdot\nabla)^2)\,\varphi_0 = \varGamma^2\varphi_0 = 2\hat{\mathbf{k}}\cdot\nabla\times\mathscr{U}_B = 4, \quad \text{at } z = 0,\\ \varGamma^2\varphi_0 = 0, \quad \text{at } z = 1.\end{aligned}\right\} \tag{2.21.15}$$

Furthermore, the conditions on the temperature are equivalent to the following:

$$\left.\begin{aligned}\frac{\partial\varphi_0}{\partial z} &= \mathscr{T}_B(r,\theta) \quad \text{on } z = 0,\\ \frac{\partial\varphi_0}{\partial z} &= \mathscr{T}_T(r,\theta) \quad \text{on } z = 1.\end{aligned}\right\} \tag{2.21.16}$$

Thus, there are two constraints at each plate which is the proper number required by the 'biharmonic' equation, (2.21.11).

To complete the specification of the boundary value problem, two independent conditions at the sidewall, $r = a$, must be found. This is achieved by analysing the vertical boundary layer using the stretched co-ordinate $\zeta = E^{-\frac{1}{3}}(a-r)$. It is readily shown that in this layer

$$\tilde{u}_0 = 0, \quad \tilde{v}_0 = 0, \quad \tilde{\varphi}_0 = 0, \tag{2.21.17}$$

and

$$-\frac{\partial\tilde{u}_1}{\partial\zeta}+\frac{\partial\tilde{w}_0}{\partial z} = 0, \tag{2.21.18}$$

$$2\tilde{u}_1 = \frac{\partial^2\tilde{v}_1}{\partial\zeta^2}, \tag{2.21.19}$$

$$0 = \tilde{T}_0+\frac{\partial^2\tilde{w}_0}{\partial\zeta^2}, \tag{2.21.20}$$

$$0 = \frac{\sigma_P}{f_R}\tilde{w}_0-\frac{\partial^2\tilde{T}_0}{\partial\zeta^2}. \tag{2.21.21}$$

The last two equations have exactly the same structure as those for the Ekman layer and their integration, subject to $\tilde{w}_0 = 0$ on $\zeta = 0$, yields

$$\left.\begin{aligned} \tilde{w}_0 &= A(\theta, z)\exp\left[-\left(\frac{\sigma_P}{4f_R}\right)^{\frac{1}{4}}\zeta\right]\sin\left(\frac{\sigma_P}{4f_R}\right)^{\frac{1}{4}}\zeta, \\ \tilde{T}_0 &= \left(\frac{\sigma_P}{f_R}\right)^{\frac{1}{2}} A(\theta, z)\exp\left[-\left(\frac{\sigma_P}{4f_R}\right)^{\frac{1}{4}}\zeta\right]\cos\left(\frac{\sigma_P}{4f_R}\right)^{\frac{1}{4}}\zeta. \end{aligned}\right\} \quad (2.21.22)$$

Components $\tilde{u}_1$ and $\tilde{v}_1$ are then determined by integrating (2.21.18) and (2.21.19).

The function $A(\theta, z)$ cannot be fully determined without considering the detailed structure of the corner layers, $|r-a| < E^{\frac{1}{2}}$, $|z| < E^{\frac{1}{2}}$, or $|z-1| < E^{\frac{1}{2}}$, where the vertical and horizontal layers overlap. (These regions are unimportant in the motion of a homogeneous fluid because the vertical layers there, are much wider than $E^{\frac{1}{2}}$. In that case, the conditions at the end plates are those obtained from the analysis of the Ekman layer, (2.17.3).) However, it is possible, surprisingly, to obtain sufficient information to complete the specification of the interior problem for φ_0.

The boundary conditions on the velocity components, together with (2.21.12), imply that

$$\frac{\partial \varphi_0}{\partial \theta} = 0, \quad \frac{\partial \varphi_0}{\partial r} = 0 \quad \text{on } r = a; \qquad (2.21.23)$$

the temperature boundary condition there is

$$\left(\frac{\sigma_P}{f_R}\right)^{\frac{1}{2}} A(\theta, z) = \mathcal{T}_L(\theta, z) - \frac{\partial \varphi_0}{\partial z}. \qquad (2.21.24)$$

The result of averaging the continuity equation, (2.21.18), across the sidewall layer is

$$\frac{\partial}{\partial z}\int_0^\infty \tilde{w}_0\, d\zeta = u_1(a, \theta, z),$$

because $u_1 + \tilde{u}_1 = 0$ at $r = a$. Since

$$u_1 = -\frac{1}{2r}\frac{\partial \varphi_1}{\partial \theta},$$

it follows that

$$\int_0^{2\pi} d\theta \int_0^\infty \tilde{w}_0\, d\zeta = c. \qquad (2.21.25)$$

In the simply-connected domain of a single cylinder (i.e., $r = a$ is the only vertical wall), the *net* vertical mass transport must be nil, in which case, $c = 0$. Upon substituting (2.21.22) for $\tilde{w}_0$, we conclude that

$$\int_0^{2\pi} A(\theta, z)\, d\theta = 0.$$

The function $A(\theta, z)$ may now be eliminated from (2.21.24) and the result is

$$\frac{\partial \varphi_0}{\partial z} = \frac{1}{2\pi}\int_0^{2\pi} \mathscr{T}_L(\theta, z)\, d\theta \quad \text{on } r = a. \qquad (2.21.26)$$

(Note that $\dfrac{\partial \varphi_0}{\partial \theta} = 0$ by (2.21.23).) *Only the horizontally averaged wall temperature has an effect on the interior motion.* Moreover, the pressure levels at the two end plates are related by the averaged sidewall temperature:

$$\varphi_0(a, 1) - \varphi_0(a, 0) = \frac{1}{2\pi}\int_0^1 dz \int_0^{2\pi} \mathscr{T}_L(\theta, z)\, d\theta.$$

The complete boundary value problem for φ_0 consists then of (2.21.11) with boundary conditions (2.21.15), (2.21.16), (2.21.23) and (2.21.26). It is not an easy task to find the solution and in this respect, the approximate techniques developed for similar problems in the theory of elasticity may be of some practical benefit. No attempt will be made here to solve any particular problem and we conclude this discussion with some general observations on the motion of stratified fluids.

In a multiply-connected geometry (see fig. 2.18), the constants which appear, like that in (2.21.25), are evaluated in terms of the circulations about the individual vertical walls. This is similar to the situation encountered in the mass flux problems in § 2.20.

Vortex line stretching is notable once again in time-dependent motions, but the mechanism is much weakened in the presence of stratification. Holton[87] studied spin-up of a stratified fluid from both a theoretical and experimental standpoint. He found that the fluid adjusts to a quasi-equilibrium distribution during the usual spin-up time but the ultimate transition to rigid rotation is accomplished by diffusion in the longer time scale, $E^{-1}\Omega^{-1}$. Pedlosky[147] has also examined this problem and some disagreement exists concerning the case of continuous stratification.

Spin-up of a multi-layered medium of homogeneous fluids differs only qualitatively from the single layer problem. The mechanisms within each stratum are the same but the layers slip upon each other. This has the effect of lengthening the characteristic time which remains, however, of the order of $E^{-\frac{1}{2}}\Omega^{-1}$.

Stability considerations cannot be overlooked in either the transient or steady motions of a stratified fluid. The establishment of Ekman layers in the initial stages of a transient motion may involve large displacements of the particles originally on the container wall. Because the formation time scale is too short for diffusive processes to alter the density appreciably, a fluid element can be shifted to an inherently unstable position. A particle heavier than its surroundings will sink and this is the reason for much of the mixing observed in transient motions. Steady forced motions can also produce instabilities in this fashion. The precise condition for mixing to develop and the extent to which it occurs have not been established.

The theory discussed here presumes that diffusion and the O(E) dynamics are the dominant processes in steady motions of a stratified fluid. Since E is very small and viscous shear is in large measure counteracted by stratification, the question naturally arises about the magnitudes of processes heretofore neglected. It becomes necessary to assess the contributions to the motion of all those terms eliminated in the linearization and this should provide a number of surprises.

CHAPTER 3

CONTAINED ROTATING FLUID MOTION: NON-LINEAR THEORIES

3.1. Introduction

The general purpose of this chapter is to examine non-linear processes in rotating flows. An attempt is made to follow the outline of the preceding chapter as closely as possible in order to facilitate a direct comparison between linear and non-linear solutions of similar problems. The desired goal is to convey enough basic information to build a reasonable fund of knowledge about non-linear effects. Hopefully, experience and familiarity with the subject can be translated into judicious approximations to make many more problems amenable to solution with the techniques currently available. As yet, results of general validity are extremely rare finds, and only a few specific non-linear examples have been solved in sufficient detail to be edifying.

3.2. Boundary layer on an infinite plate

Kármán[93] found an exact solution of the steady non-linear equations which is analogous in part to the Ekman layer solution derived and discussed in §2.3. Bödewadt[10], Batchelor[7], Stewartson[187, 189], Thiriot[207], Rogers and Lance[104, 169], Pearson[143] and Benton[8], among others, have considered modifications and generalizations of this work.

In the general problem, an infinite plane, $z = 0$, rotates with angular velocity $\Omega_w \Lambda(t)\,\hat{\mathbf{k}}$, $(|\Lambda(t)| \leqslant 1)$, in an otherwise unbounded medium. Fluid far from the plate rotates with angular velocity $\Omega_f\,\hat{\mathbf{k}}$. (The linear theory presumes that the two rotation speeds are only slightly different at any time.) If the maximum angular velocity,

$$\Omega_{\text{max}} = \max\,[\Omega_w, \Omega_f],$$

is used to characterize the time, then the *dimensionless* equations of motion in an *inertial frame* (see (1.2.7) et seq.) are

$$\left.\begin{aligned}\frac{\partial}{\partial t}\mathbf{q}+\mathbf{q}\cdot\nabla\mathbf{q} &= -\nabla p+E\nabla^2\mathbf{q},\\ \nabla\cdot\mathbf{q} &= 0.\end{aligned}\right\} \tag{3.2.1}$$

Here the Rossby number is taken as unity, the characteristic velocity is $\Omega_{max} L$, and the boundary conditions are

$$\mathbf{q} = \frac{\Omega_w \Lambda(t)}{\Omega_{max}} r\hat{\boldsymbol{\theta}} \quad \text{on } z = 0,$$

and
$$\lim_{z\to\infty}\left(\hat{\boldsymbol{\theta}}\cdot\mathbf{q}-\frac{\Omega_f}{\Omega_{max}}r\right) = 0, \quad \lim_{z\to\infty}\hat{\mathbf{r}}\cdot\mathbf{q} = 0.$$

The vertical velocity component cannot be prescribed at infinity because the boundary layer at the plate induces a circulation to sustain a mass transport. This flux persists throughout the infinite realm.

Since the motion is axially symmetric, it is convenient to use cylindrical co-ordinates in which case the equations reduce to

$$\frac{\partial u}{\partial t}+u\frac{\partial u}{\partial r}-\frac{v^2}{r}+w\frac{\partial u}{\partial z} = -\frac{\partial p}{\partial r}+E\left(\frac{\partial^2 u}{\partial r^2}+\frac{1}{r}\frac{\partial u}{\partial r}+\frac{\partial^2 u}{\partial z^2}-\frac{u}{r^2}\right), \tag{3.2.2}$$

$$\frac{\partial v}{\partial t}+u\frac{\partial v}{\partial r}+\frac{uv}{r}+w\frac{\partial v}{\partial z} = E\left(\frac{\partial^2 v}{\partial r^2}+\frac{1}{r}\frac{\partial v}{\partial r}+\frac{\partial^2 v}{\partial z^2}-\frac{v}{r^2}\right), \tag{3.2.3}$$

$$\frac{\partial w}{\partial t}+u\frac{\partial w}{\partial r}+w\frac{\partial w}{\partial z} = -\frac{\partial p}{\partial z}+E\left(\frac{\partial^2 w}{\partial r^2}+\frac{1}{r}\frac{\partial w}{\partial r}+\frac{\partial^2 w}{\partial z^2}\right), \tag{3.2.4}$$

$$\frac{1}{r}\frac{\partial}{\partial r}(ru)+\frac{\partial w}{\partial z} = 0. \tag{3.2.5}$$

Kármán's similarity solution is of the form

$$u = rF(z,t), \quad v = rG(z,t), \quad w = H(z,t), \tag{3.2.6}$$

with
$$p = K(t)\,r^2+L(z,t).$$

Substitution of these expressions into the system (3.2.2)–(3.2.5), and the elimination of the pressure, yields

$$\left.\begin{aligned} \frac{\partial^2 F}{\partial t \partial z} + 2F\frac{\partial F}{\partial z} - 2G\frac{\partial G}{\partial z} + \frac{\partial}{\partial z}\left(H\frac{\partial F}{\partial z}\right) &= E\frac{\partial^3 F}{\partial z^3}, \\ \frac{\partial G}{\partial t} + 2FG + H\frac{\partial G}{\partial z} &= E\frac{\partial^2 G}{\partial z^2}, \\ 2F &= -\frac{\partial H}{\partial z}. \end{aligned}\right\} \quad (3.2.7)$$

The assumed functional forms are entirely consistent with the prescribed boundary conditions and

$$F(0, t) = 0, \quad G(0, t) = \frac{\Omega_w}{\Omega_{max}}\Lambda(t), \quad H(0, t) = 0;$$

$$F(\infty, t) = 0, \quad G(\infty, t) = \frac{\Omega_f}{\Omega_{max}}.$$

Although rather arbitrary initial conditions on F, G and H can be set, the motion is usually thought to start from a state of rest or rigid rotation.

The lack of a definite characteristic distance, in the case of an infinite fluid bounded by a single plate, means that no parameter having a length scale need appear explicitly in the basic formulation. The Ekman number E can be removed through a redefinition of variables which is, in essence, a boundary layer scaling.

Let

$$\zeta = E^{-\frac{1}{2}}z, \quad F = F(\zeta, t), \quad G = G(\zeta, t), \quad H = E^{\frac{1}{2}}H(\zeta, t).$$

By eliminating F from the problem, the equations can be reduced to the following fifth order system,

$$\left.\begin{aligned} \frac{1}{2}\frac{\partial^2 H}{\partial t \partial \zeta} - \frac{1}{4}\left(\frac{\partial H}{\partial \zeta}\right)^2 + G^2 + \frac{1}{2}H\frac{\partial^2 H}{\partial \zeta^2} &= \frac{1}{2}\frac{\partial^3 H}{\partial \zeta^3} + \left(\frac{\Omega_f}{\Omega_{max}}\right)^2, \\ \frac{\partial G}{\partial t} + H\frac{\partial G}{\partial \zeta} - G\frac{\partial H}{\partial \zeta} &= \frac{\partial^2 G}{\partial \zeta^2}, \end{aligned}\right\} \quad (3.2.8)$$

with $$G - \frac{\Omega_w}{\Omega_{max}}\Lambda = H = \frac{\partial H}{\partial \zeta} = 0 \quad \text{on } \zeta = 0,$$

$$G - \frac{\Omega_f}{\Omega_{max}} = \frac{\partial H}{\partial \zeta} = 0 \quad \text{at } \zeta = \infty, \text{ and } G(\zeta, 0), H(\zeta, 0) \text{ prescribed.}$$

Kármán's original problem is a special case of this formulation which corresponds to steady motion with $\Lambda(t) = 1$ and $\Omega_f = 0$. Bödewadt's problem also concerns steady motion but has $\Omega_w = 0$ and $\Omega_{max} = \Omega_f$.

Cochran's [34] numerical solution of Kármán's problem is displayed in fig. 3.1. The profiles show no oscillations in the boundary layer and, in this respect, are quite untypical of the general steady

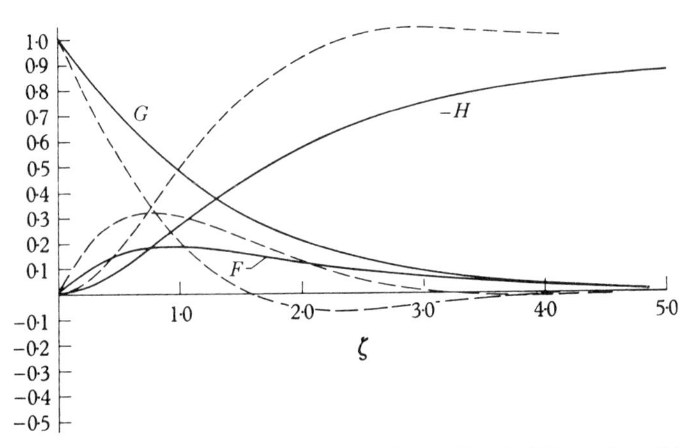

Fig. 3.1. Solid curves are the similarity velocity profiles for Kármán's problem. Dashed curves are the solutions of the corresponding linear problem, p. 32.

problem. In fact, Rogers and Lance [169] carried out the numerical integrations for a large number of parameter settings ranging from

$$\frac{\Omega_f}{\Omega_w} = 0 \quad \text{to} \quad \frac{\Omega_w}{\Omega_f} = 0,$$

and they found that Kármán's solution is the *only* solution which does not exhibit oscillatory behaviour in the boundary layer. Any fluid rotation at infinity, is sufficient to produce such fluctuations.

Pearson [143] and Benton [8] computed the time-dependent evolution of Kármán's solution with $\Lambda = \mathcal{S}(t)$, (an impulsive start), and zero initial conditions. Benton's results for the azimuthal velocity profile are exhibited in fig. 3.2. As in the linear theory, the steady state is almost completely established within the first few revolutions of the plate, $t \doteq 2$. After this initial phase, steady boundary layer theory applies accurately, although small inertial oscillations do persist for a much longer period.

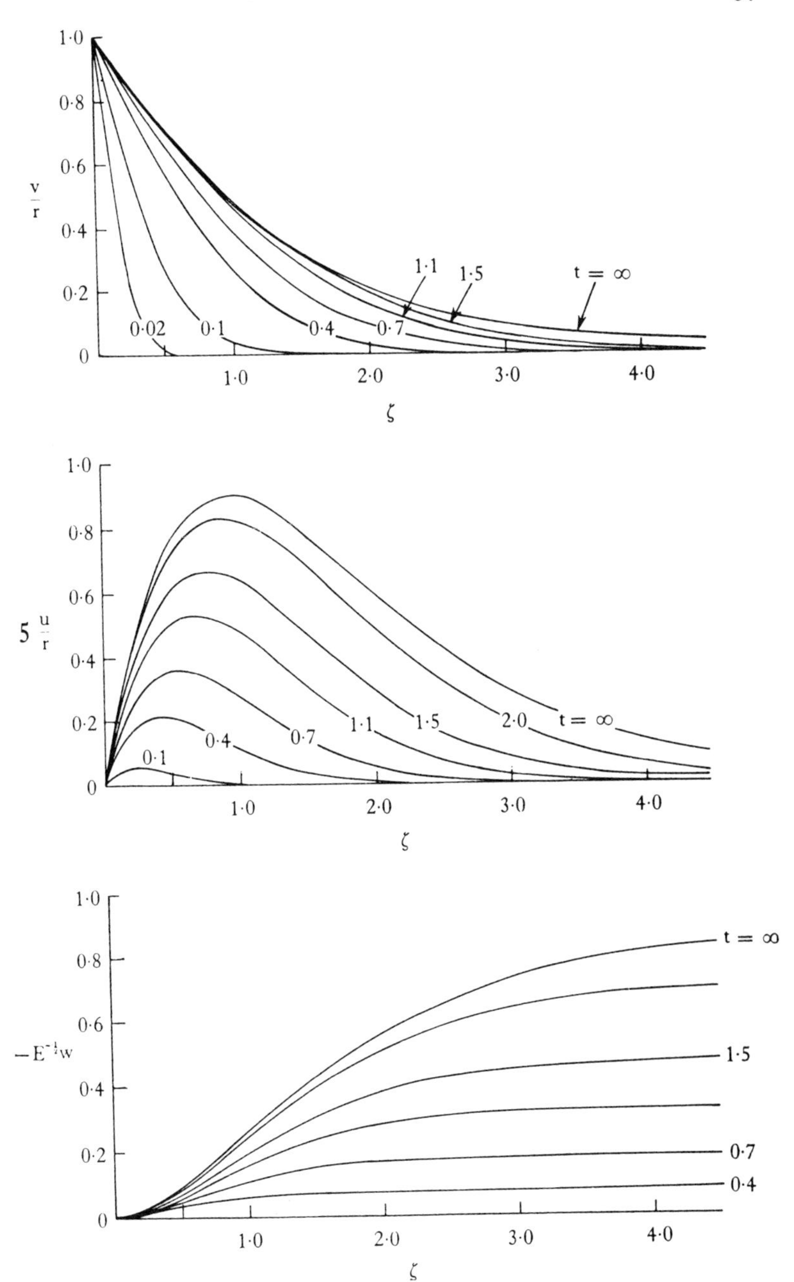

Fig. 3.2. The transient evolution of Kármán's solution, [8].

Bödewadt's solution is shown in fig. 3.3. A great deal of controversy exists about the realization of this idealized motion. Much of this stems from the fact that the flow begins at infinity and, thenceforth, emerges from the boundary layer to feed the interior. Moreover, a fluid element moving inward along the plate has to lose memory of its original state and fall completely under the control of local inviscid flow field outside the layer. It is difficult to accept

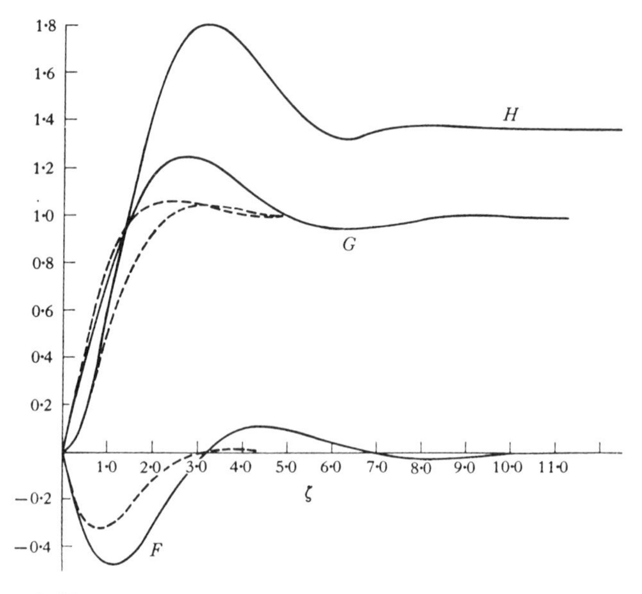

Fig. 3.3. Solid curves are the similarity velocity profiles for Bödewadt's problem. The dashed curves are the solutions of the corresponding linear problem.

that all this happens without boundary layer separation and flow instability being involved.

From an analytical point of view, the solution has a firm foundation. The investigations of the analogous finite disk problem show that the similarity profiles are indeed excellent approximations over the inner half of the plate. Rott and Lewellen's[177] study of the question of terminal similarity, which is based on momentum integral methods (§3.6), and the associated experiments of Maxworthy[127] leave no doubt that the Bödewadt solution can be used effectively and accurately in the determination of interior flows.

This is certainly the case in all the linear problems that have been experimentally verified.

Disagreement still exists on the matter of the stability of the boundary layer. Of course, a definitive set of experiments would resolve the dispute. Such experiments have not been made, but it seems likely that the layer will prove stable under restrictive conditions, namely, a small enough value of a boundary layer Reynolds number.

The calculations of Rogers and Lance[169] for the general steady problem show a continuous transition in the velocity profiles as the ratio Ω_f/Ω_w changes. Fig. 3.4 exhibits the vertical component of velocity at infinity induced by mass transport in the boundary layer versus Ω_f/Ω_w. The information provided in this diagram will be extremely useful in developing approximate methods of analysis (as in § 3.6).

An unusual feature of the numerical solution is that the axial transport is not a monotonic function of Ω_f/Ω_w. Slight fluid rotation at infinity actually increases the magnitude of the normal flux. This is quite possibly related to the oscillations established in the boundary layer as a result of the rotational motion at infinity, but no clear cause and effect mechanism has been given.

3.3. Boundary layer on a finite plate

The question of applicability and pertinence of similarity solutions to real flows is ever present. In what circumstances, and for which boundary conditions, will these solutions represent an adequate and reliable approximation? This and the next section will attempt to throw some light on these issues.

Rogers and Lance[170] investigated the boundary layer in a rotating fluid over a stationary, finite disk in order to examine the validity and limitations of the Bödewadt solution. Their procedure is to compute numerically sufficiently many terms of the series expansion given by Stewartson[189], which is valid near the outer edge of the disk, so that an adequate representation of the flow is obtained at smaller radii, say $r = \frac{1}{2}$. This series solution is then compared to the similarity solution for the infinite plate problem which presumably holds at least near the very centre of the disk. Hopefully, the two should attach to each other and indeed they

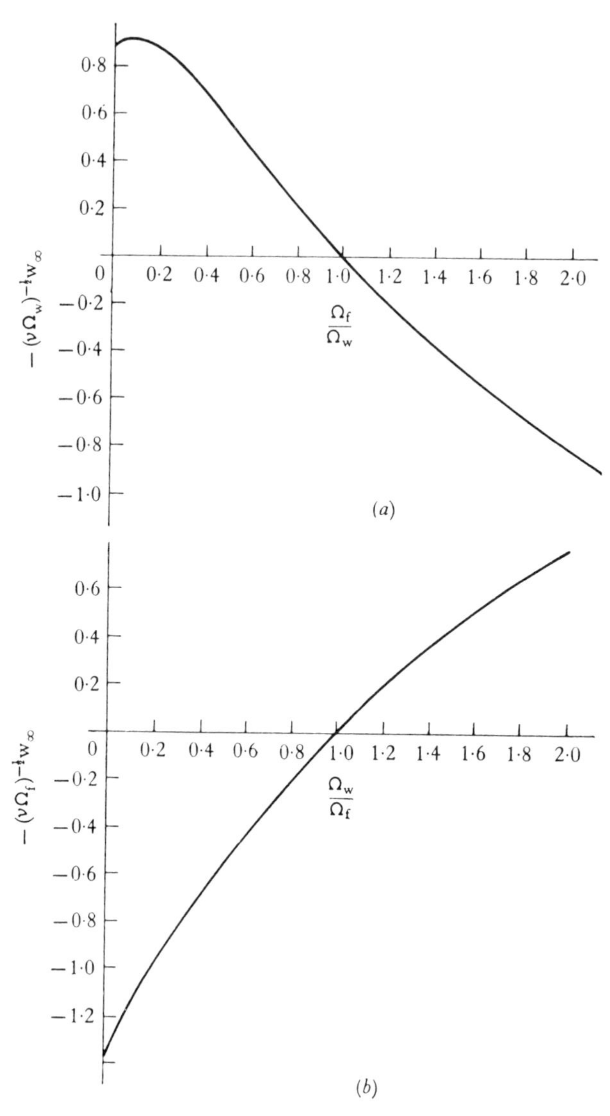

Fig. 3.4. The induced vertical velocity at a flat plate as a function of the rotation speeds of the wall, Ω_w, and the fluid outside the boundary layer, Ω_f [169].

seem to; in fact, Bödewadt's solution is found to be a good approximation over the inner half of the disk. It may be concluded that the boundary layer flow in the central region depends essentially on the local inviscid flow field and has completely 'forgotten' the manner in which it originated at the edge of the disk.

The configuration studied consists of a fluid filled semi-infinite cylinder, of radius a and rotation rate Ω_f, which is capped with a *stationary* disk. Far from the end plate, the fluid rotates rigidly with its sidewalls and the pressure gradient exactly balances the centrifugal force. Flow near the disk is confined to a viscous boundary layer and the solution of the non-linear boundary layer equations should be reasonably accurate over most of the end plate surface—except, perhaps, in the immediate vicinity of the corners. Here, radial shear terms can be expected to become increasingly important, but these effects are ignored in this first attempt. As a consequence, the boundary layer equations, which are based solely on the vertical derivatives, are found to imply infinite axial velocities at the outer radius although the other components do remain finite. This singularity could be corrected by a proper sidewall boundary layer calculation, but the net effect on the central flow, the question at hand, would most probably be small and not really worth the effort from this standpoint.

The approximate, *dimensionless* equations in an *inertial* frame that govern the boundary layer region are, according to § 1.6:

$$u\frac{\partial u}{\partial r}+w\frac{\partial u}{\partial z}-\frac{v^2}{r} = -r+E\frac{\partial^2 u}{\partial z^2}, \tag{3.3.1}$$

$$u\frac{\partial v}{\partial r}+w\frac{\partial v}{\partial z}+\frac{uv}{r} = E\frac{\partial^2 v}{\partial z^2}, \tag{3.3.2}$$

$$\frac{\partial}{\partial r}(ru)+\frac{\partial}{\partial z}(rw) = 0. \tag{3.3.3}$$

The pressure is essentially constant through the layer and equal to its value at 'infinity'; the radius a and $\Omega_f\, a$ are the characteristic scales. The boundary conditions are simply

$$u = v = w = 0 \quad \text{on } z = 0,\ 0 \leqslant r \leqslant 1,$$

$$u \to 0, \quad v \to r \quad \text{as } z \to \infty,\ 0 \leqslant r < 1,$$

and

$$u = 0, \quad v = 1 \quad \text{at } r = 1, \text{ for all } z.$$

The introduction of a stream function,

$$u = \frac{1}{r}\frac{\partial\psi}{\partial z}, \quad w = -\frac{1}{r}\frac{\partial\psi}{\partial r},$$

facilitates the solution procedure.

Stewartson's expansion method employs the variables

$$\mathfrak{x} = 1 - r, \quad \eta = E^{-\frac{1}{2}} z \mathfrak{x}^{-\frac{1}{4}}, \tag{3.3.4}$$

and the series

$$\psi = E^{\frac{1}{2}} \mathfrak{x}^{\frac{3}{4}} \sum_{n=0}^{\infty} \psi_n(\eta)\, \mathfrak{x}^n, \quad v = \sum_{n=0}^{\infty} V_n(\eta)\, \mathfrak{x}^n. \tag{3.3.5}$$

These expansions are substituted into (3.3.1)–(3.3.3) and the coefficients of like powers of $\mathfrak{x}$ equated to obtain an infinite number of problems for the unknown functions. The first is

$$\left.\begin{aligned} \frac{d^3\psi_0}{d\eta^3} - \frac{3}{4}\psi_0 \frac{d^2\psi_0}{d\eta^2} + \frac{1}{2}\left(\frac{d\psi_0}{d\eta}\right)^2 &= 1 - V_0^2, \\ \frac{d^2V_0}{d\eta^2} - \frac{3}{4}\psi_0 \frac{dV_0}{d\eta} &= 0, \end{aligned}\right\} \tag{3.3.6}$$

with

$$V_0(0) = \psi_0(0) = \frac{d}{d\eta}\psi_0(0) = 0 \quad \text{and} \quad V_0(\infty) - 1 = \frac{d}{d\eta}\psi_0(\infty) = 0.$$

The numerical solutions of these equations are compared to the Bödewadt solution in two ways. At position $z = E^{\frac{1}{2}}$, which is close to the outer edge of the boundary layer, the velocity profiles obtained from both sources are plotted together versus radial distance. Fig. 3.5 shows the results; the curves seem to join at about $r = \frac{1}{2}$. Secondly, the profiles are compared at a definite radius, $r = \frac{1}{2}$, as functions of the axial co-ordinate $zE^{-\frac{1}{2}}$, and this is illustrated in fig. 3.6. Reasonable agreement is attained.

Presumably, calculations of still higher order problems would make the fit even smoother, as suggested by the inclusion of third order terms, but the appearance of a reverse circulation can cause trouble. The axial velocity is not shown because, although qualitative agreement does exist, the similarity is not as close for reasons mentioned previously. The results indicate that the similarity solution is reasonably accurate in the region $r < \frac{1}{2}$. A similar conclusion was drawn by King and Lewellen[96] from a momentum-integral solu-

A matched expansion procedure joining series solutions valid mall r and $1-r$ remains to be completed, but some discussion of approach is given by Rott and Lewellen[177].

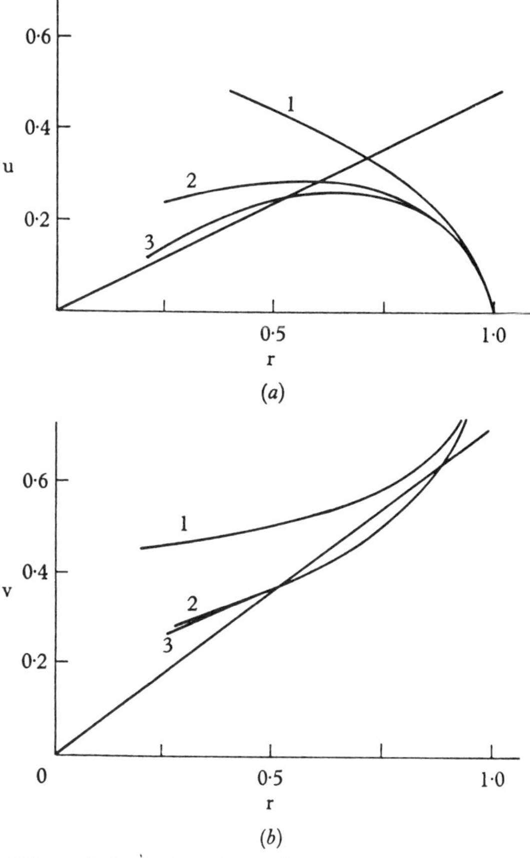

Fig. 3.5. (*a*) The radial velocity at $\zeta = 1$ for a finite disk rotating in a stationary fluid[170]. Curves 1, 2 and 3 show the effect of increasing the number of terms in the 'edge' expansion (3.3.5). (*b*) The azimuthal velocity at $\zeta = 1$ for the same problem.

It is no more difficult to solve the more general problem in which the end plate is also allowed to rotate. This is a particularly interesting situation when the disk and sidewalls rotate exactly counter to each other. Although the equations (3.2.8) seem to hold in this situation too, no acceptable similarity solution for the single, infinite

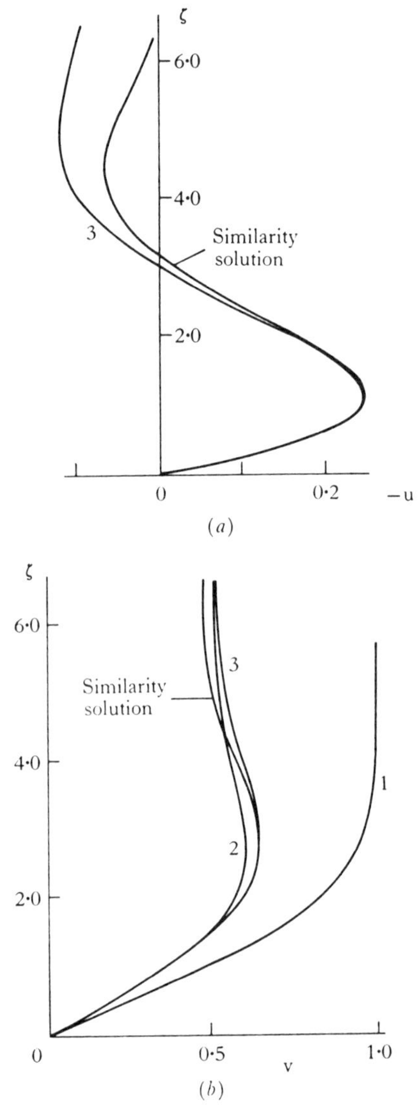

Fig. 3.6. (*a*) The radial velocity at $r = \frac{1}{2}$ as a function of the boundary layer co-ordinate ζ. The curve labelled 3 corresponds to three terms of the 'edge' expansion. (*b*) The azimuthal velocity at $r = \frac{1}{2}$.

plate in an unbounded fluid has been found. The computations based on the edge expansion for the finite plate indicate that a counter-rotating plate delays the onset of mass efflux from the boundary layer. For example, when $\Omega_w = 0$, $r = 0 \cdot 84$ is the position at which the axial velocity reverses sign and the boundary layer begins to exude fluid into the interior. However, at $\Omega_w = -0{\cdot}9\ \Omega_f$, the corresponding value is $r = 0{\cdot}66$. Furthermore, there seems to be a region near the centre where the fluid rotates faster than the disk (in the same direction). Pearson[143] found the same type of behaviour in an analogous situation discussed in the next section. It should be noted that these conclusions are, in many respects, incomplete and more work is required to elevate them to the status of absolute acceptance.

3.4. Motion between concentric plates

Flow between infinite, concentric, rotating disks was first discussed by Batchelor[7], later by Stewartson[187], Lance and Rogers[104] and by Pearson[143]. The first two of these are theoretical analyses, whereas the latter two are numerical investigations of the steady and time-dependent motions respectively.

The simplest case has both disks rotating with the same instantaneous angular velocity and for this and many other conditions, the flow is describable as a similarity solution of equation (3.2.7). A solution of this form presumes implicitly that the sidewall entry conditions at infinite radius are unimportant, so that only the local relationship between the internal motion and viscous boundary layers is of any consequence. Of course, the relevance of such similarity solutions to the finite disk configuration must be closely scrutinized and their limitations realistically assessed.

When the plates rotate steadily in the same direction and the Ekman number is small, the flow field consists of a boundary layer at each plate and an interior state of rigid rotation at an intermediate rate that is compatible with flux requirements. The calculations of Rogers and Lance[169], reported in §3.2, are already sufficient to describe the motion quite accurately.

Let the rotation speed of one disk be maintained at the constant value $\Omega_w = \Omega_1$, while the angular velocity of the other may be set at any value between zero and Ω_1. At the faster disk, there must be a net influx of mass into the boundary layer in a manner similar to that described in Kármán's problem. Likewise, at the slower disk, there must be a mass efflux according to Bödewadt's similarity solution. The interior rotates rigidly at the exact intermediate rate for which the local mass efflux from one boundary layer matches identically, at each radial position, with that required by the other.

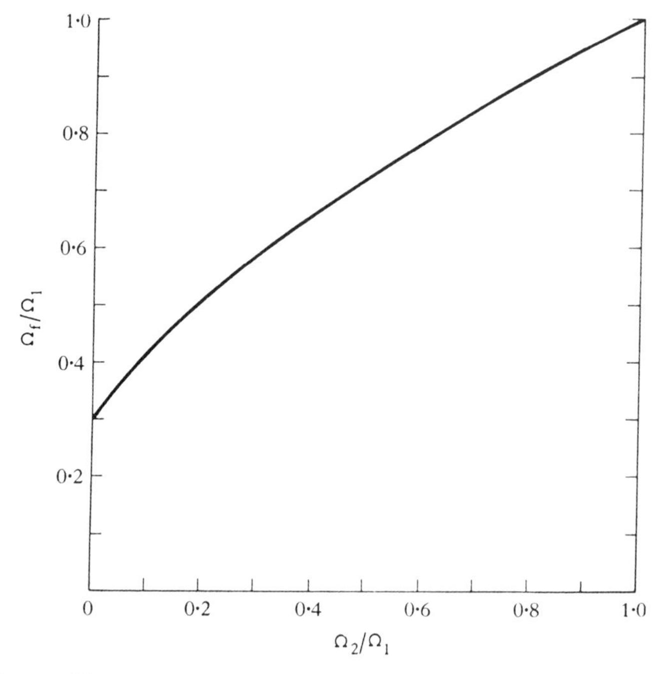

Fig. 3.7. Two concentric plates rotate with frequencies Ω_1 and Ω_2. Ω_f is the angular velocity of the contained fluid for which the vertical mass transports induced by the boundary layers at each plate balance exactly, [104].

Calling the rotation rate of the internal fluid Ω_f, ($\Omega_f r$ is the velocity at 'infinity' as seen from each layer), then the relationship of Ω_1, Ω_2 and Ω_f dictated by this local flux or transport balance is shown in fig. 3.7. The azimuthal velocity in the interior varies from $\Omega_1 r$, when the two plates rotate rigidly in unison, to the value of $0{\cdot}3\Omega_1 r$, when one plate is stationary. This solution assumes implicitly that the transient evolution of the steady flow is one in which viscous diffusion eventually permeates the fluid to affect all fluid particles. For infinite plates, this is undoubtedly true; the assumption is not necessarily valid in many real problems when the inflow from lateral regions can be controlled.

This analysis does not apply to the case of counter-rotating disks, where a substantial departure from an internal state of rigid rotation is observed.

Pearson[143] numerically integrated the time-dependent equations, (3.2.7), for a variety of boundary conditions, but with the fluid always in an initial state of rest. The angular velocity of one or both plates is impulsively increased and the calculations are carried out until the final steady state is achieved. The steady solutions obtained this way are in overall agreement with the direct calculations of the time-*in*dependent equations by Lance and Rogers[104] in every case presented except one, which is discussed shortly.

One example of a transient numerical solution for small Ekman number, $E = 10^{-3}$, is shown in fig. 3.8. In this problem, $\Omega_1 = S(t)$, $\Omega_2 = 0$. Especially noteworthy is the fact that two flow regimes are in evidence. In the interval of spin-up, $t \doteq E^{-\frac{1}{2}}$, $(5 < t < 150)$, the flow consists of a single, almost steady, boundary layer on the surface of the moving disk. Elsewhere, the fluid motion is very small and essentially inviscid; most of the fluid particles are as yet unaffected by viscous diffusion. Vorticity diffusion radically alters the pattern in the longer characteristic time scale, $t \doteq 1000 \doteq E^{-1}$, when viscous processes permeate the interior. By that time, all fluid particles have experienced some viscous control and the ultimate steady state predicted by the steady boundary layer calculations is obtained.

It seems clear that if, in real circumstances, the internal fluid can be affected by viscous diffusion, either directly or by recirculation, then the flow should be similar to the final steady state pictured. If, however, the configuration is arranged so that the main body of fluid is unaffected by viscosity, then the flow established by the boundary layer alone in the spin-up time is more relevant.

Numerical results for the case of counter-rotating disks are shown in fig. 3.9. The Ekman number there is only 10^{-2}, not truly small, and the boundary layers are very broad. Steady and unsteady calculations agree when $\Omega_2 = -\Omega_1$ and $E = 10^{-2}$, but this is not the case at smaller values of the Ekman number. At $E = 10^{-3}$, Rogers and Lance[169] compute a symmetric pattern about the mid-plane, but Pearson's results show a marked asymmetry with an internal angular velocity of greater magnitude than that of either plate. The source of this discrepancy is not clear but it seems possible that there might be some inherent numerical instability associated with the internal flow when the effects of each plate tend to cancel. Since

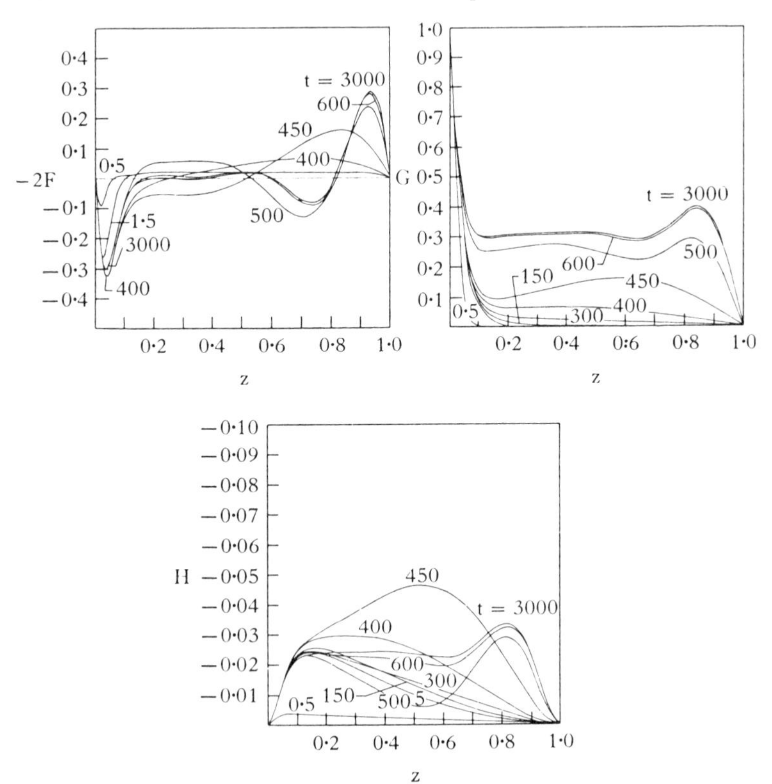

Fig. 3.8. The transient fluid motion between two plates, one of which is started impulsively; $\Omega_1 = S(t)$ and $\Omega_2 = 0$. $E = 10^{-3}$ and functions G, H and F are defined in (3.2.6), [143].

neither plate exercises predominant control, the initial setting, and conditions at the sidewalls, may also become crucial factors in the selection of the final state.

The boundary conditions at the lateral walls naturally enter into the determination of which solution occurs in the case of finite rotating disks. Two configurations have received considerable attention from an experimental standpoint (Schultz-Grunow[179], Picha and Eckert[152], Stewartson[187]). The first involves an open arrangement of two centred disks of equal size; one disk is *stationary* and the other rotates uniformly. Observations show that the main body of fluid or gas between the disks remains essentially quiescent. There is motion in the boundary layer on the rotating plate, but little elsewhere. On the other hand, if the disks are surrounded or shrouded by a close fitting cylindrical container, there is

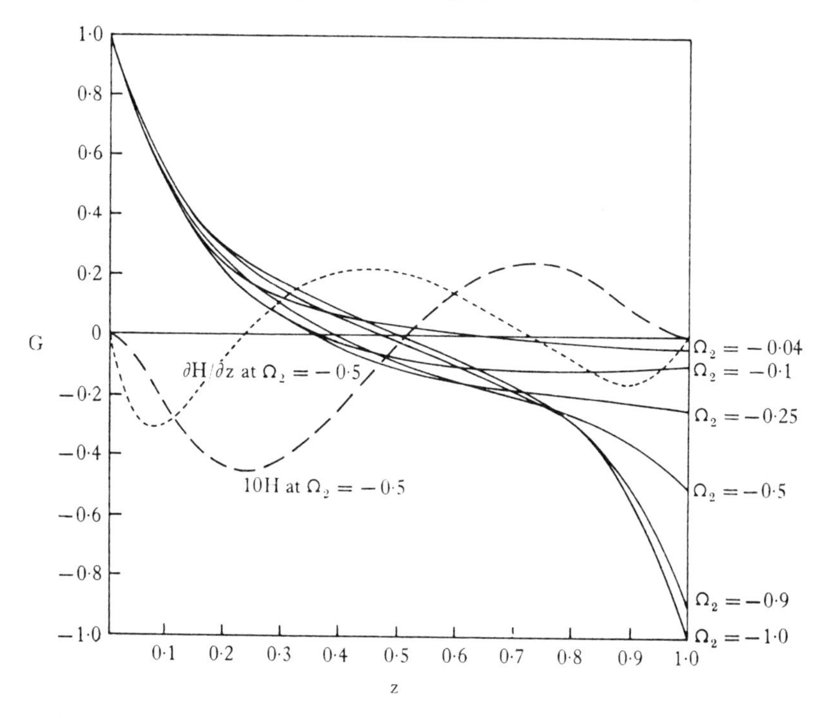

Fig. 3.9. The final steady motion between counter-rotating disks with $\Omega_1 = 1$, $\Omega_2 = -1, -0{\cdot}9, -0{\cdot}5, -0{\cdot}25, -0{\cdot}1, -0{\cdot}04$ and $E = 10^{-2}$, [143].

motion of the internal fluid which closely resembles the flow between infinite plates. The azimuthal velocity is then about three-tenths the speed of the turning plate, over most of the central core. The experimental results regarding counter-rotating disks show no circular motion of the internal fluid and, hence, support the numerical results of Rogers and Lance.

The different flows can be explained on the basis of Pearson's numerical work. The boundary layer on the rotating disk in the unshrouded or free configuration acts upon an inexhaustible supply of inviscid fluid fed continually by the infinite, external reservoir. A particle entering the region between the plates remains there for the order of the spin-up time, $E^{-\frac{1}{2}}$, before it is ejected from the boundary layer back into the reservoir by centrifugal forces. Thus, viscous diffusion is always counteracted and can never permeate the entire fluid domain. The resultant motion is that established directly by boundary layer action (the profiles shown in fig. 3.8 which correspond to the spin-up time, $t \doteq E^{-\frac{1}{2}}$). The major flow is confined to the boundary layer, and motion elsewhere is of a much smaller magnitude. However, for the shrouded disks, the interior is affected by viscosity because the fluid between plates is contained and the same fluid particles are acted upon continually. The ultimate state then corresponds to the infinite disk problem wherein diffusion plays a similar role.

When both disks of an *unshrouded* configuration revolve (not necessarily at the same frequency), a central core of the internal fluid is also set into rotation at a comparable rate, but elsewhere the almost quiescent interior conditions are unperturbed. The reasons for this may lie in the fact that any rotation whatsoever, acquired by an interior fluid particle, is markedly accentuated as it approaches the centre line because of the conservation of angular momentum.

3.5. Spherical boundaries

Motion produced by a rapidly rotating sphere in an otherwise quiescent fluid of infinite extent was studied by Howarth[89]. Fox[53] and Banks[3] computed his solution in greater detail; the former author generalized the approach to solve analogous problems pertaining to many other body shapes. In most respects, the flow pattern resembles that established by a single rotating plate, the identification being strongest at the polar regions. Boundary layers form at these positions and mass, drawn in from the outside, is expelled along the surface towards the equatorial latitudes. Near the equator, the flows within the modified Kármán boundary layers on both hemispheres impinge upon each other. The interaction produces a radial jet of fluid at the equator, shown in fig. 3.10 which is taken from the work of Bowden and Lord[13], (see also Kreith *et al.*[99]). A theoretical analysis of the motion at the equator is given by Stewartson[189].

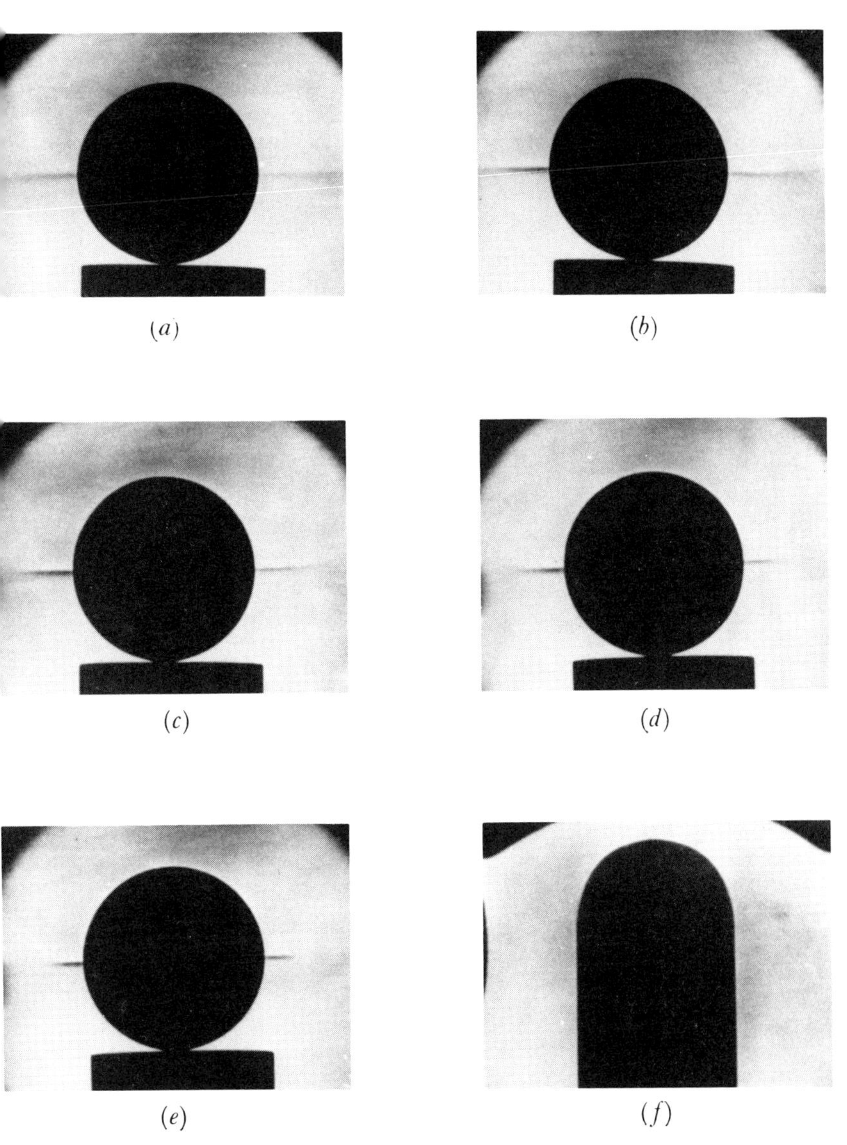

Fig. 3.10. Schlieren photographs, by Bowden and Lord [13], of the boundary layer on a rotating sphere. A radial jet is clearly visible and appears to grow shorter as the equatorial speed is increased from 6 m/s in (*a*) to 18 m/s in (*f*), possibly because of some turbulence. No jet is present in the case of a rotating half body shown in (*f*).

Although ordinary boundary layer theory is incapable of describing the collision, it seems to apply extremely well over the rest of the spherical surface. The disturbance created near the equator has a minor influence on the incoming boundary layer flows when the Ekman number is small. The numerical solutions discussed shortly corroborate this conclusion.

The steady axisymmetric boundary layer equations (the *dimensionless* form of (1.6.3)–(1.6.5) with $\Omega = 0$) written in terms of spherical co-ordinates

$$(\xi_1, \xi_2, \xi_3) = (\Theta, \theta, \mathfrak{r} - 1),$$
$$(h_1, h_2, h_3) = (1, \sin\Theta, 1),$$
$$(q_1, q_2, q_3) = (u, v, w),$$

are

$$\frac{1}{\sin\Theta}\frac{\partial}{\partial\Theta}(u\sin\Theta) + \frac{\partial w}{\partial \mathfrak{r}} = 0, \tag{3.5.1}$$

$$u\frac{\partial u}{\partial\Theta} + w\frac{\partial u}{\partial\mathfrak{r}} - v^2\cot\Theta = E\frac{\partial^2 u}{\partial\mathfrak{r}^2}, \tag{3.5.2}$$

$$u\frac{\partial v}{\partial\Theta} + w\frac{\partial v}{\partial\mathfrak{r}} + uv\cot\Theta = E\frac{\partial^2 v}{\partial\mathfrak{r}^2}. \tag{3.5.3}$$

Here, the pressure outside the layer is a constant and the usual characteristic scales, $[\![\mathfrak{r}_0, \Omega^{-1}, \mathfrak{r}_0\Omega]\!]$, have been used in the dimensionalization.

The method of solution is based on a series expansion in the polar angle Θ;

$$\left.\begin{aligned} u &= \Theta F_1(\zeta) + \Theta^3 F_3(\zeta) + \ldots, \\ v &= \Theta G_1(\zeta) + \Theta^3 G_3(\zeta) + \ldots, \\ w &= E^{\frac{1}{2}}[H_1(\zeta) + \Theta^2 H_3(\zeta) + \ldots], \end{aligned}\right\} \tag{3.5.4}$$

where

$$\zeta = E^{-\frac{1}{2}}(\mathfrak{r} - 1).$$

A system of ordinary differential equations for the unknown functions results from the substitution of these series into (3.5.1)–(3.5.3), and equating like powers of Θ. The first is

$$\left.\begin{aligned} F_1^2 + H_1\frac{dF_1}{d\zeta} - G_1^2 &= \frac{d^2F_1}{d\zeta^2}, \\ 2F_1G_1 + H_1\frac{dG_1}{d\zeta} &= \frac{d^2G_1}{d\zeta^2}, \\ 2F_1 + \frac{dH_1}{d\zeta} &= 0; \end{aligned}\right\} \tag{3.5.5}$$

with $F_1(\infty) = G_1(\infty) = 0, \quad F_1(0) = H_1(0) = 0, \quad G_1(0) = 1.$

The first seven functions of each set were computed by Fox and Banks and the numerical values seem to imply good overall convergence in non-equatorial latitudes. Furthermore, there is no indication of either separation or premature mass efflux from the boundary layer as predicted by Nigam[135, 136]. The equatorial

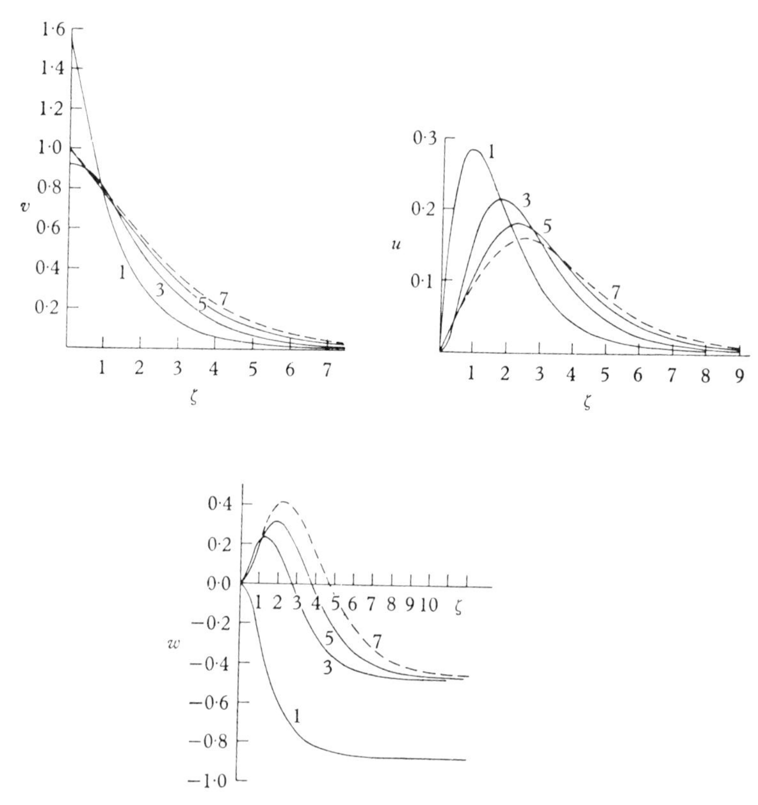

Fig. 3.11. The azimuthal, radial and meridional components of velocity at the equator, $\Theta = \frac{1}{2}\pi$, [3]. The number labels indicate the highest order function used in the series in (3.5.4).

velocity components, given by Banks, are shown in fig. 3.11; the numbers on the graphs indicate the highest order function included in the computation.

Linear motions between concentric spheres, first treated by Proudman[155], were discussed in the last chapter; flow within a narrow gap spherical annulus receives further attention in chapter 5.

Here, we will describe some of the numerical results obtained by Pearson[144] for the time-dependent problem of axially symmetric motion between concentric spheres. The spheres rotate about the same axis but with different rotation rates. (The narrow gap annulus is one particular case of the general problem and there is no indication that special difficulties are associated with it.)

The completed calculations are, at present, too few to be able to draw many inferences apart from the important one that the program works well. Some information is available about the flow in the annulus whose gap width is the radius of the inner sphere. In this example, the inner sphere is kept stationary while the outer one is started impulsively. The transient evolution and the ultimate steady circulation are shown in figures 3.12 and 3.13. The formation of an internal shear is quite evident and the flow pattern near the pole is similar to that which develops between concentric disks. With a little effort, it should be possible to determine the structure of the free shear layers and their dependence on the Rossby number. The behaviour of the fluid near the equator would be of significant interest when the inner sphere spins too. A number of standing questions will be resolved now that an accurate numerical technique is available. That new questions will arise seems inevitable.

3.6. Momentum-integral methods

A much more empirical approach to the solution of the equations of motion is based on the momentum-integral methods introduced by Kármán[94] in a study of both laminar and turbulent boundary layers on a rotating plate. Later, Schultz-Grunow[179] analysed the flow between finite rotating disks in this manner and more recently the same methods have been applied to problems of this sort by Mack[122, 123] and King and Lewellen[96], among others. Rott and Lewellen[177] review the applications of the theory dealing with rotating fluids.

A great difficulty with relatively crude procedures is that to succeed they depend very much on hindsight and the ingenuity, experience, and knowledgeability of the particular researcher. One can never be sure in completely new situations that gross approximations are really adequate and will yield trustworthy results. (For example, is a monotonic boundary layer profile always an accurate approximation?) However, in many situations, especially those involving turbulent boundary layers and other strong interactions, there are essentially no other theoretical techniques available and

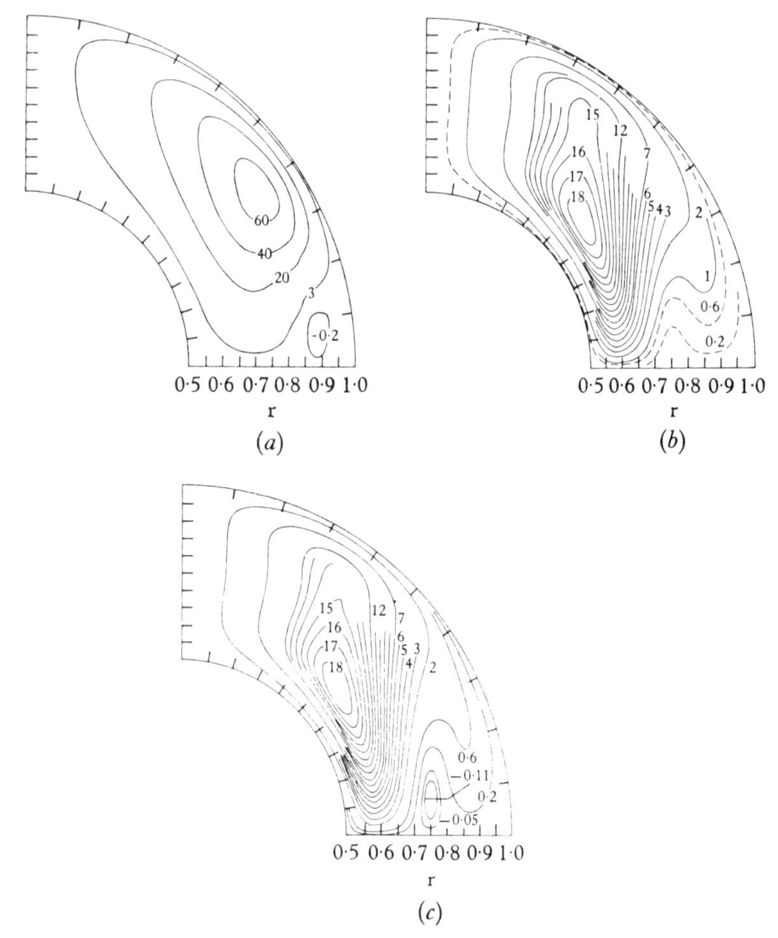

Fig. 3.12. The inner sphere is held fixed and the outer sphere is started impulsively, [144]. Contour lines of $10^4\psi$, the axially symmetric stream function defined on page 101, are shown at varying times: (*a*) $t = 8$, (*b*) $t = 47$, (*c*) $t = 223$. In all cases, $E = 10^{-3}$.

in these cases, momentum-integral methods become a most practical means of analysis. The presentation in this section is based on the work of Rott and Lewellen[177].

Consider steady, axially symmetric fluid motions in a container with the same geometrical restrictions. The analysis of a boundary layer formed adjacent to the wall may be facilitated through the introduction of the arc length, s, which *increases* in the direction of boundary layer growth. The surface itself is described parametrically in terms of this co-ordinate by

$$r = R(s), \quad z = \mathscr{Z}(s); \quad 1 = \left(\frac{dR}{ds}\right)^2 + \left(\frac{d\mathscr{Z}}{ds}\right)^2.$$

According to the notation of § 1.6, let

$$(\xi_1, \xi_2, \xi_3) = (s, \theta, \mathfrak{y}); \, (h_1, h_2, h_3) = (1, R(s), 1);$$

and

$$(q_1, q_2, q_3) = (q_s, v, q_{\mathfrak{y}}).$$

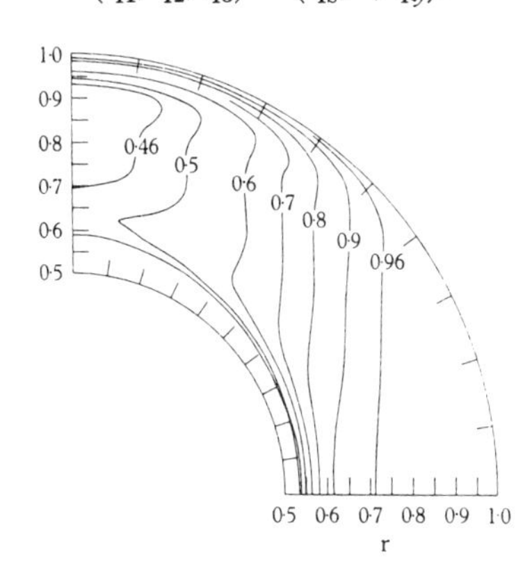

Fig. 3.13. Contour lines of angular velocity at t = 223, E = 10^{-3}. Conditions are the same as for fig. 3.12.

The *unscaled* boundary layer equations are then

$$q_s \frac{\partial q_s}{\partial s} + q_{\mathfrak{y}} \frac{\partial q_s}{\partial \mathfrak{y}} - \frac{v^2}{R}\frac{dR}{ds} = -\frac{v_I^2}{R}\frac{dR}{ds} + \frac{\partial \pi_s}{\partial \mathfrak{y}}, \tag{3.6.1}$$

$$\frac{q_s}{R}\frac{\partial \Gamma}{\partial s} + \frac{q_{\mathfrak{y}}}{R}\frac{\partial \Gamma}{\partial \mathfrak{y}} = \frac{\partial \pi_\theta}{\partial \mathfrak{y}}, \tag{3.6.2}$$

$$\frac{1}{R}\frac{\partial}{\partial s}(Rq_s) + \frac{\partial q_{\mathfrak{y}}}{\partial \mathfrak{y}} = 0. \tag{3.6.3}$$

Here, $\mathfrak{y}$ is the co-ordinate normal to the boundary; π_s, π_θ are stress components;

$$\Gamma = Rv;$$

and conditions outside the boundary layer, which are assumed to be known, are denoted by the subscript I. A *dimensional* notation is used because several different scales will enter into the following analysis; the co-ordinate system here is an *inertial* or *laboratory frame*.

Upon integrating each of these equations in the normal direction from zero to 'infinity', it follows that

$$\frac{d}{ds}\left(R\int_0^\infty q_s^2\,d\eta\right) - \int_0^\infty \frac{1}{R^2}\left(\frac{dR}{ds}\right)(\Gamma^2-\Gamma_I^2)\,d\eta = -R\pi_s^{(0)}, \quad (3.6.4)$$

$$\frac{d}{ds}\left(R\int_0^\infty q_s(\Gamma-\Gamma_I)\,d\eta\right) + \left(\frac{d\Gamma_I}{ds}\right)\int_0^\infty Rq_s\,d\eta = -R^2\pi_\theta^{(0)}, \quad (3.6.5)$$

$$q_\eta\big|_{\eta=\infty} = q_{\eta_I} = -\frac{1}{R}\frac{d}{ds}\left(R\int_0^\infty q_s\,d\eta\right), \quad (3.6.6)$$

with $\pi_s^{(0)} = \pi_s\big|_{\eta=0}$.

Approximations are now made to reduce these equations to a system of readily soluble, *ordinary* differential equations for certain meaningful averages that describe the boundary layer structure. Two quantities are most significant—the meridional mass transport within the layer, $\mathcal{M}(s)$, and the boundary layer thickness, $\delta(s)$. The former is, by definition,

$$\mathcal{M}(s) = \int_0^\infty R(s)\,q_s\,d\eta, \quad (3.6.7)$$

but the latter concept is less precise. Experience with boundary layer phenomena indicates that very little change occurs beyond a certain distance from the wall, $\delta(s)$. Moreover, within this distance, the dependent variables are very often of a similarity form, since the length scale characterizing the total geometry is not relevant to local considerations. An approximation based on these observations is

$$q_s = \frac{U}{F(1)}\frac{d}{d\zeta}F(\zeta) = \frac{\mathcal{M}}{\delta R\,F(1)}\frac{d}{d\zeta}F(\zeta), \quad (3.6.8)$$

$$\Gamma = \Gamma_W + (\Gamma_I - \Gamma_W)\,G(\zeta), \quad (3.6.9)$$

with $\zeta = \eta/\delta(s)$.

Here, both F and G are identically zero for $\zeta \geqslant 1$, and subscript W indicates evaluation at the wall. For $\zeta < 1$, the functions are chosen freely to model the actual flow and to simplify the mathematics. In other words, it is explicitly assumed that the boundary conditions at 'infinity' can be satisfied to a high degree of precision at a finite position, and that the velocity profiles are of the prescribed form. The conditions $G(1) = 1$, $G(0) = 0$, $F(0) = 0$, hold.

The substitution of these expressions into (3.6.4) and (3.6.5) leads to a system of ordinary differential equations for $\mathscr{M}$ and δ:

$$\chi_2 \frac{d}{ds}\left(\frac{\mathscr{M}^2}{R\delta}\right) + \frac{\delta}{R^2}\frac{dR}{ds}[(\Gamma_I - \Gamma_W)^2 \chi_3 + \Gamma_W(\Gamma_I - \Gamma_W)\chi_4] = -R\pi_s^{(0)}, \tag{3.6.10}$$

$$\frac{d}{ds}\Big((\Gamma_I - \Gamma_W)\mathscr{M}\Big) - \chi_1 \mathscr{M}\frac{d}{ds}\Gamma_I = \chi_1 R^2 \pi_\theta^{(0)}, \tag{3.6.11}$$

where

$$\left.\begin{aligned} \chi_1^{-1} &= \int_0^1 \frac{1}{F(1)}\frac{dF}{d\zeta}(1 - G(\zeta))\,d\zeta, \quad & \chi_2 &= \int_0^1 \left(\frac{1}{F(1)}\frac{dF}{d\zeta}\right)^2 d\zeta, \\ \chi_3 &= \int_0^1 (1 - G^2(\zeta))\,d\zeta, & \chi_4 &= 2\int_0^1 (1 - G(\zeta))\,d\zeta. \end{aligned}\right\} \tag{3.6.12}$$

The remaining equation is used to calculate the normal flux:

$$q_\eta = -\frac{1}{R}\frac{d}{ds}\mathscr{M}(s), \quad \text{at } \eta = \infty. \tag{3.6.13}$$

The forms of the stress functions, $\pi_\theta^{(0)}$ and $\pi_s^{(0)}$, have deliberately been left indefinite so that a general formula can be used that is appropriate for both laminar and turbulent flow. The shear laws at $\eta = 0$ employed by Rott and Lewellen[177] are

$$\left.\begin{aligned} \pi_\theta^{(0)} &= \chi_5 \frac{(\Gamma_I - \Gamma_W)^2}{R^2}\left(\frac{\nu R}{(\Gamma_I - \Gamma_W)\delta}\right)^\mu, \\ \pi_s^{(0)} &= \chi_6 \frac{\mathscr{M}(\Gamma_I - \Gamma_W)}{R^2\delta}\left(\frac{\nu R}{(\Gamma_I - \Gamma_W)\delta}\right)^\mu, \end{aligned}\right\} \tag{3.6.14}$$

where $\mu = 1$ for laminar flow, and $\mu = \frac{1}{4}$ for turbulent flow (the Blasius value). The constants are

$$\chi_5 = \frac{d}{d\zeta}G(0), \quad \chi_6 = \frac{1}{F(1)}\frac{d^2}{d\zeta^2}F(0),$$

in the former case, and 0·0225 and $\chi_5/F(1)$, in the latter. Note that the forms of $\pi_\theta^{(0)}$ and $\pi_s^{(0)}$ for laminar flow follow directly from the basic assumptions.

In ordinary examples, the flow conditions at the plate, Γ_W, and in the inviscid interior, Γ_I, are usually assumed known, as is the direction of boundary layer growth. Reasonable guesses are made for $F(\zeta)$ and $G(\zeta)$ and the parameters, χ_i, are calculated. Equations (3.6.10) and (3.6.11) are then solved, numerically if need be, for the unknown functions, $\mathscr{M}(s)$, $\delta(s)$, subject to proper boundary conditions at the origin of the viscous layer, say, $\mathscr{M}(0) = 0$, $\delta(0) = 0$. If the choices are made judiciously, the resultant solutions can be excellent approximations.

In rotating flow problems, the swirl, $\Gamma_I(r)$, is most often an unknown. In fact, the determination of $\Gamma_I(r)$ may be the prime objective, in which case it is necessary to obtain approximate analytical solutions of the boundary layer equations that retain this function in an explicit manner. Further simplifications are required for this purpose.

Consider the flow above a stationary plate, $\Gamma_W = 0$, and let $\Gamma_I(r) = \Gamma(r)$. The solution of (3.6.10) and (3.6.11) near $s = 0$, $(\delta(0) = \mathscr{M}(0) = 0)$, can be obtained in the form of a series. If

$$\left.\begin{aligned} \delta(s) &= s^{\gamma}(\delta_0 + s\delta_1 + \dots), \\ \mathscr{M}(s) &= s^{\sigma}(\mathscr{M}_0 + s\mathscr{M}_1 + \dots), \end{aligned}\right\} \tag{3.6.15}$$

and

$$\Gamma(s) = \Gamma_0 + s\Gamma_1 + \dots,$$

$$R(s) = R_0 + sR_1 + \dots,$$

then

$$\left.\begin{aligned} \delta_0 &= \left(\frac{\ell(\chi_1\chi_5)^2}{\sigma}\right)^{\frac{1}{2}(2\sigma-1)} R_0(-R_0R_1)^{\frac{1}{2}(1-2\sigma)}\left(\frac{\nu}{\Gamma_0}\right)^{2(1-\sigma)}, \\ \mathscr{M}_0 &= (\chi_1\chi_5)^{2\sigma-1}\ell^{\sigma-1}\sigma^{-\sigma}(-R_0R_1)^{1-\sigma}\Gamma_0^{2\sigma-1}\nu^{2(1-\sigma)}, \end{aligned}\right\} \tag{3.6.16}$$

with

$$\sigma = \frac{2+\mu}{2(1+\mu)}, \quad \gamma = \frac{1}{2(1+\mu)}, \quad \ell = \frac{3+2\mu}{2+\mu}\frac{\chi_2}{\chi_3} + \frac{\chi_6}{\chi_1\chi_3\chi_5}.$$

In the neighbourhood of the origin, the boundary layer thickness is related to the flux by

$$\delta^{2+\mu} = \frac{\chi_1\chi_5\ell}{-R_1}\nu^{\mu}\left(\frac{R_0}{\Gamma_0}\right)^{1+\mu}\mathscr{M}. \tag{3.6.17}$$

This relationship is taken by Rott and Lewellen to apply throughout the entire boundary layer domain with the initial constants, R_0, Γ_0,

R_1, replaced by their local values, R, Γ_I, $\frac{dR}{ds}$; therefore

$$\delta^{2+\mu} = \frac{\chi_1 \chi_5 \ell}{-\frac{dR}{ds}} \nu^{\mu} \left(\frac{R(s)}{\Gamma_I(s)}\right)^{1+\mu} \mathcal{M}. \qquad (3.6.18)$$

It has been found that this approximation is the only power law consistent with both initial and terminal similarity.

Equation (3.6.11) can now be solved in closed form using (3.6.18), which, in effect, replaces (3.6.7). Thus,

$$\mathcal{M} = \sigma^{-\sigma} (\chi_1 \chi_5)^{2\sigma-1} \ell^{\sigma-1} \nu^{2(1-\sigma)} \Gamma_I^{\chi_1 - 1}$$
$$\times \left\{ \int_0^s \Gamma_I^{2-(\chi_1/\sigma)} \left(-R \frac{dR}{ds}\right)^{(1-\sigma)/\sigma} ds \right\}^{\sigma}. \qquad (3.6.19)$$

This equation is still too complicated for practical use and a simpler approximate relationship between the local value of $\mathcal{M}$ and Γ_I is desirable. A manageable formula can be obtained for laminar flow over a flat plate by interpolation between two extreme cases. Let

$$\frac{1}{F(1)} \frac{dF}{d\zeta} = 12\zeta(1-\zeta)^2, \quad G(\zeta) = 2\zeta - \zeta^2, \qquad (3.6.20)$$

so that $\chi_1 = 2{\cdot}5$, $\chi_2 = 1{\cdot}375$, $\chi_3 = 0{\cdot}467$, $\chi_4 = 0{\cdot}667$. $\chi_5 = 2$, $\chi_6 = 12$. The *potential vortex*, Γ_I = constant ($= \Gamma_0$), then has an associated mass flux, according to (3.6.19),

$$\mathcal{M} = 1{\cdot}26(\nu\Gamma_0)^{\frac{1}{2}} (R_0^{\frac{4}{3}} - R^{\frac{4}{3}})^{\frac{3}{4}}, \qquad (3.6.21)$$

whereas for *rigid rotation*, $\Gamma_1 = \Omega R^2$,

$$\mathcal{M} = 1{\cdot}26\nu^{\frac{1}{2}} \frac{\Omega^{\frac{1}{2}} R^2}{R_0} (R_0^{\frac{4}{3}} - R^{\frac{4}{3}})^{\frac{3}{4}}. \qquad (3.6.22)$$

A formula that includes both of these results as limiting cases is

$$\mathcal{M} = 1{\cdot}26 \left(\frac{\nu}{\Gamma_0}\right)^{\frac{1}{2}} (R_0^{\frac{4}{3}} - R^{\frac{4}{3}})^{\frac{3}{4}} \Gamma_I, \qquad (3.6.23)$$

and this will be used as the basic relationship between $\mathcal{M}$ and Γ_I, when a rotating flow persists above a stationary plate.

The analysis of the problem in which the plate rotates faster than the fluid is similar and will not be presented here. As a rule, $\Omega r^2 - \Gamma$

must not change sign for the approximations to remain valid. If $\Gamma_W = \Omega R^2$ and $R = s$, then the laminar flux calculation leads to the following approximation;

$$\mathcal{M} = 0{\cdot}55\left(\frac{\nu}{\Omega}\right)^{\frac{1}{2}}(\Omega R^2 - \Gamma_I). \tag{3.6.24}$$

These results, (3.6.23) and (3.6.24), may be combined to solve the problem of flow between shrouded disks, one of which rotates while the other remains stationary. The reasoning is based on the fact that mass conservation requires the *net* radial flux between plates to be zero. The transport in either boundary layer is $O(E^{\frac{1}{2}})$, but the interior flow contribution is only $O(E)$. (This is a rather simple result and it is established by a formal analysis in the next section.) Therefore, to lowest order, the boundary layer fluxes must balance *each other* at every radial position. This local matching and detailed balancing determine the possible interior flow, Γ_I. Equating the expressions for $\mathcal{M}$ in the preceding equations yields

$$\Gamma_I(r) = \frac{0{\cdot}55\,\Omega r^2}{0{\cdot}55 + 1{\cdot}26\left(1 - \left(\frac{r}{a}\right)^{\frac{4}{3}}\right)^{\frac{3}{4}}}, \tag{3.6.25}$$

where $R_0 = a$ is the radius of the tank. This formula implies that $v_I = 0{\cdot}3\Omega r$ near the central axis, which is in exact agreement with the analysis of the infinite disk problem, and it compares favourably at all positions with the experimental data of Maxworthy[127], fig. 3.14.

Further development of these procedures, their improvement and modification, may be found in the cited references.

3.7. Spin-up

The small convective circulation established by viscous boundary layers is the most important process controlling spin-up even in the non-linear regime. The predominance of this mechanism makes certain that the spin-up time scale, $E^{-\frac{1}{2}}\Omega^{-1}$, is characteristic in all circumstances. The principal difference between container spin-up from one rotation rate to another and spin-up from rest is the role of angular momentum. In the latter case, the main body of fluid is not rotating initially and must be completely flushed, spun-up from rest and returned to the interior by the boundary layers. This is probably the more interesting situation at this time because of the new features exhibited and it will be discussed first for this reason. Spin-down to rest is not considered because the flow is unstable and becomes turbulent.

If the fluid is already revolving, then comparatively little motion is required to effect a change of rotation rate. The conservation of angular momentum requires a particle to spin faster as it is convected inward and the concomitant stretching of vortex lines by the induced boundary layer flux increases the total vorticity. The

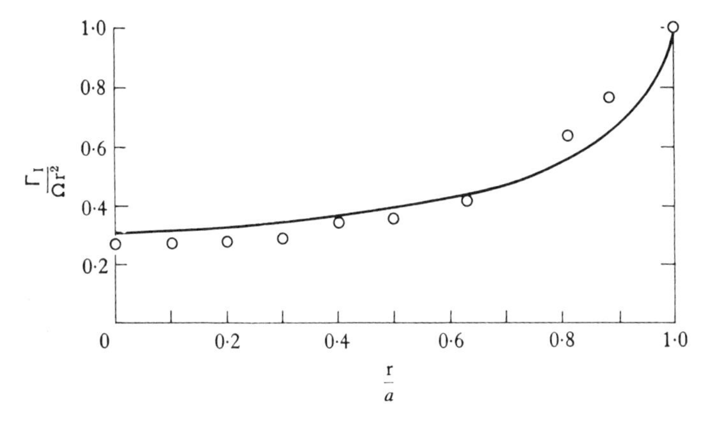

Fig. 3.14. The angular velocity of the interior fluid between two finite shrouded disks, one fixed and the other rotating, as given by (3.6.25). The circles are experimental points of Maxworthy[127].

particle moves just so far until its speed and vorticity are at their required levels. Obviously, this non-linear problem is a direct extension of the linear problem and only quantitative changes are expected.

Consider a very simple problem of spin-up from rest. A symmetrical, fluid-filled container is initially in a state of rest and the container is then impulsively set into rotation about its symmetry axis with constant angular velocity Ω. An analytical description is desired of the approach to the final state of rigid rotation.

From a physical standpoint, the flow development is as follows. Boundary layers form at the container wall within the first few revolutions following the impulsive start. These layers remain essentially steady thereafter, throughout the critical phases of motion. Their main function is to produce a secondary flow to draw in non-rotating fluid, spin it up and return it to the interior. Rotating fluid emerging from the boundary layers is convected across the breadth of the container and the process concludes when the original fluid has been completely flushed and replaced.

The *dimensionless* equations of motion, scaled by $[\![L, \Omega^{-1}, \Omega L]\!]$, are, in terms of a cylindrical, *inertial* co-ordinate system,

$$E^{\frac{1}{2}}\frac{\partial u}{\partial \tau}+u\frac{\partial u}{\partial r}+w\frac{\partial u}{\partial z}-\frac{v^2}{r} = -\frac{\partial P}{\partial r}+E\left(\frac{\partial^2 u}{\partial r^2}+\frac{\partial}{\partial r}\frac{u}{r}+\frac{\partial^2 u}{\partial z^2}\right), \tag{3.7.1}$$

$$E^{\frac{1}{2}}\frac{\partial v}{\partial \tau}+u\left[\frac{\partial v}{\partial r}+\frac{v}{r}\right]+w\frac{\partial v}{\partial z} = E\left(\frac{\partial^2 v}{\partial r^2}+\frac{\partial}{\partial r}\frac{v}{r}+\frac{\partial^2 v}{\partial z^2}\right), \tag{3.7.2}$$

$$E^{\frac{1}{2}}\frac{\partial w}{\partial \tau}+u\frac{\partial w}{\partial r}+w\frac{\partial w}{\partial z} = -\frac{\partial P}{\partial z}+E\left(\frac{\partial^2 w}{\partial r^2}+\frac{1}{r}\frac{\partial w}{\partial r}+\frac{\partial^2 w}{\partial z^2}\right), \tag{3.7.3}$$

$$\frac{\partial}{\partial r}(ru)+\frac{\partial}{\partial z}(rw) = 0. \tag{3.7.4}$$

The Rossby number, ε, is taken as unity and the container height, L, characterizes the length scale. Certain terms have been eliminated because, by symmetry, all functions are independent of the azimuthal angle. The boundary conditions are

$$u = v-r = w = 0 \tag{3.7.5}$$

on the container surface, and $u = v = w = 0$ initially.

The solution is once again representable as a combination of a boundary layer flow and a nearly inviscid, interior motion. Only the motion in the inviscid interior shall be described in detail. Since the induced secondary circulation must be $O(E^{\frac{1}{2}})$, the internal velocity components and pressure are expressible as

$$\left.\begin{aligned} u &= E^{\frac{1}{2}}u_1(r,z,\tau)+\dots,\\ v &= v_0(r,z,\tau)+E^{\frac{1}{2}}v_1(r,z,\tau)+\dots,\\ w &= E^{\frac{1}{2}}w_1(r,z,\tau)+\dots,\\ P = p &= p_0(r,z,\tau)+E^{\frac{1}{2}}p_1(r,z,\tau)+\dots.\end{aligned}\right\} \tag{3.7.6}$$

The substitution of these expansions in the foregoing equations shows that

$$\left.\begin{aligned} \frac{v_0^2}{r} &= \frac{\partial p_0}{\partial r}, \\ o &= \frac{\partial p_0}{\partial z}. \end{aligned}\right\} \tag{3.7.7}$$

Thus, the azimuthal velocity is independent of height to this order,

$$v_0 = v_0(r, \tau). \tag{3.7.8}$$

Furthermore, it follows from (3.7.2) that

$$\frac{\partial v_0}{\partial \tau} + u_1 \left(\frac{\partial v_0}{\partial r} + \frac{v_0}{r} \right) + w_1 \frac{\partial v_0}{\partial z} = o,$$

whence

$$\frac{\partial}{\partial \tau}(r v_0) + u_1 \frac{\partial}{\partial r}(r v_0) = o, \tag{3.7.9}$$

by virtue of (3.7.8). The radial velocity is also independent of the vertical co-ordinate, z. (For steady motions, $u_1 = o$, is an immediate consequence of this relationship and proves, in particular, that $u = O(E)$ in the interior. This fact was used in the last section in the discussion preceding (3.6.25).) In order to solve (3.7.9), an explicit relationship between u_1 and v_0 is needed and this is provided by a detailed boundary layer analysis. The exact relationship can be found only under severe restrictions—moderate spin-up from one speed to another—and the theory in this case is developed later in this section. It has not been possible to analyse spin-up from rest in this fashion, but a very successful approximate procedure was introduced by Wedemeyer[223] which in spirit and execution is similar to the momentum-integral methods of the last section.

Information accumulated in previous chapters shows clearly that the boundary layer processes are nearly steady throughout the essential phases of certain transient fluid motions. Many results derived from an analysis of steady boundary layers apply almost as well to time-dependent situations—after the first few revolutions. Therefore, the crucial step is to regard all boundary layers as steady and to use the established formulas connecting mass transport in the viscous layers with the inviscid secondary circulation. In this manner, an approximate formula is derived relating u_1 and v_0. The cylindrical container is considered first since this is the configuration about which most is known. The analysis is then extended in the same spirit to include arbitrary axially symmetric geometries.

The *net* radial transport of fluid in the boundary layers and inviscid interior must be zero at any time, and at any position $r = \text{constant}$. This is expressed by

$$\int_0^1 u\,dz = 0,$$

which is the result of the vertical integration of (3.7.4) from one plate to the other. The radial velocity component, like all the others, is, in essence, composed of three parts—an inviscid, interior component and representations for two boundary layers;

$$u = u_I(r, \tau) + \tilde{u}_B + \tilde{u}_T.$$

Subscripts refer to location. In this particular geometry, the boundary layer flows are identical, $\tilde{u}_B = \tilde{u}_T$, and the replacement of u, above, by this decomposition yields

$$u_I(r, \tau) + 2\int_0^\infty \tilde{u}_B(r, \zeta, \tau)\,d\zeta = 0. \tag{3.7.10}$$

However, the transport within the boundary layer is related to the normal mass flux there, and the integration of the continuity equation through the boundary layer implies

$$\frac{\partial}{\partial r} r \int_0^\infty \tilde{u}\,d\zeta = -r\,\tilde{w}(r, 0, \tau).$$

For the configuration consisting of two infinite concentric plates,

$$\tilde{w}(r, 0, \tau) = \tilde{w}(0, \tau),$$

and the radial integration of the preceding equation in this special case results in

$$\int_0^\infty \tilde{u}(r, \zeta, \tau)\,d\zeta = +\frac{r}{2}\tilde{w}(0, \tau). \tag{3.7.11}$$

It is assumed next that the real transient boundary layer can be locally approximated by the *steady* layer having the same instantaneous conditions in the interior and at the plate. The steady computations of Rogers and Lance[169], summarized in fig. 3.4 (*a*), can then be used to relate the normal velocity at any radial position to the difference in rotation rates there of the internal fluid, Ω_I, and of the plate, Ω; (in the previous notation, $\Omega_f = \Omega_I$, $\Omega_w = \Omega$). This dependency is nearly linear and it will be taken as such;

$$\tilde{w}(0, \tau) = \varkappa E^{\frac{1}{2}}\left(1 - \frac{\Omega_I}{\Omega}\right)$$

with the constant $\varkappa$ to be chosen appropriately. Hence, (3.7.11) is approximately

$$\int_0^\infty \tilde{u}(r, \zeta, \tau)\, d\zeta = \tfrac{1}{2} r \varkappa E^{\frac{1}{2}}\left(1 - \frac{\Omega_I}{\Omega}\right).$$

Another major simplification is now introduced. Following Wedemeyer[223], the preceding formula is taken as a local approximation in the general time-dependent situation as well, where Ω_I is also a function of r, and τ. The interior rotation rate, Ω_I, is interpreted in terms of the internal azimuthal velocity by

$$\frac{\Omega_I}{\Omega} = \frac{v_I(r, \tau)}{r}.$$

The result of combining all these approximations is the formula

$$u_I(r, \tau) = -\varkappa E^{\frac{1}{2}}(r - v_I),$$

or, in view of (3.7.6), $\quad u_1 = -\varkappa(r - v_0). \qquad (3.7.12)$

This equation expresses the desired connection of the secondary circulation with the primary swirl as established by the viscous boundary layers.

Substitution of (3.7.12) for u_1 in (3.7.9) yields a first order partial differential equation for the azimuthal velocity:

$$\frac{\partial v_0}{\partial \tau} - \varkappa\left(1 - \frac{v_0}{r}\right)\frac{\partial}{\partial r}(r v_0) = 0. \qquad (3.7.13)$$

Wedemeyer takes $\varkappa$ to be 0·886, but Rott and Lewellen use 1·10. For reasons that will be apparent shortly, we shall set $\varkappa = 1$. The solution of this equation subject to the boundary condition,

$$v_0 = a$$

at the sidewall, $r = a\ (= r_0/L)$, and the initial condition, $v_0 = 0$, is

$$\left.\begin{aligned} v_0 &= \frac{r\, e^{2\tau} - \dfrac{a^2}{r}}{e^{2\tau} - 1}, \quad r \geqslant a e^{-\tau}, \\ v_0 &= 0, \qquad\qquad\quad r \leqslant a e^{-\tau}. \end{aligned}\right\} \qquad (3.7.14)$$

The locus $r = ae^{-\tau}$ represents the position of a column of fluid particles originally at the sidewall as it is convected inward. This surface separates the rotating fluid which is injected into the interior, from the non-rotating fluid that is drained into the boundary layers. The azimuthal velocity is shown as a function of time in fig. 3.15. The characteristic time scale of spin-up remains $E^{-\frac{1}{2}}\Omega^{-1}$.

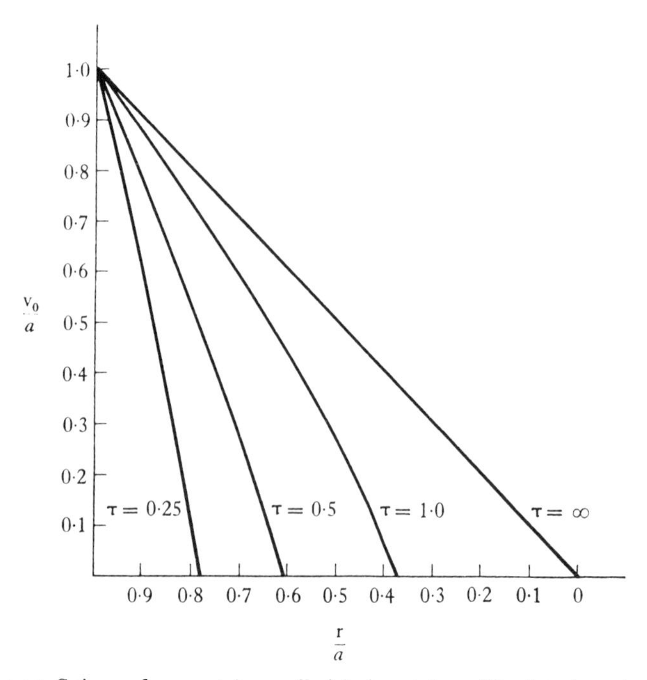

Fig. 3.15. Spin-up from rest in a cylindrical container. The interior azimuthal velocity is shown as a function of position with time as a parameter.

The *linear* problem corresponding to an impulsive but infinitesimal change in the angular velocity of the system was solved completely in § 2.6. Under present conditions of axially symmetric spin-up, the *dimensionless* radial velocity u_1 in the *rotating* frame can be determined directly from (2.6.17):

$$u_1 = -\frac{1}{2}\frac{\partial v_0}{\partial \tau}.$$

The azimuthal component (see (2.6.7), (2.6.23) and (2.8.9)) for the most general axisymmetric container configuration, is

$$\left.\begin{aligned} v_0(r,\tau) &= r[1-\exp(-\Upsilon(r)\tau)]; \\ \Upsilon(r) &= \frac{\left(1+\left(\frac{df}{dr}\right)^2\right)^{\frac{1}{4}}+\left(1+\left(\frac{dg}{dr}\right)^2\right)^{\frac{1}{4}}}{f+g}. \end{aligned}\right\} \qquad (3.7.15)$$

Hence,
$$u_1 = -\tfrac{1}{2}\Upsilon(r)(r-v_0), \qquad (3.7.16)$$

and this expresses the radial velocity in terms of the instantaneous and infinitesimal differential between the rotation rates of the fluid and the container. In the right circular cylinder given by $f = 1$, $g = 0$,

$$u_1 = -(r-v_0),$$

and this relationship between the *dimensionless velocity components* is *identical* with (3.7.12) if $\varkappa = 1$. As a working hypothesis, the equation for the *scaled* radial velocity component and the rotation differential that is obtained from the linear theory, (3.7.16), is also assumed to hold in the non-linear range, for a large Rossby number. Therefore, the introduction of (3.7.16) into (3.7.9) results in

$$\frac{\partial}{\partial\tau}(rv_0) - \tfrac{1}{2}\Upsilon(r)(r-v_0)\frac{\partial}{\partial r}(rv_0) = 0. \qquad (3.7.17)$$

The initial condition is one of quiescence but the spatial condition must be suitably modified. If the container has a lateral sidewall, as, for example, a circular cylinder with a conical bottom, then the condition is unchanged. However, in other cases, typified by the sphere, the condition $v = r$ applies at the position of maximum radius only. The boundary layers pump liquid outwards and the fluid having acquired rotation returns to the interior in the most lateral regions of the container, where the bottom and top boundary layers interact. Observations of spin-up in the sphere show an initial turbulent out-flow near the equator which very shortly thereafter resolves itself into a laminar motion of the type predicted by the theory. This is also the region of return flow in the linear case.

The general solution of (3.7.17) is

$$\tau = 2\int_r^a \frac{\xi}{\Upsilon(\xi)(\xi^2 - rv_0(r,\tau))}\,d\xi, \qquad (3.7.18)$$

and the lead particle path of the rotational flow regime is obtained from this by setting $v_0 = 0$. The complete integration of (3.7.18) need not be trivial but the specific formula for the sphere

$$f(r) = g(r) = (1-r^2)^{\frac{1}{2}}, \quad a = 1,$$

is found to be

$$\tau = -\tfrac{4}{3}(1-r^2)^{\frac{3}{4}} + (1-rv_0)^{\frac{3}{4}} \ln\left[\frac{(1-rv_0)^{\frac{1}{4}} + (1-r^2)^{\frac{1}{4}}}{(1-rv_0)^{\frac{1}{4}} - (1-r^2)^{\frac{1}{4}}}\right] - 2(1-rv_0)^{\frac{3}{4}} \tan^{-1}\left(\frac{1-r^2}{1-rv_0}\right)^{\frac{1}{4}}. \quad (3.7.19)$$

There are other geometries for which it is feasible to derive an implicit functional relationship for τ and v, and the cylinder fitted with a conical bottom is one that is perhaps particularly suited for experimental purposes.

Wedemeyer reported experimental results in reasonable agreement with the theoretical predictions for the cylindrical container. This was true for turbulent boundary layers too, but this aspect will not be discussed here.

In § 1.1, a rather elementary visual demonstration of the core flow was described (see fig. 1.4). The photograph shows that after the container is spun-up, a light front moves from the outer wall across the width of the tank. The 'front' corresponds to the motion of the lead particle, and is convincing evidence that the core flow is strongly independent of the vertical co-ordinate. The boundary layers top and bottom are shown as the thin, dark horizontal regions adjacent to each plate.

Non-linear spin-up from an initial state of rigid rotation can be analysed in much the same manner. It is convenient to think of the rotation rate of the container being changed from $(1-\varepsilon)\Omega$ to Ω (with the dimensionalization based on $[\![L, \Omega^{-1}, L\Omega]\!]$). Equation (3.7.13) with $\varkappa = 1$ is then still the approximate formula for the azimuthal velocity in the cylindrical tank but the boundary conditions are now $\quad v_0(r, 0) = (1-\varepsilon)\, r, \quad v_0(a, \tau) = a.$

The solution, in the fluid domain of particles which are originally in the interior and remain there after spin-up, is

$$v_0 = \frac{r}{1 + \dfrac{\varepsilon}{1-\varepsilon} e^{-2\tau}}.$$

A better analytical approximation of the curve in fig. 3.4 than that of (3.7.12) would lead to a more accurate formula for v_0.

A different approach to non-linear spin-up was developed by Greenspan and Weinbaum [76]. This treatment presumes a moderately small change of angular velocity from an established state of rigid rotation and represents the direct extension of the linear theory. The analysis is exact, in a formal sense, but spin-up from rest is not a problem that can be successfully analysed in this manner.

The problem considered involves a filled, axially symmetric container initially revolving about its symmetry axis with angular speed Ω. Subsequently, the rotation rate is impulsively changed a modest amount to $(1+\varepsilon)\Omega$ and the fluid ultimately adjusts to the new state of rigid rotation.

It is advantageous to formulate this problem with respect to the *original rotating* co-ordinate system whose angular velocity is $\Omega\hat{\mathbf{k}}$. In this frame, the *dimensionless* equations of motion, $[\![L, \Omega^{-1}, \varepsilon\Omega L]\!]$, are

$$\left.\begin{aligned} \nabla\cdot\mathbf{q} &= 0, \\ E^{\frac{1}{2}}\frac{\partial}{\partial\tau}\mathbf{q}+\varepsilon\mathbf{q}\cdot\nabla\mathbf{q}+2\hat{\mathbf{k}}\times\mathbf{q} &= -\nabla p+E\nabla^2\mathbf{q}.\end{aligned}\right\} \tag{3.7.20}$$

The introduction of a stream function by

$$\mathbf{q} = E^{\frac{1}{2}}\frac{\partial\chi}{\partial z}\hat{\mathbf{r}}+v\hat{\boldsymbol{\theta}}-E^{\frac{1}{2}}\frac{1}{r}\frac{\partial}{\partial r}(r\chi)\hat{\mathbf{k}} \tag{3.7.21}$$

and the elimination of the pressure lead to two equations for χ and v:

$$\left(E^{\frac{1}{2}}\mathscr{G}-\frac{\partial}{\partial\tau}\right)v-\frac{\varepsilon}{r}\left[\frac{\partial\chi}{\partial z}\frac{\partial}{\partial r}(rv)-\frac{\partial v}{\partial z}\frac{\partial}{\partial r}(r\chi)\right]-2\frac{\partial\chi}{\partial z} = 0, \tag{3.7.22}$$

$$E\left(E^{\frac{1}{2}}\mathscr{G}-\frac{\partial}{\partial\tau}\right)\mathscr{G}\chi-\varepsilon\left[E\frac{\partial\chi}{\partial z}\left(\frac{\partial}{\partial r}(\mathscr{G}\chi)-\frac{1}{r}\mathscr{G}\chi\right)\right.$$
$$\left.-\frac{E}{r}\frac{\partial}{\partial z}(\mathscr{G}\chi)\frac{\partial}{\partial r}(r\chi)-2\frac{v}{r}\frac{\partial v}{\partial z}\right]+2\frac{\partial v}{\partial z} = 0, \tag{3.7.23}$$

where $$\mathscr{G} = \frac{\partial}{\partial r}\frac{1}{r}\frac{\partial}{\partial r}r+\frac{\partial^2}{\partial z^2}.$$

The solution procedure is a familiar one by now. Each dependent variable is a composition of boundary layer terms and representations valid in the inviscid core:

$$v = v_I + \tilde{v}_B + \tilde{v}_T,$$

$$\chi = \chi_I + \tilde{\chi}_B + \tilde{\chi}_T.$$

All functions are expanded as power series in $E^{\frac{1}{2}}$ and a series of non-linear problems evolve which must be solved sequentially. Actually, only terms of order $E^{\frac{1}{2}}$ are ever of interest, and, practically speaking, two terms of the series suffice. The basic interior motion $(E = 0)$ satisfies

$$\frac{\partial v_I}{\partial \tau} + \frac{\varepsilon}{r}\left(\frac{\partial \chi_I}{\partial z}\frac{\partial}{\partial r}(rv_I) - \frac{\partial v_I}{\partial z}\frac{\partial}{\partial r}(r\chi_I)\right) + 2\frac{\partial \chi_I}{\partial z} = 0, \quad (3.7.24)$$

$$\left(1 + \frac{\varepsilon v_I}{r}\right)\frac{\partial v_I}{\partial z} = 0, \quad (3.7.25)$$

from which it may be deduced that

$$v_I = v_I(r, \tau),$$

and

$$\frac{\partial \chi_I}{\partial z} = -\frac{\dfrac{\partial v_I}{\partial \tau}}{2 + \dfrac{\varepsilon}{r}\dfrac{\partial}{\partial r}(rv_I)}. \quad (3.7.26)$$

The last two equations are the direct analogues of (3.7.8) and (3.7.9) in the new co-ordinate system.

From this point on, the analysis becomes very complicated and only a general description of procedure, the aims, methods and results can be given here. In principle, the analysis follows the same steps as in the linear theory, except that now the Rossby number must always be included in a manner that maintains uniform validity through the spin-up phase. Solutions for the boundary layer functions must also be uniformly valid in their spatial dependence.

The boundary layer equations are steady to lowest order but involve radial derivatives. Hence, a solution of a system of non-linear partial differential equations is required and for this reason, it is assumed that ε is moderately small. The analysis leads to six transcendental equations in six unknowns (two for the core and two for each viscous layer) from which a single equation is obtained that relates the azimuthal velocity with its time derivatives:

$$\frac{\partial v_I}{\partial \tau} = F(r, \tau, v_I).$$

The requirement that the solution be uniformly valid necessitates the introduction of a stretched time co-ordinate

$$\tau = t + \varepsilon\tau_1(r, t) + \dots$$

and use of techniques developed by Poincaré [153] and Lighthill [107]. A consistent expansion procedure is carried out in this fashion and the final result is a uniformly valid representation for each variable in the separate regions. Results valid in the core are perhaps the most important and the solution, accurate to $O(\varepsilon^2)$ uniformly in time through spin-up, is

$$v_I(r, \tau) = r[1 - \exp(-\Upsilon(r)\, t)] + O(\varepsilon^2), \qquad (3.7.27)$$

$$\tau = t + \varepsilon \left\{ -\frac{t}{2} + \frac{13}{40} \frac{r\dfrac{d}{dr}\Upsilon(r)}{\Upsilon(r)} \left(1 - \frac{v}{r}\right) t + \left[\frac{4}{5\Upsilon(r)} + \frac{47}{280} \frac{r\dfrac{d}{dr}\Upsilon(r)}{\Upsilon^2(r)} + \frac{69}{140} \frac{r}{\Upsilon(r)} \frac{\dfrac{df}{dr} + \dfrac{dg}{dr}}{f+g} \right] \frac{v}{r} \right\}; \qquad (3.7.28)$$

$\Upsilon(r)$ is given in (3.7.15). The solution for the stream function is given here to $O(\varepsilon)$ only, because the next term is excessively complicated in the general case (see [76]):

$$\chi_I(r, \tau) = -\frac{\Upsilon(r)\, rz}{2} e^{-\Upsilon t} + \frac{r}{2} \left[\frac{f\left(1 + \left(\dfrac{dg}{dr}\right)^2\right)^{\frac{1}{2}} - g\left(1 + \left(\dfrac{df}{dr}\right)^2\right)^{\frac{1}{2}}}{f+g} \right] e^{-\Upsilon t} + O(\varepsilon). \qquad (3.7.29)$$

Flow between the disks, $f = 1, g = 0$, has the particular form

$$\left.\begin{aligned} v_I &= r(1 - e^{-2t}), \\ \chi_I &= -r(z - \tfrac{1}{2})\,[1 - \varepsilon(\tfrac{1}{2} - \tfrac{1}{5}e^{-2t})]\, e^{-2t}, \\ \tau &= t + \varepsilon(-\tfrac{1}{2}t + \tfrac{2}{5}(1 - e^{-2t})). \end{aligned}\right\} \qquad (3.7.30)$$

A plot of the azimuthal velocity versus time for $E = 0{\cdot}001$ is shown in fig. 3.16 together with Pearson's transient numerical solution for the same configuration [143]. Obviously, both approaches are in excellent agreement and this tends to confirm the validity of the boundary layer approximation and the accuracy of the numerical

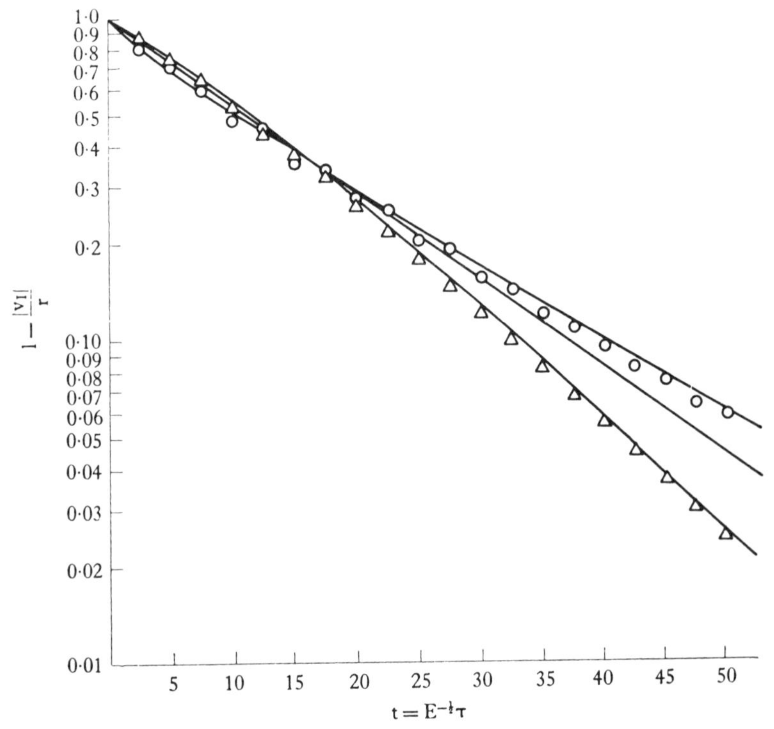

Fig. 3.16. The deviation from rigid rotation in the non-linear spin-up between concentric infinite plates. The solid curves result from boundary layer theory, (3.7.30); the circles are taken from Pearson's transient numerical solution of the complete equations [143]. The parameters are $E = 0{\cdot}001$, $\varepsilon = 0, \pm 0{\cdot}5$.

calculations. The computations do indicate the occurrence of small inertial oscillations, a feature not detectable by this boundary layer theory.

Evidently, a significant variation of the Rossby number produces little effective change in the fluid motion during spin-up. This somewhat negative conclusion really means that non-linear effects are of minor importance compared to basic viscous process *in this particular problem.*

Another interesting conclusion is that the e-folding time, τ_c, of the decay is essentially a constant, independent of ε in the range $|\varepsilon| < 0.5$. (At $\tau = \tau_c, v_I = r(1 - e^{-1})$.) Note carefully that this statement is true only when the time is made dimensionless with respect to the *original rotation* rate, Ω, of the container. An appropriate scale factor, which is generally a function of ε, is involved when another characteristic time is used.

Although the experiments of Greenspan and Howard [75] qualitatively corroborate these conclusions, no exact measurements have been made. It should not be difficult to check theory versus experiment for any suitable container shape and the findings displayed in fig. 3.16 seem especially suited for this purpose.

3.8. Some experiments with non-linear phenomena

A few experiments are described in this section which specifically probe into the effects of non-linear processes on Taylor–Proudman columns and vertical shear layers.

Hide and Ibbetson [84] extended Taylor's original column experiment in order to acquire more quantitative data on column formation. The important objectives of this program are the delineation of the parameter space for which pillars form, the dependence of the shape and structure of the column on the Rossby number and its relationship to the geometry of the moving body that produces the disturbance. Unfortunately, only a preliminary report of the results attained is available and, as a consequence, discussion of all these points must be brief and incomplete.

In one experiment, a circular cylinder of height H and diameter L moves at a uniform speed U across the base of a large rotating tank (radius r_0). As the Rossby number, $\varepsilon = U/Lr_0$, increases, the flow changes from an almost purely columnar state to one in which the fluid penetrates the region above the cylinder, as shown in fig. 3.17. The drawing was made from streak lines observed visually at a position above the object, near the middle of the tank. The streamlines are, for the most part, essentially two-dimensional.

As the Rossby number is increased further, non-linear effects tilt the column and alter the streamlines significantly by producing a strong cross-flow over the obstacle. Surprisingly, a flow pattern of this general nature, fig. 3.18, was predicted by Stewartson [186] who

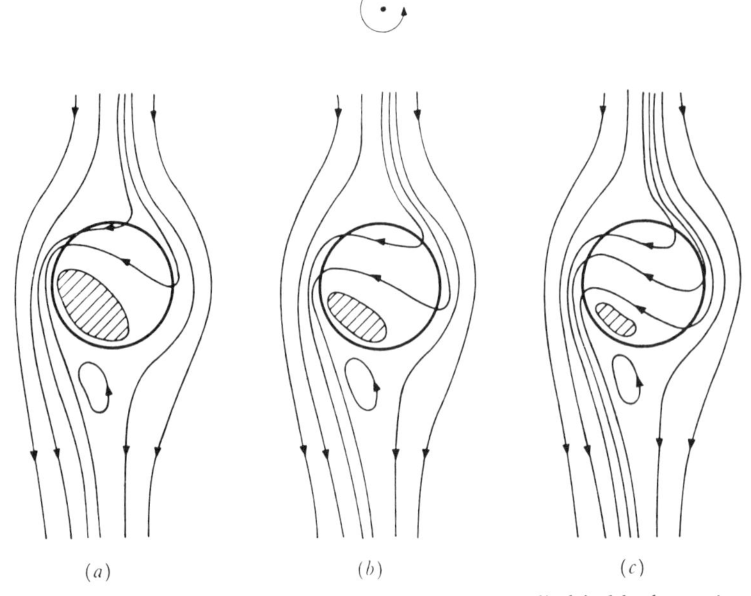

Fig. 3.17. Drawing of the relative fluid motion past a cylindrical body moving perpendicular to the rotation axis, [84]. The values of the Rossby number are: (a) $\varepsilon = 3 \cdot 10^{-3}$, (b) $\varepsilon = 1 \cdot 4 \; 10^{-2}$, (c) $\varepsilon = 2 \cdot 3 \; 10^{-2}$.

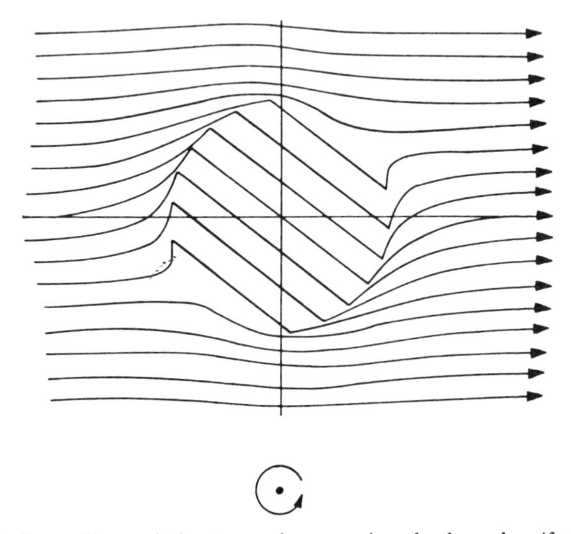

Fig. 3.18. Streamlines relative to a sphere moving slowly and uniformly in a direction perpendicular to the rotation axis, [186]. The horizontal plane shown is any one which does not intersect the object.

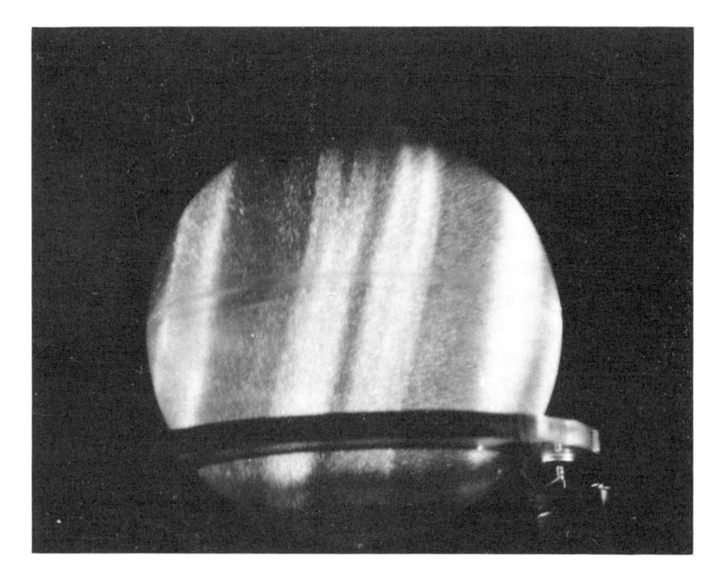

Fig. 3.19. A photograph by W. V. R. Malkus showing free shear layers in a slowly precessing, rotating sphere.

studied the slow, time-dependent motion of an ellipsoid moving perpendicular to $\mathbf{\Omega}$ in an unbounded fluid. This similarity of a distinctly non-linear motion and the limiting flow, as $t \to \infty$, of a linear theory is probably accidental. But a relationship of this kind does exist when the body moves along the rotation axis because the governing equations can be shown to be analogous (see §4.6).

The occurrence of a column also depends very critically on the height of the obstacle, but there is insufficient information on this aspect of the problem to be very definite. Eventually, it should be possible to determine, experimentally and theoretically, the different flow regimes in which columns are or are not observed. One continuing difficulty is the lack of a precise experimental criterion for columnar flow.

The experiments of Malkus[125] display a number of unusual effects that arise from non-linear wave interactions in a precessing sphere. The spin-over problem, in which the rotation axis of a fluid-filled sphere is impulsively turned through a small angle α, was analysed in §2.12 (see equation (2.12.13) et seq.). The agreement between experiment and the linear theory is good if the angle change is only a few degrees. As α is increased to about 8°, a new phenomenon appears in the fluid which persists and grows stronger at still larger values of the inclination angle. The new effect, shown in fig. 3.19, consists, in part, of free shear layers within the flow which are almost perfectly straight. The same phenomenon develops when the rotation axis of a sphere is precessed about a fixed direction in space. Precession can be viewed as a sequence of small impulsive changes in angle α in the manner just described. (Actually, the photograph is one of a precessing configuration.)

Measurements of the velocity in the precessing sphere indicate that a steady two-dimensional zonal current is established which changes direction several times in $0 \leqslant r \leqslant 1$. The magnitude of this circulation seems consistent with the hypothesis that it has its origin in a non-linear mechanism. Non-linear terms can couple two inertial waves with frequencies λ_1 and λ_2 to produce additional waves at frequencies $\lambda_1 \pm \lambda_2$. In particular, a single wave can interact with itself to provide a steady zonal current, and an oscillation at twice the basic frequency. Interactions within the boundary layers at the critical latitudes may also be important (see the discussion on p. 179).

Malkus[124] conjectures that instabilities forming on these shear layers could be of great significance in geophysical problems related to the magnetic field of the earth. Further work will be required to substantiate this view. Some experiments on free shear layer instabilities are described in chapter 6.

3.9. Large angle precession

The motion within a precessing spheroid was studied in §2.13. Here, the problem is considered anew, but without the restrictions presumed by a linear theory. The experiments of Malkus with precessing spheres are described in the last section.

The configuration is that shown in fig. 2.9, p. 71. The rotation axis of the spheroid is made to precess about its original direction in space by tilting the bottom turntable of the apparatus through an angle α. The inclination angle need not be small but otherwise conditions are identical with those detailed earlier.

The co-ordinate system selected to describe the motion is one for which the symmetry and precession axes, $\boldsymbol{\Omega}$ and $\boldsymbol{\omega}$, are fixed. This reference frame, designated as system $\mathscr{E}$, rotates with angular velocity $\boldsymbol{\omega}$, but has the z axis aligned with the main rotation axis $\boldsymbol{\Omega}$ of the body. This is an especially useful set of axes when the body is symmetrical about $\boldsymbol{\Omega}$, but then, the shell appears to slip on the surface of the contained fluid mass.

In terms of the velocity vector $\mathbf{u}$ defined by

$$\mathbf{q}_{\text{inert}} = \mathbf{u} + (\boldsymbol{\omega} + \boldsymbol{\Omega}) \times \mathbf{r},$$

the *inviscid dimensional* equations of motion in $\mathscr{E}$ are

$$\frac{\partial}{\partial t}\mathbf{u} + 2(\boldsymbol{\omega} + \boldsymbol{\Omega}) \times \mathbf{u} + (\nabla \times \mathbf{u}) \times (\boldsymbol{\Omega} \times \mathbf{r})$$
$$+ (\nabla \times \mathbf{u}) \times \mathbf{u} = -\frac{1}{\rho}\nabla p + \mathbf{r} \times (\boldsymbol{\omega} \times \boldsymbol{\Omega}), \qquad (3.9.1)$$

$$\nabla \cdot \mathbf{u} = 0, \qquad (3.9.2)$$

where $\quad \boldsymbol{\Omega} = \Omega\hat{\mathbf{k}} \quad$ and $\quad \boldsymbol{\omega} = \omega(-\sin\alpha\,\hat{\mathbf{j}} + \cos\alpha\,\hat{\mathbf{k}}) = \omega\hat{\boldsymbol{\omega}}.$

The initial condition corresponding to an impulsive change in the direction of $\boldsymbol{\omega}$, ($\boldsymbol{\omega}_+ = \boldsymbol{\omega}$ at $t = 0+$, $\boldsymbol{\omega}_- = \omega\hat{\mathbf{k}}$ at $t = 0-$) is, in terms of the vorticity,

$$\mathfrak{B} = \nabla \times \mathbf{u} = \pm 2\omega[\sin\alpha\,\hat{\mathbf{j}} + (1 - \cos\alpha)\,\hat{\mathbf{k}}]. \qquad (3.9.3)$$

Furthermore, the inviscid boundary condition is $\mathbf{u}\cdot\hat{\mathbf{n}} = 0$ at the surface

$$\frac{1}{a^2}(x^2+y^2)+\frac{1}{b^2}z^2 = 1.$$

(The vorticity is unaffected by an impulsive shift in container attitude although the velocity must be immediately corrected by a potential flow that arises to maintain the boundary conditions. All this was discussed previously on p. 72.)

Since the initial vorticity vector has no spatial dependence, Hocking[85] assumed simply that thereafter, the components of $\mathfrak{B}$ remain functions only of time:

$$\mathfrak{B} = (a^2+b^2)\,A(t)\,\hat{\imath}+(a^2+b^2)\,B(t)\,\hat{\jmath}+2a^2C(t)\,\hat{\mathbf{k}}. \qquad (3.9.4)$$

Vorticity solutions of this type have a long history and a comprehensive discussion of early work appears in Lamb[103]. The corresponding form for $\mathbf{u}$ turns out to be

$$\mathbf{u} = [-a^2yC(t)+a^2zB(t)]\,\hat{\imath} + [a^2xC(t)-a^2zA(t)]\,\hat{\jmath}+[-b^2xB(t)+b^2yA(t)]\,\hat{\mathbf{k}}. \qquad (3.9.5)$$

The continuity equation and the spatial boundary condition are then automatically satisfied. By taking the curl of (3.9.1) and substituting for $\mathbf{u}$, the following equations result for the unknown time coefficients:

$$(a^2+b^2)\frac{dA}{dt} = 2\omega\Omega\sin\alpha+((a^2-b^2)\,\Omega+2\omega a^2\cos\alpha)\,B + 2a^2\omega\sin\alpha\, C+a^2(a^2-b^2)\,BC, \qquad (3.9.6)$$

$$(a^2+b^2)\frac{dB}{dt} = -((a^2-b^2)\,\Omega+2\omega a^2\cos\alpha)\,A - a^2(a^2-b^2)\,AC, \qquad (3.9.7)$$

$$a^2\frac{dC}{dt} = -\omega b^2\sin\alpha\, A. \qquad (3.9.8)$$

The initial conditions are

$$A(0) = 0, \quad B(0) = -\frac{2\omega}{a^2+b^2}\sin\alpha, \quad C(0) = -\frac{\omega}{a^2}(1-\cos\alpha). \qquad (3.9.9)$$

The velocity $\mathbf{u}$ also satisfies the complete equations of motion, including viscous terms, but the no-slip boundary conditions are violated.

The solution of the foregoing system can be obtained in exact form, but it is more convenient to integrate the equations numerically, having chosen a particular parameter setting. However, certain special cases warrant comment.

In the linear range, with α small, C(t) is effectively zero and all product terms, BC etc., are negligible. The integrations can be performed easily and the solution is the inviscid version of that quoted earlier in § 2.13, when the different co-ordinate systems are related. Furthermore, if the factor $(a^2-b^2)\Omega+2\omega a^2\cos\alpha$ is zero, it is apparent from (3.9.6) that at early times

$$A(t) = \frac{2\omega\Omega\sin\alpha}{(a^2+b^2)}t;$$

the growth is then proportional to time and a linear resonance is produced. The inclusion of non-linear terms restricts the amplitudes to finite values and the solutions are oscillatory in time. Thus, no unbounded solutions exist in the non-linear inviscid range when all the terms are properly taken into account.

The solution is particularly simple for that value of α for which

$$A(t) = 0, \quad B(t) = B(0), \quad C(t) = C(0).$$

The motion observed is then completely steady in this system of co-ordinates, and is given by (3.9.5). The vorticity vector is a constant in this case.

Fig. 3.20 shows the coefficient A(t) versus time for various parameter settings including 'resonance'. The other coefficients show similar behaviour and a typical set of curves is illustrated in fig. 3.21.

The omission of viscous effects in this analysis severely restricts the usefulness of the solution. For example, at small eccentricities, for which $a-b = O(E^{\frac{1}{2}})$, the viscous boundary layers are as important in producing motion as the normal pressure forces exerted by the casing. In fact, the Ekman layer is the only driving mechanism in the precessing spherical container; the inviscid, non-linear solution for this geometry represents rigid rotation about the original rotation axis. The absence of dissipation also makes it impossible to describe the eventual development of a steady motion although the initial transient phase is correctly given.

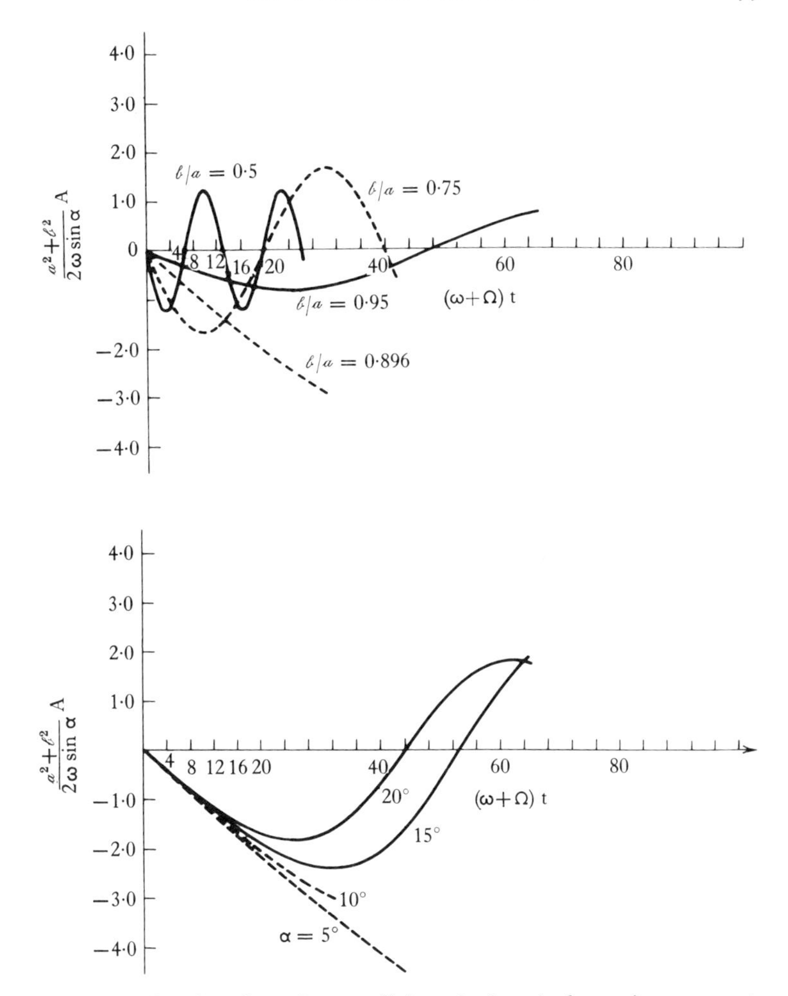

Fig. 3.20. The time-dependent coefficient A, (3.9.4), for various parameter settings. (*a*) Four values of the axes ratio b/a at $\omega/\Omega = -0{\cdot}1$ and $\alpha = 10°$ including the 'resonance' setting, $b/a = 0{\cdot}896$. (*b*) Four values of the inclination angle at $\omega/\Omega = 0{\cdot}1$, $b/a = 0{\cdot}896$, including the 'resonance' setting.

Experiments with the precessing spheroid, and the linear theory of §2.13 reveal that a steady, constant vorticity flow component (relative to system $\mathscr{E}$) does indeed arise in the interior. An investigation of non-linear effects in the steady, viscous, precession problem was initiated by Busse[22], whose approach consists of a systematic perturbation expansion in powers of the Ekman and

Rossby numbers, similar to that in § 3.7. Although the interior flow is assumed to be one of nearly constant vorticity, the magnitude and direction of the vorticity vector are determined as part of the analysis. In this manner, corrections are made for the location of the

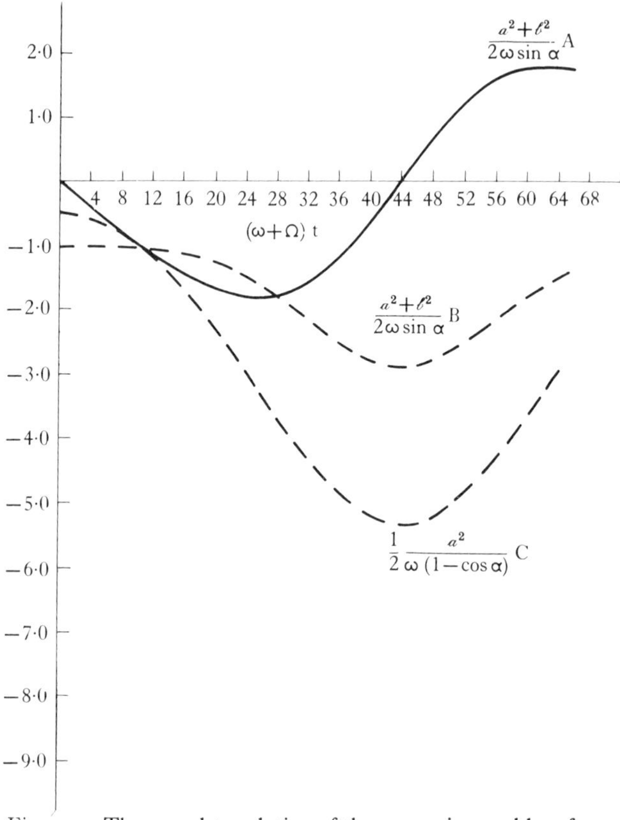

Fig. 3.21. The complete solution of the precession problem for $a/\ell = 0{\cdot}896$, $\alpha = 20°$, $\omega/\Omega = 0{\cdot}1$.

critical latitudes, which are found to be symmetrically placed with respect to the main rotation axis of the fluid and not that of the shell. The analysis is quite complicated but the formula obtained for the vorticity vector reduces correctly to the linear result and seems to agree well with the preliminary experimental evidence of Malkus. A detailed comparison is yet to appear.

3.10. Vortex flows and similarity solutions

Meteorological and oceanographic phenomena motivate the search for exact solutions of the Navier–Stokes equations that represent intense vortices. Although the literature on vortex flows and similarity solutions is extensive, most often it is very difficult to attach real physical meaning to the particular solutions that have been found. This is so because the effects produced by spatial boundaries are rarely included in special solutions. These effects, boundary layers and secondary circulations, are truly basic in rotating fluid phenomena and the significance of any analysis of a homogeneous fluid problem that takes no account of such fundamental processes must be seriously questioned.

There are certain outstanding exceptional cases and some of these, Kármán's solution for example, have already occupied a central position in previous sections. However, this particular approach to the subject will not be pursued here with any special vigour. Reference is made to the comprehensive review of Donaldson [38] and the treatise of Truesdell [209] for those who wish to explore this direction more thoroughly. The few examples presented in this section seem typical, and the most relevant of the entire class to the study and understanding of strong vortices occurring in nature.

The potential flow,

$$\mathbf{q} = -\Xi r\,\hat{\mathbf{r}} + \frac{\Gamma}{2\pi r}\,\hat{\boldsymbol{\theta}} + 2\Xi z\hat{\mathbf{k}}, \tag{3.10.1}$$

is an exact solution of the complete *dimensional* equations of motion in *inertial* space ((1.2.1) and (1.2.2) with $\boldsymbol{\Omega} = 0$) and represents an axisymmetric, stagnation point flow. The parameter Ξ, (sec^{-1}), characterizes the disturbance. Although many vortices may be of this type, the solution is not completely satisfactory because there is a singularity at $r = 0$. Burgers [21] and Rott [175] found the following solution that corrects this difficulty:

$$\mathbf{q} = -\Xi r\,\hat{\mathbf{r}} + \frac{\Gamma}{2\pi r}\left[1 - \exp\left(-\frac{\Xi r^2}{2\nu}\right)\right]\hat{\boldsymbol{\theta}} + 2\Xi z\hat{\mathbf{k}}. \tag{3.10.2}$$

This shows that in equilibrium, swirling fluid is carried toward the axis and the motion is essentially inviscid outside a central core of radius

$$r = \left(\frac{2\nu}{\Xi}\right)^{\frac{1}{2}}.$$

Of course, viscosity is also an important factor at the surface $z = 0$ and (3.10.2) is enlightening only in a restrictive interpretation. Information is given on the nature and extent of the viscous core which prevents singular behaviour at the axis of the incompressible vortex. Rott[176] found a class of time-dependent solutions which include thermal diffusion and have (3.10.2) as a limiting case.

A more systematic procedure for the analysis of vortex flows was developed by Lewellen[106] who managed to put many earlier efforts within one consistent framework. His method involves perturbations about a vortex flow that can have a more general radial dependence than that corresponding to rigid rotation. However, a similarity assumption imposes some difficulty in satisfying arbitrary boundary conditions.

Lewellen makes the problem dimensionless using L, Γ_∞, $\mathscr{M}$ to characterize the length, swirl, $\Gamma = rv$, and the axisymmetric stream function, ψ, defined by

$$u = \frac{1}{r}\frac{\partial\psi}{\partial z}, \quad w = -\frac{1}{r}\frac{\partial\psi}{\partial r}.$$

Here, $\mathscr{M}$ is the radial mass flux and it is not necessarily small or related in magnitude to the Ekman number. The radial velocity, u, is scaled differently from v to model atmospheric conditions.

The *dimensionless* theory, in an *inertial* frame, is

$$\varepsilon E\mathscr{D}^2\psi - \varepsilon^2\left\{r\frac{\partial\psi}{\partial z}\frac{\partial}{\partial r}\left(\frac{1}{r^2}\mathscr{D}\psi\right) - \frac{1}{r}\frac{\partial\psi}{\partial r}\frac{\partial}{\partial z}\mathscr{D}\psi\right\} + \frac{2}{r^2}\Gamma\frac{\partial\Gamma}{\partial z} = 0, \tag{3.10.3}$$

$$E\mathscr{D}\Gamma - \frac{\varepsilon}{r}\left(\frac{\partial\psi}{\partial z}\frac{\partial\Gamma}{\partial r} - \frac{\partial\psi}{\partial r}\frac{\partial\Gamma}{\partial z}\right) = 0, \tag{3.10.4}$$

where

$$\mathscr{D} = \frac{\partial^2}{\partial z^2} + r\frac{\partial}{\partial r}\frac{1}{r}\frac{\partial}{\partial r}$$

and

$$\varepsilon = \mathscr{M}/L\Gamma_\infty, \quad E = \nu/\Gamma_\infty.$$

All functions can now be expanded in powers of ε; however, Lewellen first sets

$$E = \varepsilon/R_E$$

and takes the radial Reynolds number, $R_E = \mathscr{M}/\nu L$, and not E, as an independent parameter. To proceed with the derivation of a particular class of solutions, let $\eta = r^2$, and set

$$\Gamma = \sum_{n=0}^{\infty} \Gamma_n(\eta, z)\,\varepsilon^{2n},$$

$$\psi = \sum_{n=0}^{\infty} \psi_n(\eta, z)\,\varepsilon^{2n}.$$

It follows upon substituting these expansions into (3.10.3) and (3.10.4) that

$$\Gamma_0 \frac{\partial \Gamma_0}{\partial z} = 0 \tag{3.10.5}$$

and
$$2R_E^{-1}\eta \frac{d^2\Gamma_0}{d\eta^2} - \frac{\partial \psi_0}{\partial z}\frac{d\Gamma_0}{d\eta} = 0. \tag{3.10.6}$$

Therefore, the stream function must be linear in z:

$$\psi_0 = A_{00}(\eta) + zA_{01}(\eta). \tag{3.10.7}$$

(The solution of Burgers and Rott corresponds to $\psi_0 = -z\eta$.) It is even possible to prescribe boundary conditions at two axial positions, say $z = 0$ and $z = 1$. These might be of the general form

$$\psi_0(\eta, 0) = \mathscr{\psi}_0(\eta) \quad (\mathscr{\psi}_0(0) = 0),$$

$$\psi_0(\eta, 1) = \mathscr{\psi}_1(\eta) \quad (\mathscr{\psi}_1(0) = 0),$$

which can be used as the appropriate conditions exterior to Ekman layers. Thus, (3.10.7) and (3.10.6) become

$$\psi_0 = \mathscr{\psi}_1(\eta) + [\mathscr{\psi}_0(\eta) - \mathscr{\psi}_1(\eta)]\,z,$$

$$2R_E^{-1}\eta \frac{d^2\Gamma_0}{d\eta^2} - [\mathscr{\psi}_0(\eta) - \mathscr{\psi}_1(\eta)]\frac{d\Gamma_0}{d\eta} = 0. \tag{3.10.8}$$

The integration of the last equation subject to conditions

$$\lim_{\eta \to \infty} \Gamma_0 = \Gamma_\infty, \quad \Gamma_0(0) = 0$$

and the arbitrary prescription of functions $\mathscr{\psi}_0$ and $\mathscr{\psi}_1$ yields one particular solution for Γ_0. Models for many phenomena can be constructed on this basis.

As many higher order terms can be computed as desired, and the theory can be extended [106] to include viscous boundary layers at container walls. The centrifuge problem, (see § 2.19), with large mass injection was analysed in this way as a perturbation about an established vortex flow that is *not* solid body rotation. However, the qualitative character of the flow is found to be much the same as before.

Lewellen has shown that it is not possible to join a boundary layer to an internal motion whose swirl, Γ_0, decreases with increasing radius. The potential flow, $\Gamma_0 = \text{constant}$, has the weakest radial dependence for which a match to an Ekman layer can still be made, but what happens when Γ_0 is forced to have an unacceptable radial variation is a matter for speculation.

CHAPTER 4

MOTION IN AN UNBOUNDED ROTATING FLUID

4.1. Classification

The separation of rotating fluid motions into two classes was made in § 2.1. As the title implies, this chapter deals with motions produced by bodies moving through unbounded rotating fluids. Before plunging into the analysis of specific problems, it is well to pause and reflect upon the validity of the basic idealization—an unbounded rotating fluid. Certainly the approximation is sound in many circumstances and it is perhaps most appropriate in problems of steady, viscous flows, or transient motions at early times. However, there are situations in which it can be the source of difficulties and there is serious trouble in the theory of *inviscid*, *steady* motions, either linear or non-linear, when the container walls are placed at infinity. Too many conceptual limit processes, $\nu \to 0, t \to \infty, L \to \infty$, etc. are invoked and whether or not the final results of an analysis make sense may depend strongly on the ordering of the limits. For these reasons, problems of this type may not even be well-posed.

The control exerted on the main inviscid flow by the boundary layers at container walls has been definitely established in the preceding chapters. The extent of this influence is one of the most interesting and important features of the subject, so much so, that caution is required whenever the bounding walls are to be excluded from consideration.

4.2. Plane inertial waves

The fact that small disturbances in a uniformly rotating, incompressible fluid can propagate as wave motions has long been established, an early discussion of the theory having been given by Kelvin[95]. Among the important theoretical and experimental investigations that have appeared since, are those of Fultz[60], Long[113], Chandrasekhar[30], Phillips[151]; some of this material was discussed previously in connection with contained inertial waves and the general theory of chapter 2. However, many important and particularly revealing features of wave propagation remain to be studied and the present section is an attempt to complete, or at least bring to date, our knowledge of wave motion in a rotating field. The presentation here is based on that of Phillips.

The *dimensionless*, linear, inviscid theory

$$\left.\begin{aligned} \nabla\cdot\mathbf{q} &= 0, \\ \frac{\partial}{\partial t}\mathbf{q}+2\hat{\mathbf{\Omega}}\times\mathbf{q} &= -\nabla p, \end{aligned}\right\} \tag{4.2.1}$$

has wave solutions of the form

$$\left.\begin{aligned} \mathbf{q} &= \mathscr{R}e\,\mathbf{Q}\,e^{i(\boldsymbol{\kappa}\cdot\mathbf{r}-\lambda t)}, \\ p &= \mathscr{R}e\,\Phi\,e^{i(\boldsymbol{\kappa}\cdot\mathbf{r}-\lambda t)}, \end{aligned}\right\} \tag{4.2.2}$$

where $\boldsymbol{\kappa}$ is the wave vector, $|\boldsymbol{\kappa}|$ is the wave number and λ is the frequency. (The length scale L is arbitrary.) This is entirely analogous to the form of contained inertial waves, but far simpler because $\mathbf{Q}$ and Φ are now constants. The substitution of these expressions into (4.2.1) leads to relationships among the quantities $\boldsymbol{\kappa}$, λ, $\mathbf{Q}$ and Φ. For example, the continuity equation implies

$$\mathbf{Q}\cdot\boldsymbol{\kappa} = 0; \tag{4.2.3}$$

the velocity vector is perpendicular to the propagation direction and the wave is transverse. This is the only possible type of motion in an incompressible fluid.

It follows from the momentum equation that

$$\lambda = \pm 2\hat{\mathbf{\Omega}}\cdot\hat{\boldsymbol{\kappa}} = \pm 2\cos\Theta, \tag{4.2.4}$$

where
$$\hat{\boldsymbol{\kappa}} = \boldsymbol{\kappa}/|\boldsymbol{\kappa}|$$

and Θ is the polar angle measured from the rotation axis. Thus, the frequency is dependent on the direction but not the magnitude of the wave vector. Moreover, if $\boldsymbol{\upsilon}$ is *any* vector perpendicular to $\boldsymbol{\kappa}$, then

$$\mathbf{Q} = \pm\hat{\boldsymbol{\kappa}}\times\boldsymbol{\upsilon}+i\boldsymbol{\upsilon}, \tag{4.2.5}$$

and
$$\Phi = \frac{2}{|\boldsymbol{\kappa}|}[\hat{\mathbf{\Omega}}\cdot\hat{\boldsymbol{\kappa}}\times\boldsymbol{\upsilon}\pm i\hat{\mathbf{\Omega}}\cdot\boldsymbol{\upsilon}]. \tag{4.2.6}$$

Therefore, with $\mathbf{υ}$ real,

$$\mathbf{q} = \pm\hat{\boldsymbol{\kappa}} \times \mathbf{υ} \cos(\boldsymbol{\kappa}\cdot\mathbf{r} \mp 2\hat{\boldsymbol{\Omega}}\cdot\hat{\boldsymbol{\kappa}}t) - \mathbf{υ}\sin(\boldsymbol{\kappa}\cdot\mathbf{r} \mp 2\hat{\boldsymbol{\Omega}}\cdot\hat{\boldsymbol{\kappa}}t), \quad (4.2.7)$$

$$p = \frac{2}{|\boldsymbol{\kappa}|}[\hat{\boldsymbol{\Omega}}\cdot\hat{\boldsymbol{\kappa}} \times \mathbf{υ}\cos(\boldsymbol{\kappa}\cdot\mathbf{r} \mp 2\hat{\boldsymbol{\Omega}}\cdot\hat{\boldsymbol{\kappa}}t) \mp \hat{\boldsymbol{\Omega}}\cdot\mathbf{υ}\sin(\boldsymbol{\kappa}\cdot\mathbf{r} \mp 2\hat{\boldsymbol{\Omega}}\cdot\hat{\boldsymbol{\kappa}}t)]; \quad (4.2.8)$$

a change in the definition of $\mathbf{υ}$ brings these equations to the form stated by Phillips. To obtain solutions incorporating the viscous terms as well, it is only necessary to multiply each of the foregoing by the factor

$$\exp(-E|\boldsymbol{\kappa}|^2 t).$$

In fact, the expressions are then exact solutions of the complete non-linear equations, [30], but the principle of superposition does not hold.

The phase velocity of waves moving in the direction $\boldsymbol{\kappa}$, i.e. surfaces maintaining a constant value of $\mathbf{r}\cdot\boldsymbol{\kappa} - \lambda t$ (with $\lambda = +2\hat{\boldsymbol{\Omega}}\cdot\hat{\boldsymbol{\kappa}}$), is

$$\mathbf{c}_p = 2\frac{\hat{\boldsymbol{\Omega}}\cdot\hat{\boldsymbol{\kappa}}}{|\boldsymbol{\kappa}|}\hat{\boldsymbol{\kappa}}. \quad (4.2.9)$$

The phase speed is inversely proportional to the magnitude of the wave vector and the waves are, in general, dispersive, dissipative and anisotropic. The long waves, small $|\boldsymbol{\kappa}|$, travel fastest and the short waves slowest, a condition reminiscent of surface waves in classical hydrodynamics.

The group velocity, $\mathbf{c}_g$, is the velocity of energy propagation and may be determined by calculating the energy flux. A comprehensive theory of wave and energy propagation exists (see Whitham[225]), but for present purposes only the relationship of group velocity and frequency is required. For a wave train of the form $\exp i(\boldsymbol{\kappa}\cdot\mathbf{r} - \lambda(\boldsymbol{\kappa})t)$, the group velocity is given by

$$\mathbf{c}_g = \nabla_\kappa \lambda(\boldsymbol{\kappa}) = \frac{\partial\lambda}{\partial\kappa_1}\hat{\mathbf{i}}_1 + \frac{\partial\lambda}{\partial\kappa_2}\hat{\mathbf{i}}_2 + \frac{\partial\lambda}{\partial\kappa_3}\hat{\mathbf{i}}_3, \quad (4.2.10)$$

where $\boldsymbol{\kappa} = \kappa_1\hat{\mathbf{i}}_1 + \kappa_2\hat{\mathbf{i}}_2 + \kappa_3\hat{\mathbf{i}}_3$. In a rotating fluid

$$\lambda = 2\hat{\boldsymbol{\Omega}}\cdot\hat{\boldsymbol{\kappa}} = 2\frac{\Omega_1\kappa_1 + \Omega_2\kappa_2 + \Omega_3\kappa_3}{(\kappa_1^2 + \kappa_2^2 + \kappa_3^2)^{\frac{1}{2}}}, \quad (4.2.11)$$

and it follows that

$$\mathbf{c}_g = \frac{2}{|\boldsymbol{\kappa}|}\hat{\boldsymbol{\Omega}} - \mathbf{c}_p, \quad (4.2.12)$$

or with the use of (4.2.9)

$$\mathbf{c}_g = \frac{2}{|\boldsymbol{\kappa}|}\hat{\boldsymbol{\kappa}}\times(\hat{\boldsymbol{\Omega}}\times\hat{\boldsymbol{\kappa}}). \tag{4.2.13}$$

This result shows that the energy transport is at right angles to the phase velocity. A wave appearing to move in one direction, according to the surfaces of constant phase, is actually propagating energy in a perpendicular direction.

It may be suspected now that the reflexion of these plane waves off rigid boundaries will display many odd properties. Phillips[151] showed that the reflexion involves a change of wave number and a new direction of energy transport that does not satisfy an 'equal angle' rule (Snell's law) typical of non-dispersive systems.

Let

$$\mathbf{q} = (\hat{\boldsymbol{\kappa}}\times\boldsymbol{\upsilon}+\mathrm{i}\boldsymbol{\upsilon})\exp\mathrm{i}(\boldsymbol{\kappa}\cdot\mathbf{r}-2\hat{\boldsymbol{\Omega}}\cdot\hat{\boldsymbol{\kappa}}t) \tag{4.2.14}$$

denote a wave incident upon an infinite plane whose unit normal, *pointing out of the fluid*, is $\hat{\mathbf{n}}$. It is convenient to choose $\boldsymbol{\upsilon}$ as coplanar with $\hat{\boldsymbol{\kappa}}$ and $\hat{\mathbf{n}}$, and this may be done without loss of generality because the origin of the time scale (the phase) is at our disposal.

The reflected wave is then of the form

$$\mathbf{q}' = (\pm\hat{\boldsymbol{\kappa}}'\times\boldsymbol{\upsilon}'+\mathrm{i}\boldsymbol{\upsilon}')\exp\mathrm{i}(\boldsymbol{\kappa}'\cdot\mathbf{r}\mp 2\hat{\boldsymbol{\Omega}}\cdot\hat{\boldsymbol{\kappa}}'t). \tag{4.2.15}$$

The boundary condition at the wall, $\mathbf{r}\cdot\hat{\mathbf{n}} = 0$, requires that the normal component of velocity be zero:

$$\hat{\mathbf{n}}\cdot(\mathbf{q}+\mathbf{q}') = 0.$$

Hence,

$$\hat{\mathbf{n}}\cdot(\hat{\boldsymbol{\kappa}}\times\boldsymbol{\upsilon}+\mathrm{i}\boldsymbol{\upsilon})\exp\mathrm{i}(\boldsymbol{\kappa}\cdot\mathbf{r}-2\hat{\boldsymbol{\Omega}}\cdot\hat{\boldsymbol{\kappa}}t) + \hat{\mathbf{n}}\cdot(\pm\hat{\boldsymbol{\kappa}}'\times\boldsymbol{\upsilon}'+\mathrm{i}\boldsymbol{\upsilon}')\exp\mathrm{i}(\boldsymbol{\kappa}'\cdot\mathbf{r}\mp 2\hat{\boldsymbol{\Omega}}\cdot\hat{\boldsymbol{\kappa}}'t) = 0. \tag{4.2.16}$$

Since this relationship holds at any time, and at any point on the plane, the frequencies and tangential wave numbers of both waves must be identical

$$\hat{\boldsymbol{\Omega}}\cdot\hat{\boldsymbol{\kappa}} = \pm\hat{\boldsymbol{\Omega}}\cdot\hat{\boldsymbol{\kappa}}', \tag{4.2.17}$$

and

$$\boldsymbol{\kappa}\cdot\mathbf{r} = \boldsymbol{\kappa}'\cdot\mathbf{r}.$$

The last equation can be manipulated into a better form by writing

$$\mathbf{r} = -\hat{\mathbf{n}}\times(\hat{\mathbf{n}}\times\mathbf{r}),$$

(which holds because $\mathbf{r}\cdot\hat{\mathbf{n}} = 0$ defines the plane) and it follows that

$$(\hat{\mathbf{n}}\times\mathbf{r})\cdot(\boldsymbol{\kappa}\times\hat{\mathbf{n}}-\boldsymbol{\kappa}'\times\hat{\mathbf{n}}) = 0.$$

However, $\hat{\mathbf{n}} \times \mathbf{r}$ is an arbitrary tangential vector in the bounding plane and the second vector must be identically zero:

$$\boldsymbol{\kappa} \times \hat{\mathbf{n}} = \boldsymbol{\kappa}' \times \hat{\mathbf{n}}. \tag{4.2.18}$$

The exponential factors can be eliminated from (4.2.16) to obtain conditions on the amplitudes of incident and reflected waves. These are

$$\hat{\mathbf{n}} \cdot \hat{\boldsymbol{\kappa}} \times \boldsymbol{\upsilon} = \mp \hat{\mathbf{n}} \cdot \hat{\boldsymbol{\kappa}}' \times \boldsymbol{\upsilon}'; \tag{4.2.19}$$

$$\hat{\mathbf{n}} \cdot \boldsymbol{\upsilon} = -\hat{\mathbf{n}} \cdot \boldsymbol{\upsilon}'. \tag{4.2.20}$$

Finally, the sign is chosen to insure that the energy flux of the reflected wave is directed into the fluid away from the wall.

It is a rather straightforward task to calculate $\boldsymbol{\kappa}'$ and $\boldsymbol{\upsilon}'$ from these equations, but Phillips' graphical construction, presented next, is simpler and more informative. To begin, note that the vectors $\boldsymbol{\kappa}$ and $\boldsymbol{\upsilon}$, $\boldsymbol{\kappa}'$ and $\boldsymbol{\upsilon}'$ are orthogonal:

$$\boldsymbol{\kappa} \cdot \boldsymbol{\upsilon} = 0 = \boldsymbol{\kappa}' \cdot \boldsymbol{\upsilon}'. \tag{4.2.21}$$

This follows directly from the conservation equation (4.2.3) and the amplitude relationship (4.2.5). Furthermore, $\hat{\mathbf{n}}$, $\boldsymbol{\kappa}$, $\boldsymbol{\kappa}'$ are coplanar, and $\boldsymbol{\upsilon}$ has been assumed to be in this plane, too, since the origin of the time scale is arbitrary. Equations (4.2.19) and (4.2.20) then imply that $\boldsymbol{\upsilon}'$ is also in this plane. The task is now to find $\boldsymbol{\kappa}'$, $\boldsymbol{\upsilon}'$, given $\hat{\boldsymbol{\Omega}}$, $\boldsymbol{\kappa}$, $\boldsymbol{\upsilon}$, and $\hat{\mathbf{n}}$. Let $\boldsymbol{\Omega}_P$ be the projection of $\hat{\boldsymbol{\Omega}}$ in the plane of all these vectors which, in fig. 4.1, lies along the directed line segment $\overline{AO}$. Let $\boldsymbol{\kappa}$ be $\overline{OB}$ and $\hat{\boldsymbol{\kappa}}$ the unit vector $\overline{OG}$. The construction proceeds as follows:

(i) Equation (4.2.18) implies that vectors $\boldsymbol{\kappa}$ and $\boldsymbol{\kappa}'$ must have the same projection on the reflexion plane, and this is segment $\overline{OJ}$.

(ii) Equation (4.2.17) implies that vectors $\hat{\boldsymbol{\kappa}}$ and $\hat{\boldsymbol{\kappa}}'$ have the same projected length on $\boldsymbol{\Omega}_P$, the line FA. To insure that the reflected energy flux is directed away from the wall, it is necessary to choose the minus sign in this equation; $\boldsymbol{\kappa}'$ must then be $\overline{OC}$, and $\hat{\boldsymbol{\kappa}}'$ is $\overline{OH}$.

(iii) Vectors $\boldsymbol{\upsilon}$ and $\boldsymbol{\upsilon}'$ are orthogonal to $\boldsymbol{\kappa}$ and $\boldsymbol{\kappa}'$ respectively. Equation (4.2.20) requires the normal components of both vectors to be of equal magnitude, OM. Let $\overline{DO}$ be $\boldsymbol{\upsilon}$ then $\boldsymbol{\upsilon}'$ is the directed segment $\overline{OE}$.

(iv) The group velocity is normal to $\boldsymbol{\kappa}$ and its projection in the plane of $\boldsymbol{\kappa}$ and $\hat{\mathbf{n}}$ is parallel to $\boldsymbol{\upsilon}$ and in the direction $\overline{DO}$.

(v) The corresponding projection of the group velocity of the reflected wave is in the direction $\overline{OE}$.

The process pictured here is one of backwards reflexion. The incoming and reflected flux vectors make equal angles with $\boldsymbol{\Omega}_P$, and not with the normal to the plane, $\hat{\mathbf{n}}$, as a non-dispersive wave would.

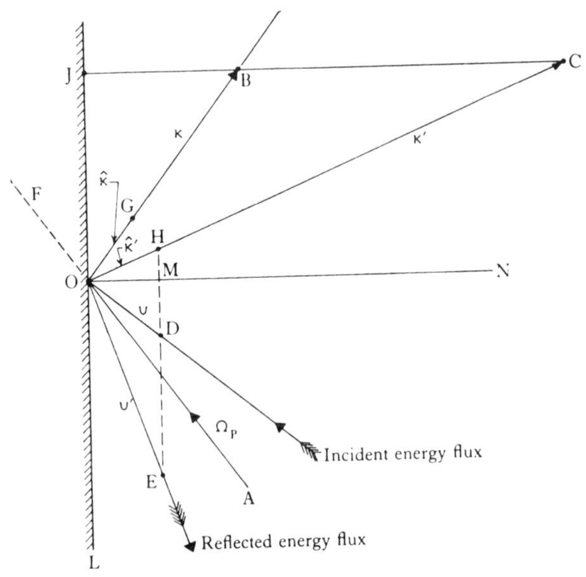

Fig. 4.1. Diagram for the graphical construction of a reflected inertial wave.

A forward reflexion occurs when the angle DOA is larger than angle AOL. If the two angles are equal, the energy of the reflected wave travels parallel to the wall and is, accordingly, much more susceptible to viscous dissipative processes. The wave vector $\boldsymbol{\kappa}'$ is generally not equal to $\boldsymbol{\kappa}$ and it is possible upon reflexion to scatter energy into different wavelengths. Incident and reflected waves do have the same frequency and the same component of $\boldsymbol{\kappa}$ in the plane of the boundary.

The similarity of triangles DOE and BOC implies the following identity:

$$\frac{|\boldsymbol{\upsilon}|}{|\boldsymbol{\kappa}|} = \frac{|\boldsymbol{\upsilon}'|}{|\boldsymbol{\kappa}'|}.$$

This, in conjunction with the relationship

$$\frac{1}{|\boldsymbol{\kappa}|^2}\,d\boldsymbol{\kappa} = \frac{1}{|\boldsymbol{\kappa}'|^2}\,d\boldsymbol{\kappa}',$$

may be used to establish the fact that the energy density per unit volume of wave number space,

$$\mathfrak{E}(\boldsymbol{\kappa}) = \frac{1}{2}\sum_{\boldsymbol{\kappa}}^{\boldsymbol{\kappa}+d\boldsymbol{\kappa}} |\mathbf{v}|^2,$$

remains unchanged in the reflexion process,

$$\mathfrak{E}(\boldsymbol{\kappa}) = \mathfrak{E}(\boldsymbol{\kappa}').$$

A study of reflexions in a viscous fluid and analysis of the boundary layer regimes, [151], leads to the conclusion that there is an $O(E^{\frac{1}{2}})$ loss in amplitude upon reflexion; here, the length scale L is the typical wavelength. Thus, a contained inertial wave packet, of unit amplitude, reverberating from one wall to the other, is dissipated after $E^{-\frac{1}{2}}$ reflexions, in the approximate time $E^{-\frac{1}{2}}\Omega^{-1}$. This has, of course, already been established, but the plane wave analysis provides a new interpretation of the dissipative process. In one case, this conclusion fails. If the reflected energy flux is parallel to the plane surface, then the energy may be totally absorbed. (Care is required in this situation because the 'inviscid' reflected amplitude is infinite.) This critical case occurs when

$$\boldsymbol{\Omega}_{\mathrm{P}}\cdot\hat{\boldsymbol{\kappa}} = \pm\,\boldsymbol{\Omega}_{\mathrm{P}}\cdot\hat{\mathbf{n}}, \tag{4.2.22}$$

which, in terms of the frequency, is

$$\lambda = \pm\,2\hat{\boldsymbol{\Omega}}\cdot\hat{\mathbf{n}}.$$

This is exactly the same condition on frequency and normal vector that identifies the location on the container surface of boundary layer eruptions which are predicted in the theory of contained inertial oscillations. Thus, the breakdown in the boundary layer at critical latitudes is connected with the total absorption of plane inertial waves in the boundary layer at these locations. A more explicit description of the erupting boundary layer in terms of inertial waves awaits an adequate theory. It would seem that there is now sufficient information about the essential mechanisms to make an attempt feasible.

4.3. Slow motion along the axis of rotation

The aim in this section is to analyse a simple, transient, rotating flow problem that explicitly exhibits the propagation of inertial waves and the evolution of a steady state. Visualization of the process by which these waves organize to form a Taylor column is the first objective; other examples are discussed later.

The particular configuration utilized in this analysis is a finite disk of radius r_0 that initially rotates rigidly with an infinite body of fluid. Subsequently, the disk is moved perpendicular to itself along the rotation axis, but its angular velocity is not changed. The movement may be one of slow forward motion with velocity $\mathfrak{U}$, or a small amplitude oscillation at frequency λ about its original position. The latter situation is explored in the next section.

The theory for the disk is easier than that corresponding to other geometries, and it is typical as well. For example, it is possible to obtain a detailed solution for slow, steady motion in a viscous medium, (Morrison and Morgan[132]); an informative, asymptotic approximation for impulsive forward motion, and even a closed form representation for quasi-steady oscillations in an inviscid fluid, (Oser [139], Reynolds [161, 162]) which has been experimentally verified, (Görtler [67], Oser [140]). Furthermore, both the transient inviscid and steady viscous solutions for slow motion agree identically in the respective limiting cases, $t \to \infty$, and $\nu \to 0$. The generalization of specific results for more arbitrary body shapes is discussed in the course of the analysis.

The relevant boundary conditions for the slow forward motion of a disk are, in the *rotating* co-ordinate system,

$$\mathbf{q} = \mathfrak{U}\hat{\mathbf{k}} \quad \text{on } r \leqslant r_0,\ z = \mathfrak{U}t, \text{ for } t \geqslant 0;$$

$$\mathbf{q} \to 0 \quad \text{as } |\mathbf{r}| \to \infty;$$

$$\mathbf{q} \equiv 0 \quad \text{for } t \leqslant 0.$$

The *dimensionless*, linear boundary value problem (infinitesimal Rossby number), utilizing the scaling rule $[\![r_0, \Omega^{-1}, \mathfrak{U}]\!]$, is

$$\frac{\partial}{\partial t}\mathbf{q} + 2\hat{\mathbf{k}} \times \mathbf{q} = -\nabla p + E\nabla^2\mathbf{q},$$

$$\nabla \cdot \mathbf{q} = 0,$$

with $\mathbf{q} = \hat{\mathbf{k}}$ on $r \leqslant 1$, $z = 0$, for $t > 0$; $\mathbf{q}$ is zero at infinity and $\mathbf{q} \equiv 0$ for $t < 0$.

The inviscid problem, $E = 0$, requires only that $\mathbf{q}\cdot\hat{\mathbf{k}} = 1$ on the disk for $t \geqslant 0$, and this is the first situation to be analysed. Let (u, v, w) be the velocity components in the cylindrical, symmetry co-ordinate system and introduce an axisymmetric stream function, ψ, by

$$u = \frac{1}{r}\frac{\partial\psi}{\partial z}, \quad w = -\frac{1}{r}\frac{\partial\psi}{\partial r}. \tag{4.3.1}$$

The Laplace transform of the linear boundary value problem, with

$$\bar{\psi} = \int_0^\infty e^{-st}\psi\,dt = \mathscr{L}\{\psi\},$$

leads to the following equation for the transformed stream function:

$$r\frac{\partial}{\partial r}\frac{1}{r}\frac{\partial\bar{\psi}}{\partial r} + \left(1 + \frac{4}{s^2}\right)\frac{\partial^2\bar{\psi}}{\partial z^2} = 0, \tag{4.3.2}$$

subject to $\quad \bar{w} = -\frac{1}{r}\frac{\partial\bar{\psi}}{\partial r} = \frac{1}{s} \quad$ on $r \leqslant 1, z = 0.$ (4.3.3)

The Hankel transform of (4.3.2) results in an ordinary differential equation in z, whose solution, symmetric in z and satisfying the requirement of no flow at infinity, is readily obtained. The general solution is

$$\bar{\psi} = r\int_0^\infty A(\kappa)\,J_1(\kappa r)\exp\left[-\kappa\,|z|\,s(s^2+4)^{-\frac{1}{2}}\right]d\kappa;$$

this follows directly from the inversion formula for the Hankel transformation. The function $A(\kappa)$ is determined by satisfying the remaining boundary conditions on the plane $z = 0$, which imply

$$\left.\begin{aligned} &\int_0^\infty \kappa A(\kappa)\,J_0(\kappa r)\,d\kappa = -\frac{1}{s}, && r \leqslant 1;\\ &\int_0^\infty \kappa A(\kappa)\,J_1(\kappa r)\,d\kappa = 0, && r > 1. \end{aligned}\right\} \tag{4.3.4}$$

These dual integral equations arise in classical potential theory and their solution is given by Titchmarsh[208], p. 339:

$$\kappa A(\kappa) = \frac{2}{\pi s}\left(\cos\kappa - \frac{\sin\kappa}{\kappa}\right). \tag{4.3.5}$$

The complete solution of the boundary value problem for $z \geqslant 0$ is then

$$\psi = \frac{2r}{\pi}\int_0^\infty \frac{J_1(\kappa r)}{\kappa}\left(\cos\kappa - \frac{\sin\kappa}{\kappa}\right)\mathfrak{K}(z,t,\kappa)\,d\kappa, \tag{4.3.6}$$

$$v = \frac{4}{\pi}\int_0^\infty J_1(\kappa r)\left(\cos\kappa - \frac{\sin\kappa}{\kappa}\right)\mathfrak{N}(z,t,\kappa)\,d\kappa, \tag{4.3.7}$$

where

$$\left.\begin{aligned}\mathfrak{K} &= \mathscr{L}^{-1}\left\{\frac{1}{s}\exp\left[-\kappa z s(s^2+4)^{-\frac{1}{2}}\right]\right\},\\ \mathfrak{N} &= \mathscr{L}^{-1}\left\{\frac{1}{s}(s^2+4)^{-\frac{1}{2}}\exp\left[-\kappa z s(s^2+4)^{-\frac{1}{2}}\right]\right\}.\end{aligned}\right\} \tag{4.3.8}$$

Of course, the task is incomplete until these integrals can be reduced to a comprehensible form and useful information extracted. To this end, consider the first of the integrals in (4.3.8) and the manipulations by which an asymptotic approximation for large time is obtained. The transform of $\mathfrak{K}$ has a simple pole at the origin and two branch points of rather severe form at $s = \pm 2i$. A finite vertical branch line between these points suffices and the contour of integration can be wrapped around this segment. If the loop integrals are carefully evaluated, then it follows that

$$\mathfrak{K} = 1 - \frac{2}{\pi}\int_0^2 \frac{\cos \mathfrak{y}t}{\mathfrak{y}}\sin\left(\frac{\kappa z\mathfrak{y}}{(4-\mathfrak{y}^2)^{\frac{1}{2}}}\right)d\mathfrak{y},$$

where $s = i\mathfrak{y}$ on the imaginary axis. The change of variable

$$\eta = \frac{\mathfrak{y}t}{(4-\mathfrak{y}^2)^{\frac{1}{2}}}$$

brings this into the form

$$\mathfrak{K} = 1 - \frac{2}{\pi}\int_0^\infty \cos\left(\frac{2\eta}{(1+(\eta/t)^2)^{\frac{1}{2}}}\right)\frac{\sin \mathfrak{b}\eta}{\eta}\frac{d\eta}{(1+\eta^2/t^2)},$$

where

$$\mathfrak{b} = \frac{\kappa z}{t} \tag{4.3.9}$$

enters as an important parameter. Therefore, in the asymptotic range of large t, large z, but moderate $\mathfrak{b}$,

$$\mathfrak{K} \sim 1 - \frac{2}{\pi}\int_0^\infty \cos 2\eta\,\frac{\sin \mathfrak{b}\eta}{\eta}\,d\eta.$$

With some simple trigonometry, the last integral can be written as two, each of the more elementary form, $\int_0^\infty \frac{\sin\eta}{\eta}\,d\eta$. The final result is then

$$\mathfrak{K} \sim \mathcal{S}(2-\mathfrak{b}), \tag{4.3.10}$$

where $\mathcal{S}$ is the unit step function (see p. 102). By an entirely equivalent procedure, it is found that

$$\mathfrak{N} \sim \tfrac{1}{2}\mathcal{S}(2-\mathfrak{b}). \tag{4.3.11}$$

Approximations for small times are easily derived but are not sufficiently informative to warrant discussion.

These formulas can be interpreted in terms of the known properties of plane waves. The requirement, $\mathfrak{b} < 2$, is interpreted best in the form

$$\frac{z}{\left(\frac{2}{\kappa}\right)} < t, \tag{4.3.12}$$

because

$$\frac{2}{\kappa} = |\mathfrak{c}_g|$$

is the group velocity of plane waves of wave number κ whose phase velocity is in a direction perpendicular to the rotation axis. The energy in these waves propagates fastest along the axis of rotation. Hence, $z/(2/\kappa)$ represents the travel time for the fastest wave disturbance of wave number κ to reach a vertical height z. The inequality (4.3.12) means simply that a certain minimum time must elapse before waves of this wave number arrive at a given vertical position.

The asymptotic solutions for stream function and velocity components are obtained by substituting the formulas for $\mathfrak{N}$ and $\mathfrak{K}$ into (4.3.6) and (4.3.7). The results are

$$\psi \sim \frac{2r}{\pi}\int_0^{2t/z} \frac{J_1(\kappa r)}{\kappa}\left(\cos\kappa - \frac{\sin\kappa}{\kappa}\right)d\kappa, \tag{4.3.13}$$

$$v \sim \frac{2}{\pi}\int_0^{2t/z} J_1(\kappa r)\left(\cos\kappa - \frac{\sin\kappa}{\kappa}\right)d\kappa, \tag{4.3.14}$$

and these expressions are particularly suited for numerical integration.

For the purposes of visualization, it is better to describe the flow in terms of the *co-ordinate system of the moving disk*. In this frame,

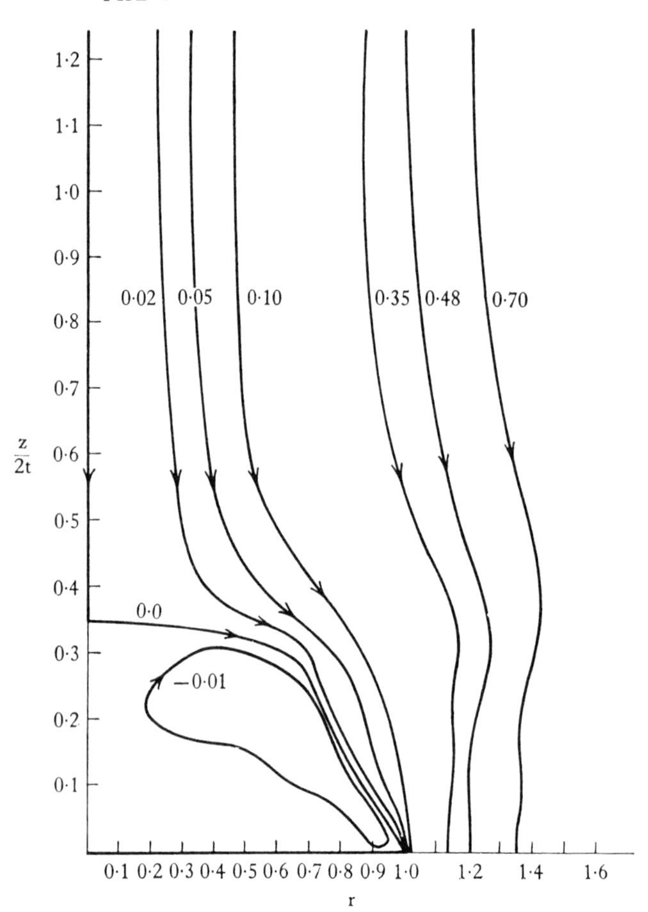

Fig. 4.2. Instantaneous streamlines relative to a disk that moves slowly along the rotation axis after an impulsive start. The plate is at $z = 0$, $r \leqslant 1$. The advancing front is the initial phase in the development of a Taylor–Proudman column.

the velocity at infinity is $-\hat{\mathbf{k}}$, the stream function is

$$\psi(r, z, t) = \tfrac{1}{2}r^2 + \psi(r, z, t), \tag{4.3.15}$$

and v is unchanged. Fig. 4.2 shows the instantaneous streamlines ψ about the disk (with respect to the observer on the disk). The vertical co-ordinate is $z/2t$, and as time increases the pattern must be continually stretched in the z direction to picture the true motion. Note the appearance of a 'stagnation point' in the flow, the broad bluff front and the reverse cellular flow behind it. A particle above the disk is eventually engulfed by the lengthening, stagnant flow regime; the major oscillations occur behind this advancing front. The net effect is similar to the flow about an imaginary obstacle the same width as the disk, but of a constantly increasing vertical dimension. This is, of course, the formation and development of a Taylor column.

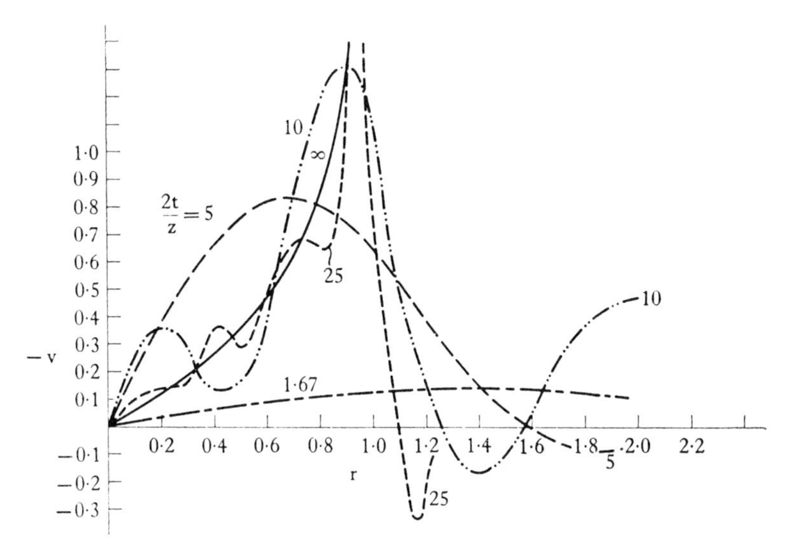

Fig. 4.3. The azimuthal velocity as a function of radial position for increasing values of the parameter $2t/z$. A discontinuity forms on the cylindrical surface $r = 1$ as the column becomes fully developed.

Fig. 4.3 illustrates the azimuthal velocity, v, as a function of radial distance for several values of $2t/z$. An obvious feature is the velocity discontinuity that forms across the vertical cylinder circumscribing the disk. It becomes clear from these two diagrams that the larger wave numbers, corresponding to dimensional wave lengths of the order of the disk diameter or smaller, carry most of the disturbance into the fluid. In other words, wave lengths approximately the size of the obstacle are the important part of the spectrum in the propagation of disturbances produced by the slow motion of a body.

The final, steady, inviscid motion, at $t = \infty$, is obtained from the asymptotic formulas, and the integrals in this limiting case are all known. The results are

$$\psi(r, z, \infty) = \psi_\infty = \begin{cases} -\dfrac{r^2}{2}, & r < 1, \\ \dfrac{1}{\pi}\left((r^2-1)^{\frac{1}{2}} - r^2 \sin^{-1}\dfrac{1}{r}\right), & r > 1; \end{cases} \tag{4.3.16}$$

$$v_\infty = -\frac{2}{\pi}\frac{r}{(1-r^2)^{\frac{1}{2}}} \mathcal{S}(1-r). \tag{4.3.17}$$

(Equation (4.3.15) is employed to find the corresponding formulas for motion relative to a reference frame fixed in the disk.)

Fluid particles in the Taylor column move with the disk, but possess an additional swirl that is independent of height. Particles outside the column have both longitudinal and azimuthal velocities but are constrained to move on a cylindrical shell because the radial velocity is zero.

The discontinuities in v and ψ across the lateral surface of the column and at the disk imply that viscous processes are important in these regions. An Ekman layer exists on the disk surface and vertical shear layers separate the inside and outside fluid domains. A complete study of the steady viscous problem (slow motion only) was given by Morrison and Morgan[132], but it is not presented here because the main features of viscous action are already well understood. One very important conclusion of their study is that the steady viscous solution agrees with the ultimate steady state of a transient inviscid analysis in the limit of zero viscosity, $E \to 0$. In this instance, the two limit processes, $\lim_{t\to\infty}$, $\lim_{E\to 0}$ are interchangeable.

Although the theory so far concerns only the motion of a disk, the results for other body shapes are entirely similar, in fact, identical when the circumscribing cylinders are the same. The important factors are the existence of a slowly moving obstacle and the shape of its shadow projection onto a $z =$ constant plane. Stewartson[185] solved the problem for a slowly moving sphere (see Grace [68, 69, 70] for an earlier attempt) and although he neglected details of the transient development, the analysis proved that the final steady state was identical to that produced by a disk of the same radius.

Reynolds'[161] extension of Morgan's[130] rapid-forcing similarity rule for axisymmetric bodies, presented in the next section, is further evidence in support of this general conclusion.

The formula following (4.2.8) shows that the characteristic decay time for a plane inertial wave is

$$t_c \doteq E^{-1}\kappa^{-2},$$

in dimensionless units. An estimate for the length of a Taylor–Proudman column in an unbounded medium can be obtained upon calculating the vertical distance travelled in time t_c by the energy in the principal part of the wave number spectrum. These are waves in the wave number range $\kappa \doteq 2$ for which the group velocity is

$$|\mathfrak{c}_g| = \frac{2}{\kappa} \doteq 1.$$

The vertical height reached in the characteristic time is then

$$|\mathfrak{c}_g|\, t_c \doteq E^{-1},$$

or dimensionally, $E^{-1}r_0$, (r_0 being the radius of the disk). The penetration distance is huge compared to the size of the obstacle. (This conclusion was also obtained by Bretherton[16] who examined another transient problem in great detail.) Non-linear effects may provide a different limitation on size.

Taylor[202], in his original investigation, remarked that the centre streamline lifted off the surface of the body, a sphere in his case, and gave a formula relating the occurrence of a column with the Rossby number of the flow. He found that it was necessary 'to make the ball move at a rate less than about one diameter per revolution of the system' to achieve columnar motion, i.e.,

$$\varepsilon = \frac{\mathfrak{U}}{r_0\Omega} < \frac{1}{\pi}.$$

The present theory, based as it is on slow motion, is certainly not capable of confirming this finding. However, a qualitative estimate can be obtained by calculating the counter-current, $\mathfrak{U}$, that would be necessary to keep the 'stagnation' point fixed on the disk. If the instantaneous stagnation point can move upstream, it is assumed that a pronounced column will ultimately form. This point moves with dimensional velocity, $0{\cdot}675 r_0 \Omega$, and, roughly speaking, $\mathfrak{U}$ should be of the same magnitude to keep it stationary. This implies that the critical Rossby number is 0·675, which is a factor of two too large, but encouraging nonetheless in view of the rough nature of the calculation. The non-linear convection terms must be taken into account in a proper theory and some progress along these lines has been made, but these matters will be reviewed in §4.6.

4.4. Oscillatory motion

We continue to use the disk as a representative body shape because it is an easy configuration to analyse in a thorough fashion. The processes illuminated this way, when divorced from specific detail, are of a fundamental nature and apply as well in cases of arbitrary geometries.

The configuration is then the same as in the preceding section but now the disk is to execute infinitesimal oscillations normal to its surface (parallel to $\mathbf{\Omega}$).

Morrison and Morgan[132] solved this problem and found integral representations for quasi-steady oscillations in a viscous fluid. Their limiting form with zero viscosity agrees with the formulas obtained later by Reynolds[162], who considered the inviscid, initial value problem. However, Reynolds' analysis leads to an exact solution of the linear theory in terms of simple known functions.

The final state of the oscillatory, inviscid flow can be determined directly by factoring the exponential time-dependence from the equations of motion and solving the reduced system. This classical procedure requires a radiation condition to make the resultant problem well-posed. The energy flux must always be directed away from the disk which is the source of the disturbance. Although this is easy to do, confusion can arise when the group and phase velocities are not in the same direction. An analysis of either the time-dependent or viscous problems circumvents the trouble by eliminating the unwanted oscillatory terms as a natural part of the solution procedure. Slow forward motion is then also a special case corresponding to a limit involving zero frequency.

The discussion of general oscillatory motion began in the introductory chapter, equation (1.3.3) et seq., and certain observations made there warrant reiteration. It will be recalled that the equation governing the spatial dependence of the pressure changes character as the frequency increases. For $\lambda < 2$, the equation is hyperbolic and the fluid can support wave motion. The characteristic surfaces are cones given by

$$r = \pm\left(\frac{4}{\lambda^2} - 1\right)^{-\frac{1}{2}} z + \text{constant.}$$

Equation (1.3.7) is elliptic when $\lambda > 2$ and it bears a strong resemblance to Laplace's equation. In fact, the identification is exact because the positive factor $1 - 4/\lambda^2$ can be absorbed into a rescaled vertical co-ordinate, a procedure used many times already. It should not be surprising then that a rapid-forcing similarity law exists which relates oscillatory solutions at $\lambda > 2$ with steady irrotational flows. Morgan[130] found such a rule for disk-like shapes and Reynolds[161] generalized the result and made it applicable to a wider class of axially symmetric bodies oscillating in a rotating fluid. But this aspect of the problem is perhaps of secondary importance, and for this reason further discussion is deferred to the end of this section, after analysis of the hyperbolic regime, $\lambda < 2$.

The mathematical formulation of the vibrating disk problem is the same as in the beginning of §4.3 except that the inviscid boundary condition on the disk, $r \leqslant 1$, $z = 0$, is replaced by

$$\mathbf{q}\cdot\hat{\mathbf{k}} = \mathscr{I}m\,\mathcal{S}(t)\exp i\lambda t. \tag{4.4.1}$$

Once again, the Laplace and Hankel transforms are applied consecutively to reduce the partial differential equations to ordinary differential equations in the vertical co-ordinate z. These are easily solved and the final oscillatory behaviour is determined from the inverse Laplace and Hankel transforms. The former is essentially a residue calculation at the simple pole $s = -i\lambda$, and no attempt is made here to describe the transient evolution as in the last section. The solution for $\lambda < 2$ is found to be

$$\psi = -\frac{2r}{\pi}\int_0^\infty \left(\frac{\sin\kappa}{\kappa} - \cos\kappa\right)\sin\left(\frac{\lambda t}{2} - \left(\frac{4}{\lambda^2} - 1\right)^{-\frac{1}{2}} z\kappa\right)\frac{J_1(\kappa r)}{\kappa}\,d\kappa, \tag{4.4.2}$$

$$v = -\frac{2}{\pi}\left(1 - \frac{\lambda^2}{4}\right)^{-\frac{1}{2}}\int_0^\infty \left(\frac{\sin\kappa}{\kappa} - \cos\kappa\right) \times \sin\left(\frac{\lambda t}{2} - \left(\frac{4}{\lambda^2} - 1\right)^{-\frac{1}{2}} z\kappa\right) J_1(\kappa r)\,d\kappa. \tag{4.4.3}$$

These integrals, or ones very similar, are tabulated, Erdelyi[42], and the resultant expressions for ψ and v are in terms of elementary functions. The formulas are easily obtained but are too lengthy to reproduce here. The instantaneous streamlines at two times are shown in fig. 4.4. Fluid motion occurs in different domains over one period of oscillation; particles immediately above the plate move with the plate in one phase while in other regions a state of rest persists simultaneously. Neither spatial variable, r nor z, is ever clearly designated as *the* time-like co-ordinate. The singular characteristic surfaces

$$r = 1 + \left(\frac{4}{\lambda^2} - 1\right)^{-\frac{1}{2}} z, \quad r = \left| 1 - \left(\frac{4}{\lambda^2} - 1\right)^{-\frac{1}{2}} z \right|,$$

demarcate the four different flow regions in the upper half plane and the adjustment of the flow from one domain to another is made through thin viscous shear layers about the discontinuity surfaces.

The particle velocity is not in phase with the disk oscillations because the energy flux must be directed outwards. Except for the appearance of singular characteristics, the flow pattern is quite complicated, a common occurrence especially when one deals with dispersive wave problems. However, the critical surfaces can be observed experimentally, as shown in fig. 1.3, and the dependence of conical apex angle on frequency measured. In this respect, agreement with theory is very good, Görtler[67].

It might seem desirable to consider point singularities instead of a body of finite size, for the formulas would surely be simpler. However, it turns out that a length scale is necessary in the inviscid problem to assure the convergence of the integrals to an acceptable final state. The realistic geometry allows a distinction to be made between waves of wavelength longer or shorter than the body dimension, and provides, thereby, a natural cut-off—a wave number beyond which the contributed effect is minimal.

As mentioned previously, the rapid forcing problem is related to irrotational flow. If $\psi_i(r, z)$ is the irrotational stream function for flow past a disk, then for any value $\lambda > 2$,

$$\psi(r, z, t) = A\psi_i\left(r, \left(1 - \frac{4}{\lambda^2}\right)^{-\frac{1}{2}} z\right) \sin \lambda t$$

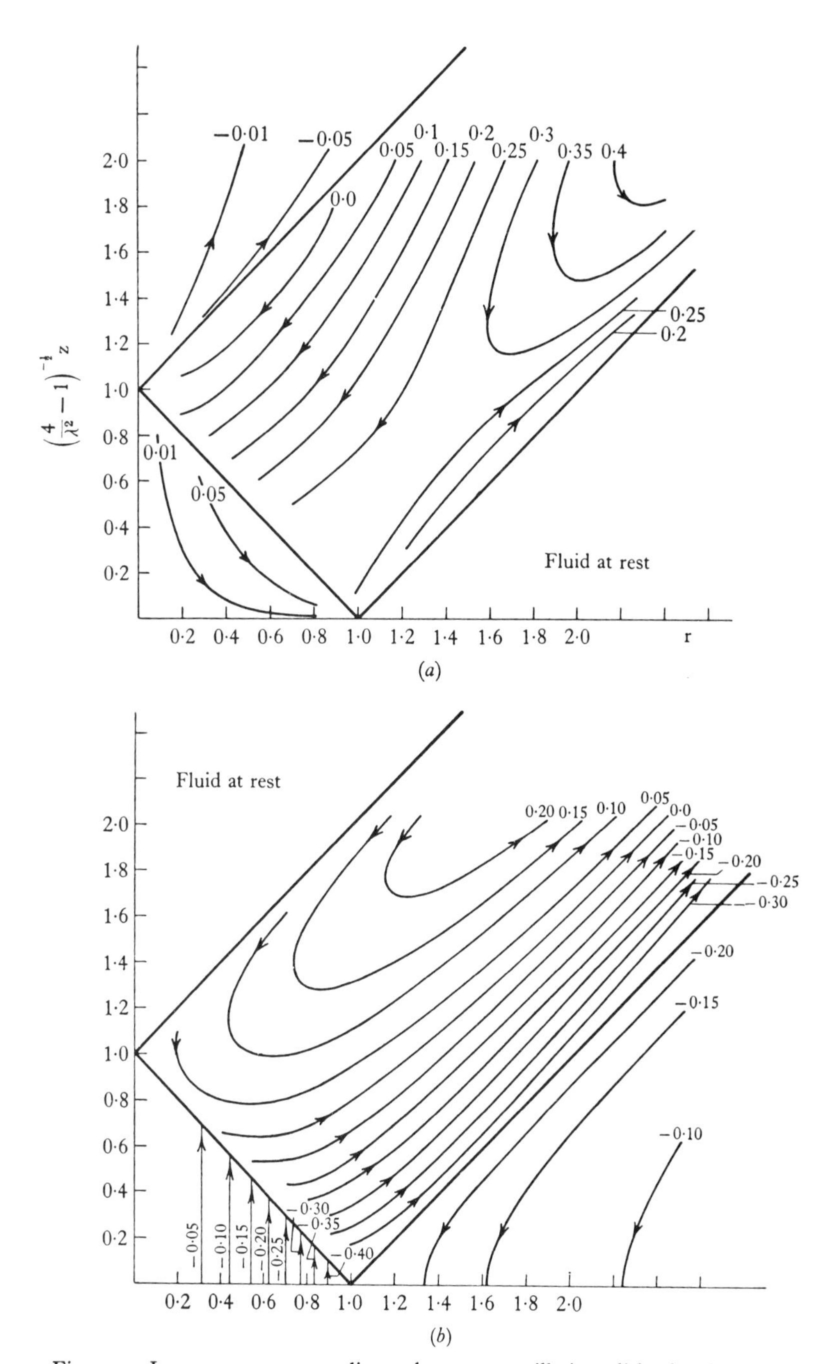

Fig. 4.4. Instantaneous streamlines about an oscillating disk showing the separation of the flow into distinct regions bounded by characteristic cones. In (*a*), $t = 0$, corresponding to the maximum downward displacement of the plate. In (*b*), $t = \frac{1}{2}\pi$ and the plate has its maximum upward velocity.

is the solution in the rotating field. Morgan showed that this result was valid for disk-like objects but Reynolds generalized the conclusion to cover any axially symmetric body undergoing small oscillations. If $z = f(r)$ specifies the actual body shape, then to apply the similarity law a related irrotational flow must be found about a distorted geometry

$$z = \left(1 - \frac{4}{\lambda^2}\right)^{-\frac{1}{2}} f(r).$$

Though this relationship precludes values $\lambda < 2$, two special cases appear to be possible exceptions. These correspond to a zero value of either multiplicative factor in the last equation. For example, $f(r) = 0$ represents the disk, an invariant shape as far as the transformation is concerned. Slow forward body motion, for which $\lambda = 0$, apparently relates any shape to a disk of the same projected cross-sectional area. Here again, there is the indication that only the projected shape on a $z =$ constant plane is important as far as steady longitudinal motion of an object is concerned. In applying these results, one must make sure that the similarity solution is physically acceptable. The energy flux must be directed away from the object at large distances and this may require careful consideration of phase relationships.

4.5. Wave propagation in a uniform current

An important problem, still unsolved in many respects, concerns the flow of a rotating fluid past a fixed body. This is equivalent to the motion of an object in an unbounded rotating fluid. Specifically, let us consider fluid motions whose far field, upstream velocity consists of rigid rotation, $\mathbf{\Omega} \times \mathfrak{r}$, and a constant, uniform current, $\mathfrak{U}$, i.e., $\mathbf{q}_\infty = \mathbf{\Omega} \times \mathfrak{r} + \mathfrak{U}$. The properties of wave motion in this field, akin to those established in §4.2 for plane inertial waves with $\mathfrak{U} = 0$, are set out here.

The linear, *dimensionless*, inviscid equations of motion, scaled with $[\![L, \Omega^{-1}, \mathfrak{U}]\!]$, were derived in §2.2, and are

$$\left.\begin{aligned} \nabla\cdot\mathbf{q} &= 0, \\ \frac{\partial}{\partial t}\mathbf{q} + \varepsilon\hat{\mathfrak{U}}\cdot\nabla\mathbf{q} + 2\hat{\mathbf{\Omega}}\times\mathbf{q} &= -\nabla p. \end{aligned}\right\} \qquad (4.5.1)$$

The Rossby number, $\varepsilon = \mathfrak{U}/\Omega L$, is not necessarily small and

$$\hat{\mathfrak{U}} = \frac{\mathfrak{U}}{\mathfrak{U}}, \quad \hat{\boldsymbol{\Omega}} = \frac{\boldsymbol{\Omega}}{\Omega}.$$

As in (4.2.1), these equations are independent of any particular cartesian co-ordinate system, although the identification $\hat{\boldsymbol{\Omega}} = \hat{\mathbf{k}}$ is made shortly, once again.

Plane waves are sought of the form

$$\mathbf{q} = \mathbf{Q} \exp i(\boldsymbol{\kappa} \cdot \mathbf{r} - \lambda t),$$
$$p = \Phi \exp i(\boldsymbol{\kappa} \cdot \mathbf{r} - \lambda t),$$

and the replacement of these expressions in (4.5.1) leads to the result

$$\lambda = \pm 2\hat{\boldsymbol{\Omega}} \cdot \hat{\boldsymbol{\kappa}} + \varepsilon \hat{\mathfrak{U}} \cdot \boldsymbol{\kappa}. \tag{4.5.2}$$

Moreover, the phase velocity is

$$\mathbf{c}_p = \left(\pm 2 \frac{\hat{\boldsymbol{\Omega}} \cdot \hat{\boldsymbol{\kappa}}}{|\boldsymbol{\kappa}|} + \varepsilon \hat{\mathfrak{U}} \cdot \hat{\boldsymbol{\kappa}} \right) \hat{\boldsymbol{\kappa}}, \tag{4.5.3}$$

and the group velocity is

$$\mathbf{c}_g = \pm \frac{2}{|\boldsymbol{\kappa}|} \hat{\boldsymbol{\kappa}} \times (\hat{\boldsymbol{\Omega}} \times \hat{\boldsymbol{\kappa}}) + \varepsilon \hat{\mathfrak{U}}. \tag{4.5.4}$$

These formulas are the appropriate generalizations of (4.2.4), (4.2.9) and (4.2.13) respectively. The meaning is clear: the wave speeds are augmented by the free stream convection; the frequency is corrected for a Doppler shift.

Suppose $\hat{\mathfrak{U}} = -\hat{\boldsymbol{\Omega}}$, then energy can be transmitted upstream, against the current, if the component of the group velocity in this direction is positive, $\mathbf{c}_g \cdot \hat{\boldsymbol{\Omega}} > 0$. This condition is

$$\frac{2}{\kappa}(1 - (\hat{\boldsymbol{\Omega}} \cdot \hat{\boldsymbol{\kappa}})^2) > \varepsilon, \quad (\kappa = |\boldsymbol{\kappa}|);$$

obviously there are always waves of sufficiently long wave length for which this inequality is true. Looked at another way, waves of wave number κ cannot propagate against the uniform flow if

$$\varepsilon > \frac{2}{\kappa}.$$

Thus, if the energy of a particular disturbance is concentrated in a limited wave number range, very little can reach far upstream when the counter-current is strong enough. In the case of a moving disk, the main part of the spectrum is $\kappa > \pi$. These waves cannot penetrate the upstream flow if

$$\varepsilon > \frac{2}{\pi},$$

an estimate which is comparable with that derived previously. Of course, this cut-off value is not precise for it can very well be decided that $\kappa > \frac{1}{2}\pi$ is a more appropriate restriction. However, it seems plausible that the oncoming flow prevents much of the disturbance from moving upstream for some value of ε of about this magnitude.

Waves, with wave vectors in the range $\boldsymbol{\kappa}$ and $\boldsymbol{\kappa}+\Delta\boldsymbol{\kappa}$, generated by a localized oscillatory disturbance of frequency λ, may be expected, asymptotically, to concentrate in the radial direction $\mathbf{r}_g$ that is parallel to their group velocity $\mathbf{c}_g$. This result can be proven in general by the arguments to be introduced, but the discussion, henceforth, is restricted to the important special case,

$$\hat{\mathbf{U}} = -\hat{\boldsymbol{\Omega}} = -\hat{\mathbf{k}},$$

for which the current is directed anti-parallel to the rotation axis. The situation corresponds to uniform motion of a body along $\boldsymbol{\Omega}$.

The general features of wave propagation can be ascertained by studying a more or less typical boundary value problem for the pressure function. This consists of the inhomogeneous, inviscid form of (2.2.4):

$$\left(\frac{\partial}{\partial t}-\varepsilon\frac{\partial}{\partial z}\right)^2 \nabla^2 p + 4\frac{\partial^2 p}{\partial z^2} = F(x,y,z)\,e^{-i\lambda t}, \tag{4.5.5}$$

with the conditions that p and all its derivatives are zero at infinity. The inhomogeneity represents some sort of local energy input and it is assumed that F decays exponentially fast as $\mathbf{r}$ becomes large in any direction. A solution is required consistent with the radiation condition. The physical basis of the inhomogeneity is left unspecified.

Let

$$p = \Phi(x,y,z)\,e^{-i\lambda t}, \tag{4.5.6}$$

so that

$$\left(i\lambda+\varepsilon\frac{\partial}{\partial z}\right)^2 \nabla^2 \Phi + 4\frac{\partial^2 \Phi}{\partial z^2} = F(x,y,z).$$

The solution of the last equation can be represented in terms of a three-dimensional Fourier integral

$$\Phi(x, y, z) = \mathscr{F}\{\bar{\Phi}\} = \iiint_{-\infty}^{\infty} \bar{\Phi}(\boldsymbol{\kappa}) \exp(i\boldsymbol{\kappa}\cdot\mathbf{r})\, dK, \qquad (4.5.7)$$

where $\boldsymbol{\kappa} = \kappa_1 \hat{\mathbf{i}} + \kappa_2 \hat{\mathbf{j}} + \kappa_3 \hat{\mathbf{k}}$, and $dK = d\kappa_1\, d\kappa_2\, d\kappa_3$. Furthermore, since

$$F = \mathscr{F}\{\bar{F}\},$$

then

$$\Phi = \mathscr{F}\left\{\frac{\bar{F}(\boldsymbol{\kappa})}{\bar{G}(\boldsymbol{\kappa})}\right\}, \qquad (4.5.8)$$

where

$$\bar{G}(\boldsymbol{\kappa}) = (\lambda + \epsilon\kappa_3)^2 (\kappa_1^2 + \kappa_2^2 + \kappa_3^2) - 4\kappa_3^2 \qquad (4.5.9)$$

is a real function of its arguments. The contour integrals in (4.5.7) are chosen in accordance with the radiation condition. Lighthill[108] has used the method of stationary phase to obtain asymptotic representations for multiple integrals of comparable type which occur in all problems of anisotropic wave motion. The results of this analysis, applied to the present problem by Nigam and Nigam[137], show that the asymptotic solution of (4.5.6) satisfying the radiation condition is

$$p \sim \frac{4\pi^2}{|\mathbf{r}|} e^{-i\lambda t} \sum \frac{A\bar{F}\, e^{i\boldsymbol{\kappa}\cdot\mathbf{r}}}{|\nabla_\kappa \bar{G}| \sqrt{\mathscr{C}}}, \qquad (4.5.10)$$

as $\mathbf{r} \to \infty$ along any specified ray, call it $\mathbf{r}_g$. The sum $\sum$ is over all points $(\kappa_1, \kappa_2, \kappa_3)$ on the surface $\bar{G} = 0$ where the normal to the surface is parallel to $\mathbf{r}$ and

$$\frac{\mathbf{r}\cdot\nabla_\kappa \bar{G}}{(\partial\bar{G}/\partial\lambda)} < 0. \qquad (4.5.11)$$

This result is valid provided that the Gaussian curvature, $\mathscr{C}$, is non-zero at each of these points, in which case $A = \pm i$ where $\mathscr{C} < 0$; $A = \pm 1$, where $\mathscr{C} > 0$ and the surface $\bar{G} = 0$ is convex to the direction of $\pm\nabla\bar{G}$. Reference is made to the cited paper for details and a complete discussion of the exceptional cases. Our interest is in the physical and geometrical interpretation of the general result and the qualitative description of the propagation process. No specific problem awaits solution.

Equation (4.5.9) for the wave number surface, when equated to zero is identical to the relationship between frequency and wave number components established previously, (4.5.2), from the consideration of plane waves. In other words, for a given frequency λ, the wave vector satisfies $\bar{G}(\kappa, \lambda) = 0.$

Furthermore, $-(\nabla_\kappa \bar{G})/(\partial\bar{G}/\partial\lambda)$ is really just the group velocity for waves of frequency λ. Since

$$\mathbf{c}_g = \frac{\partial\lambda}{\partial\kappa_1}\hat{\mathbf{i}} + \frac{\partial\lambda}{\partial\kappa_2}\hat{\mathbf{j}} + \frac{\partial\lambda}{\partial\kappa_3}\hat{\mathbf{k}},$$

this conclusion follows by implicit differentiation:

$$-\left(\frac{\partial\bar{G}/\partial\kappa_i}{\partial\bar{G}/\partial\lambda}\right) = \frac{\partial\lambda}{\partial\kappa_i}.$$

Thus, the normal to the wave number surface is parallel to the group velocity vector $\mathbf{c}_g$. The conditions underlying the asymptotic formula (4.5.10) simply state that the dominant effect in a specified radial direction $\mathbf{r}_g$ is produced by the waves whose group velocity is in the same direction. The added constraint (4.5.11) insures that the energy flux is always properly directed from the source of the disturbance.

The wave number surface is a surface of revolution about the κ_3 or z axis and its projection onto the κ_1, κ_3 plane is shown in fig. 4.5 for $\lambda < 2$. The arrows designate the direction of the group velocity which is most easily determined by mapping the wave number surface for two neighbouring values of λ. Since $\mathbf{c}_g = \nabla_\kappa \lambda$, the group velocity vector is always directed towards the surface with the larger value of λ.

Some conception of the different types of waves and the manner in which they propagate can be obtained by examining the surfaces of constant phase,

$$\boldsymbol{\kappa}\cdot\mathbf{r} - \lambda t = \mathfrak{P}, \tag{4.5.12}$$

of the asymptotic solution at different times. The wave with vector $\boldsymbol{\kappa}$, say to point $\mathfrak{p}_2$ in the figure, contributes to the asymptotic state only in the spatial direction that is parallel to its group velocity vector $\mathbf{c}_g$. At $t = t_j$, let $\mathfrak{P}_j = \mathfrak{P} + \lambda t_j$; by this time, the energy in this wave number range must have propagated a distance

$$\mathfrak{r} = \frac{\mathfrak{P}_j}{|\boldsymbol{\kappa}\cdot\hat{\mathbf{r}}|}, \tag{4.5.13}$$

in the direction dictated by $\mathfrak{c}_g$. $\boldsymbol{\kappa}\cdot\hat{\mathfrak{r}}$ is the projection of $\boldsymbol{\kappa}$ onto the line parallel to $\mathfrak{c}_g$ and is indicated by segment A_2O. The distance traversed in time t_1 is obtained by reflecting the length A_2O in the

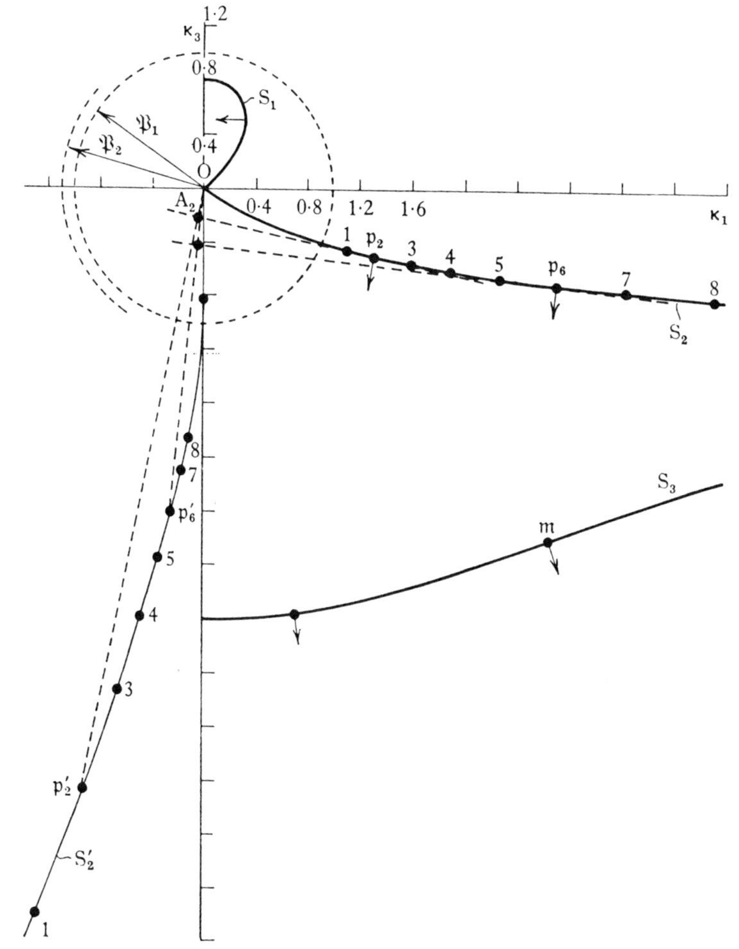

Fig. 4.5. The heavy solid lines show the three branches of the wave number surface for $\lambda = 1{\cdot}2$, $\varepsilon = 1{\cdot}0$. The construction of a surface of constant phase S_2' (light line) is indicated. Point m is an inflexion point.

sphere of radius $\sqrt{\mathfrak{P}_1}$ and is shown as $\overline{Op_2'}$. The image of the wave number surface obtained by inversion in the sphere is called the polar reciprocal surface. In this manner, the location of the major contribution from each wave is determined; the loci of all these

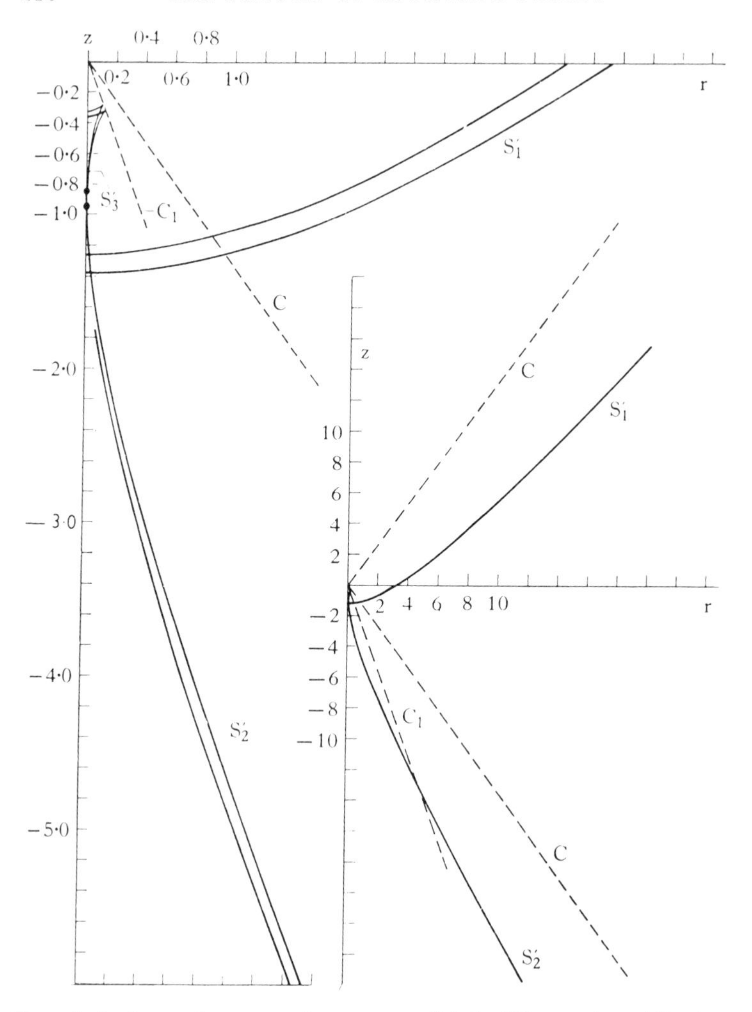

Fig. 4.6. Surfaces of constant phase at two slightly different times. The inset shows the surfaces on a contracted scale. Here $\lambda = 1{\cdot}2$, $\varepsilon = 1$, $\mathfrak{P}_1 = 1$, $\mathfrak{P}_2 = 1{\cdot}1$. The cuspal points fall on cone C_1.

points constitute the surfaces of constant phase shown in fig. 4.6 for two different times. The amplitude variation along the surface depends on the particular forcing function $F(x, y, z)$.

Three wave systems are discernible, each corresponding to a different branch of the wave number surface. Surface S_1 produces S_1' and waves that propagate both upstream and downstream, but which always lie outside the cones C, given by

$$r = \pm\left(\frac{4}{\lambda^2} - 1\right)^{-\frac{1}{2}} z.$$

These cones are the characteristic surfaces in the absence of a counter-current ($\varepsilon = 0$), and were studied in §4.3. The image of surface S_2, S_2', shows a wave system which travels downstream, completely contained by a limiting cone. Finally surface S_3, produces a cusp-shaped wave travelling downstream but contained within

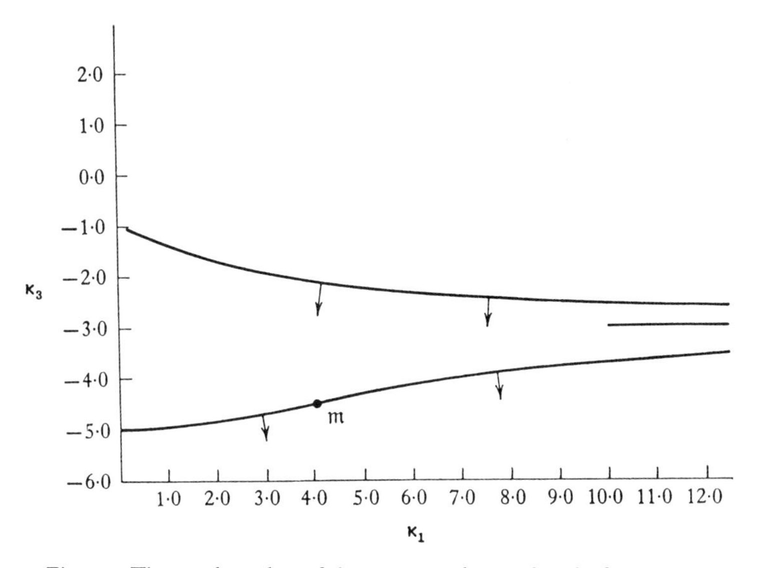

Fig. 4.7. The two branches of the wave number surface for $\lambda = 3$, $\varepsilon = 1$.

another cone C_1 which is the locus of the cuspal points. The peculiar shape of this wave arises from the fact that surface S_3 has a monoclastic or zero curvature curve whose projection in the $\kappa_2 = 0$ plane is the point m.

It is interesting to note that when no streaming motion exists and $\varepsilon = 0$, surfaces S_1 and S_2 assume their limiting shapes, the cones C, whereas S_3 degenerates to a single point at the origin.

The case $\lambda > 2$, corresponding to a frequency of excitation larger than 2Ω in absolute magnitude, may be analysed in the same fashion. The wave number surface, monoclastic curves and surfaces of constant phase are shown in figs. 4.7 and 4.8. Here, only two systems of waves exist, both cusp-shaped and both propagating in the downstream direction. As $\varepsilon \to 0$ or $\lambda \to \infty$, the two surfaces reduce to a single point—the origin; the motion, in this limit, is comparable to potential flow in an incompressible fluid.

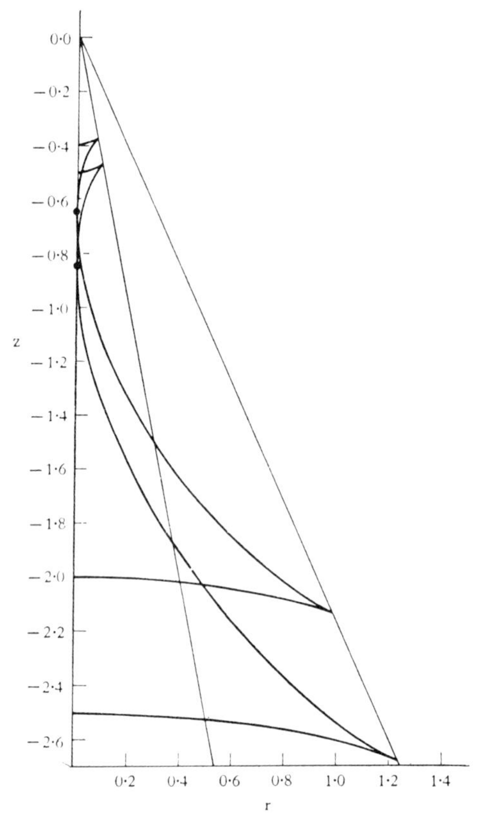

Fig. 4.8. The surfaces of constant phase (heavy lines) for $\lambda = 3$, $\varepsilon = 1$ and $\mathfrak{P}_1 = 2$, $\mathfrak{P}_2 = 2{\cdot}5$. Light lines are the loci of cuspal points. Both systems of waves propagate downstream only.

The initial value problem can be handled by a superposition of these waves over all frequencies, in other words, a Fourier integral in time. However, real problems involving bodies of finite size moving in a rotating fluid with constant forward speed, require the specification of proper boundary conditions and just what these are

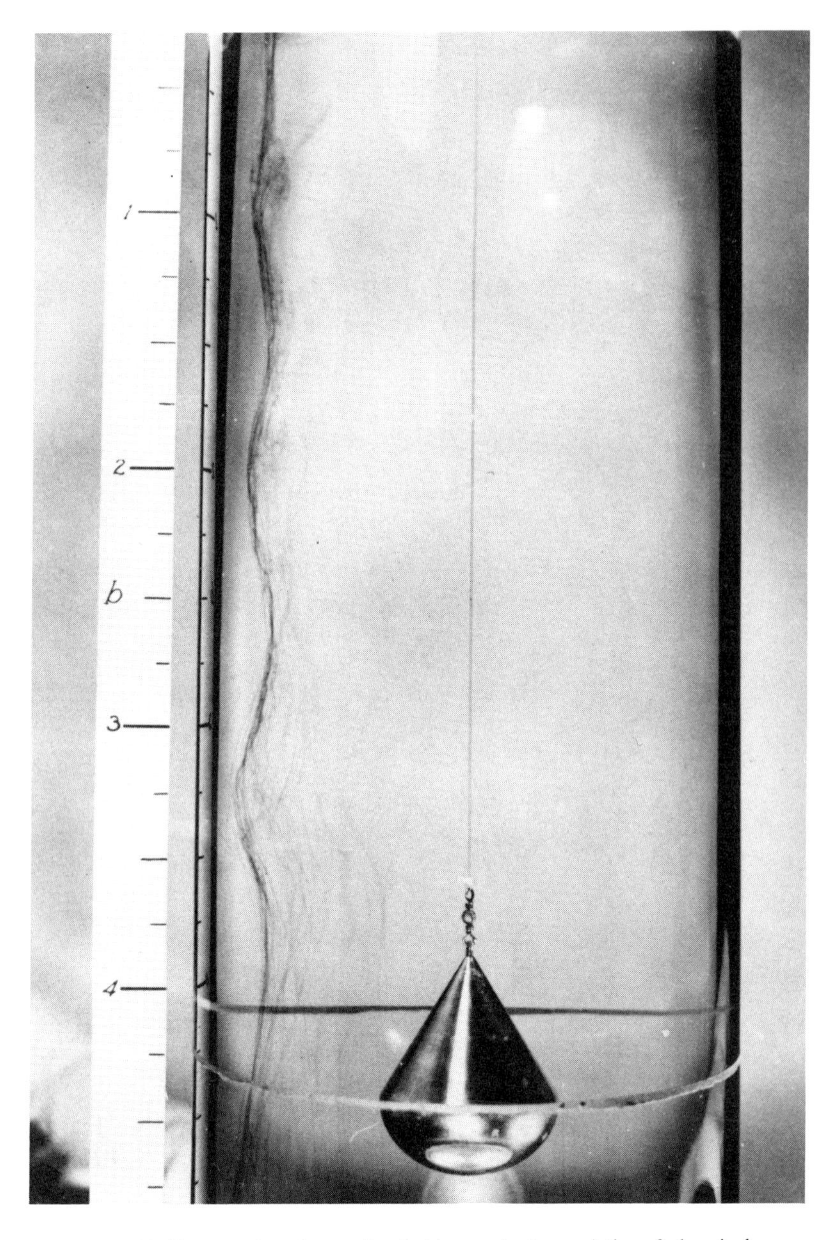

Fig. 4.9. Uniform motion of a rotating fluid past a body consisting of a hemisphere-cone combination, [115]. Dye streaks show that waves develop in the downstream direction. The radius of the tube, r_0, is 14·5 cm.; the radius of the sphere is 3·8 cm. and the upstream vertical velocity is $U = 0{\cdot}276\Omega r_0$.

in all circumstances is by no means obvious. This important issue has not been completely resolved as yet, but it receives some consideration in the next two sections.

4.6. Motion along the rotation axis at moderate speeds

Taylor[202, 203] confirmed the existence of columnar motion in a rotating fluid and also established some parametric relationships that seem to characterize this flow regime. For example, $\varepsilon < 1/\pi$ was found to be a requirement in producing slug flow as a sphere moved *along* the rotation axis. At higher values of the Rossby number, no strictly two-dimensional effects were observed. Substantially the same results were obtained later by Long[115], who used a body consisting of a hemisphere-cone combination. Long concluded that at large values of the Rossby number a definite system of waves was generated in the downstream direction; at extremely small values of ε, a forward column was observed. (Long did not find a column behind the body at any value of ε.)

Fig. 4.9 is an example of the motion, made visible by dye filaments in the fluid. These experiments motivated many theoretical efforts to determine the precise conditions separating the different states of motion. Reasonable progress has been made in understanding, at least qualitatively, the dominant processes but final success remains elusive for a number of reasons to be set forth in this section.

The slow motion problem ($\varepsilon = 0$), analysed in §4.3, revealed that in the formation of a Taylor column the main part of the disturbance is borne by waves whose wave lengths are of the order of the body diameter. Presumably, a column cannot form if these waves are convected downstream by a current flowing with a speed exceeding their group velocities. On this basis, crude initial estimates of the critical Rossby number above which columnar flow does not occur (see §§4.3 and 4.5) led to the conclusion that $\varepsilon \doteq 2/\pi$.

Wave motion of the type studied by Long does not have a theoretical basis in the 'slow motion' approximation. However, the analysis of the last section does account for convective and inertial terms in the equations of motion. It is shown there that three systems of waves develop in uniform flows; two consist of waves that propagate downstream only, while the other permits wave motion in either direction. These results lay the groundwork to explain the wave motion behind the obstacle, but they also raise questions concerning the lack of a forward disturbance at large Rossby numbers. As we have seen, waves of sufficiently long wavelengths can always penetrate the current and reach far upstream. Consequently, some disturbance must be evident in this region although the net effect can be small, perhaps negligible, if the principal waves emanating from the body are swept backwards, or if significant cancellation occurs.

The nature of the ultimate upstream conditions is one issue of central importance. Another arises in attempting to set a well-posed boundary value problem due to the number of waves proceeding downstream, but this will be considered later. One should be aware from the outset that unbounded motion is already a highly idealized state and the inviscid limit, $E = 0$, tends to aggravate this sensitive situation. End walls and Ekman layers, which were vital in all work to date, are here removed to infinity as are the lateral surfaces. The effect of these approximations must be gauged carefully for it seems very probable that the various limits involved are not interchangeable.

The assumption is often made that there is a class of steady longitudinal motions in which the flow disturbances do not appear sufficiently far from the body in the upstream direction. This hypothesis immediately precludes columnar motion. The rationalization behind it is sometimes based on two arguments. First, the experiments described earlier seem not to indicate upstream motion except at extremely small values of ε. (Later work casts doubt on this conclusion.) Secondly, there is a loose analogy with hydrodynamical surface waves produced by a travelling pressure pulse for which waves appear only downstream. We will consider the latter argument first.

Water waves on a deep ocean constitute a dispersive wave system with propagation speeds of all values. The steady motion of a disturbance leads to an indeterminate mathematical problem for which waves fore and aft are possible. The additional assumption, that no wave motion exists in the upstream direction, is sufficient to make the solution unique in a physically acceptable manner, meaning that waves trail the obstacle, as observed. The phase velocity of the main waves that are generated equals the speed of the disturbance, but their group velocity is only half this and in the same direction. This provides ample reason for the waves to fall behind the body. The solution may also be obtained as the limit of a time-dependent, initial value problem, or alternately, by introducing a little viscous dissipation into the steady equations. There is no point of disagreement and consequently, no doubt about the validity of the solution.

Waves in a rotating fluid are anisotropic as well as dispersive and energy can be channelled more effectively in the direction of the rotation axis to establish two-dimensional or columnar flows. Moreover, columnar motion is a confirmed fact showing that in certain circumstances waves do precede the body and do alter conditions at infinity. In this respect, the analogy between rotating and stratified fluids is more relevant. Furthermore, solutions based on the assumption of no upstream disturbance have not been shown to be the result of either a transient process or the limit of a viscous flow as viscosity vanishes. On the contrary, the analysis of an unsteady flow by Trustrum[211], to be discussed shortly, indicates that the final motion does indeed incorporate a two-dimensional geostrophic component extending to infinity in both directions. Its structure depends on initial conditions and the transient development. The discrepancy between these theoretical and experimental findings requires explanation; in this respect, it must be noted that precise measurements have not been made over the entire flow field. Long does not fully report upon the introduction of dye directly above the object where the columnar effects are most prominent. Clearly, a more definitive set of experiments is needed and recent unpublished results by T. Brooke Benjamin do indicate column formations in both fore and aft directions.

It seems likely that the hypothesis of no upstream disturbance is not strictly correct and that a body introduced into the stream will always affect the entry conditions at infinity to some degree. However, the assumption may be, and indeed probably is, an excellent *approximation* over a wide range of parameter settings because the amount of energy appearing upstream can be a small fraction of the total generated. Solutions derived on this basis can then be meaningful and significant.

The theory developed by Taylor[202], Long[115], Fraenkel[54], and Squire [181] concerns *steady, non-linear* rotational flows and it becomes advantageous, for the time being, to use the *dimensional* equations of motion written in an *inertial co-ordinate system.* A dimensionless notation is re-employed shortly; Squire's development is followed here.

Let (ξ_1, ξ_2, ξ_3) be a general orthogonal co-ordinate system with metric coefficients h_1, h_2, h_3 and let q_1, q_2, q_3 denote the corresponding velocity components. Consider now, only steady axially symmetric motions and take ξ_2 to represent the azimuthal angle about the symmetry axis so that $h_2 = r$. All variables are then independent of the angle ξ_2 and a general stream function, Ψ, can be defined by

$$h_2 q_1 = \frac{1}{h_3}\frac{\partial \Psi}{\partial \xi_3}, \quad h_2 q_3 = -\frac{1}{h_1}\frac{\partial \Psi}{\partial \xi_1}. \tag{4.6.1}$$

It is also convenient, as in § 3.6, to introduce the swirl,

$$\Gamma = h_2 q_2. \tag{4.6.2}$$

The substitution of these expressions into the basic inviscid equations of motion (i.e., (1.2.2) with $\mathbf{\Omega} = 0$) leads directly to several important conclusions. For example, it follows from the ξ_2 component of the steady momentum equation that

$$\frac{\partial(\Psi, \Gamma)}{\partial(\xi_1, \xi_3)} = 0.$$ [1]

Therefore, under the stated conditions, the swirl depends only on the stream function,

$$\Gamma = \Gamma(\Psi). \tag{4.6.3}$$

The azimuthal component of the steady vorticity equation can be written as

$$2\Gamma\frac{\partial(\Gamma, h_2)}{\partial(\xi_1, \xi_3)} - h_2\frac{\partial(\Psi, \mathfrak{D}\Psi)}{\partial(\xi_1, \xi_3)} + 2\mathfrak{D}\Psi\frac{\partial(\Psi, h_2)}{\partial(\xi_1, \xi_3)} = 0, \tag{4.6.4}$$

where
$$\mathfrak{D}\Psi = \frac{h_2}{h_1 h_3}\left[\frac{\partial}{\partial \xi_1}\left(\frac{h_3}{h_1 h_2}\frac{\partial \Psi}{\partial \xi_1}\right) + \frac{\partial}{\partial \xi_3}\left(\frac{h_1}{h_2 h_3}\frac{\partial \Psi}{\partial \xi_3}\right)\right]; \tag{4.6.5}$$

[1] The Jacobian is defined as follows;

$$\frac{\partial(F, G)}{\partial(x, y)} = \frac{\partial F}{\partial x}\frac{\partial G}{\partial y} - \frac{\partial F}{\partial y}\frac{\partial G}{\partial x}.$$

the individual components of vorticity are

$$\mathfrak{B}_1 = -\frac{1}{h_2 h_3}\frac{\partial \Gamma}{\partial \xi_3}, \quad \mathfrak{B}_2 = \frac{1}{h_2}\mathfrak{D}\Psi, \quad \mathfrak{B}_3 = \frac{1}{h_1 h_2}\frac{\partial \Gamma}{\partial \xi_1}.$$

Considerable simplification is gained by choosing ξ_1 = constant to be a surface of constant Ψ; in other words

$$\Psi = \Psi(\xi_1). \tag{4.6.6}$$

Γ and Ψ are then independent of ξ_3, q_1 is zero by definition and (4.6.4) reduces to

$$\frac{\partial}{\partial \xi_3}\mathfrak{D}\Psi - \frac{2}{h_2}\frac{\partial h_2}{\partial \xi_3}\left(\mathfrak{D}\Psi + \Gamma\frac{d\Gamma}{d\Psi}\right) = 0.$$

Upon integrating this equation, we obtain

$$\mathfrak{D}\Psi + \Gamma\frac{d\Gamma}{d\Psi} = r^2 A(\Psi) = r\mathfrak{B}_2 + \Gamma\frac{d\Gamma}{d\Psi}, \tag{4.6.7}$$

where A is an arbitrary function.

Another important result follows from consideration of the ξ_3 component of the momentum equation:

$$q_2\mathfrak{B}_1 - q_1\mathfrak{B}_2 = \frac{1}{h_3}\frac{\partial \mathfrak{H}}{\partial \xi_3},$$

where

$$\mathfrak{H} = \frac{p}{\rho} + \tfrac{1}{2}(q_2^2 + q_3^2)$$

denotes the total 'pressure'. In terms of the stream function, and with use of (4.6.6), it follows that

$$\mathfrak{H} = \mathfrak{H}(\Psi). \tag{4.6.8}$$

Thus, the total 'pressure' is a function of Ψ, which is, after all, Bernoulli's theorem.

If the swirl is very small, (4.6.7) reduces to the equation governing symmetric flows with vorticity (potential flows if the vorticity is zero). This is discussed in Lamb[103].

On the other hand, when the swirl is very large, two distinct possibilities arise. $\Gamma\, d\Gamma/d\Psi$ can be large compared to $r\mathfrak{B}_2$, in which case (4.6.7) is approximately

$$\Gamma\frac{d\Gamma}{d\Psi} = r^2 A(\Psi).$$

Since both Γ and A depend solely on Ψ, this expression implies that

$$\Psi = \Psi(r). \tag{4.6.9}$$

The stream function does not depend on the vertical co-ordinate and columnar motion is predicted. Of course, this is just the Taylor–Proudman theorem described in a laboratory co-ordinate system. The second possibility occurs when the terms are of comparable magnitude. Equation (4.6.7) is then very similar to the Helmholtz equation of classical wave theory, and wave motion in the rotating fluid is to be expected.

The evaluation of the arbitrary function, $A(\Psi)$, depends on the specific problem under discussion. Let the fluid far upstream consist of a constant uniform velocity along the line of symmetry plus a rigid rotation about this axis. (Attention is drawn, once again, to the possibility that this assumption may be untenable.) Asymptotically,

$$(\xi_1, \xi_2, \xi_3) \sim (r, \theta, z),$$

and

$$\mathbf{q} \sim \Omega\hat{\mathbf{k}} \times \mathbf{r} - \mathfrak{U}\hat{\mathbf{k}},$$

so that

$$(q_1, q_2, q_3) \sim (0, \Omega r, -\mathfrak{U}).$$

These upstream conditions may be expressed as

$$\Gamma = \Omega r^2, \quad \Psi = \tfrac{1}{2}\mathfrak{U}r^2, \quad p = p_0 + \tfrac{1}{2}\rho\Omega^2 r^2,$$

or, equivalently

$$\Gamma = 2\frac{\Omega}{\mathfrak{U}}\Psi, \quad \mathfrak{H} = \frac{p_0}{\rho} + \tfrac{1}{2}\mathfrak{U}^2 + 2\frac{\Omega^2}{\mathfrak{U}}\Psi. \tag{4.6.10}$$

This establishes the relationships between Γ, $\mathfrak{H}$ and Ψ along all streamlines originating at infinity in the upstream direction. It is convenient now to abandon the general co-ordinate system in favour of cylindrical co-ordinates. Equation (4.6.7) can then be written as

$$\mathscr{D}\Psi + 4\frac{\Omega^2}{\mathfrak{U}^2}\Psi = 2r^2\frac{\Omega^2}{\mathfrak{U}},$$

where

$$\mathscr{D} = r\frac{\partial}{\partial r}\frac{1}{r}\frac{\partial}{\partial r} + \frac{\partial^2}{\partial z^2}.$$

The deviation from uniform flow is measured more appropriately by the perturbed stream function,

$$\psi = \Psi - \frac{\mathfrak{U}r^2}{2},$$

which satisfies

$$\mathscr{D}\psi + 4\frac{\Omega^2}{\mathfrak{U}^2}\psi = 0.$$

It is remarkable that with the stated hypotheses a system of nonlinear partial differential equations reduce to a linear equation for the stream function alone. It becomes clear why this theory deserves and receives much attention.

The *dimensionless* form of the foregoing is

$$\mathscr{D}\psi + \frac{4}{\varepsilon^2}\psi = 0, \tag{4.6.11}$$

where the usual characteristic dimensions have been used,

$$[\![L, \Omega^{-1}, \mathfrak{U}]\!],$$

and $\varepsilon = \mathfrak{U}/\Omega L$. In particular, $\frac{1}{2}\mathfrak{U}L^2$ scales the stream function itself.

An inconsistency obviously arises as $\varepsilon \to 0$ because the upstream boundary conditions are no longer compatible with the existence of an obstacle in the fluid. It is interesting to note that although no approximations were made in the derivation of (4.6.11), it is nevertheless identical with the equation obtained from a linearization based on the Oseen replacement. This fact lends credence to analyses based on that kind of substitution.

The wave motion exhibited in fig. 4.9 has its theoretical analogue in the solutions of (4.6.11) appropriate to the cylindrical container. If the dimensionless radius of the tube is $a = r_0/L$ (L is the body radius), then separable solutions can be found of the form

$$rJ_1\left(\xi_{n1}\frac{r}{a}\right)\exp\left[i\left(\frac{4}{\varepsilon^2} - \left(\frac{\xi_{n1}}{a}\right)^2\right)^{\frac{1}{2}} z\right],$$

with

$$J_1(\xi_{n1}) = 0, \quad n = 1, 2, \ldots.$$

For values

$$\frac{4}{\varepsilon^2} < \left(\frac{\xi_{11}}{a}\right)^2,$$

the solutions grow exponentially as $|z| \to \infty$ in one direction, and decay in the other. In this case, the superposition of all such modes to solve the problem of the flow about an arbitrary body must yield a fluid motion that is essentially similar to potential flow. However, if

$$\frac{4}{\varepsilon^2} > \left(\frac{\xi_{n1}}{a^2}\right)^2,$$

then a certain number of these natural modes represent wave motion in the longitudinal direction. The first instance of wave motion occurs when

$$\varepsilon \leqq \frac{2a}{\xi_{11}}$$

with $\xi_{11} = 3{\cdot}8317$. If the Rossby number satisfies this inequality, the wavelength of the disturbance is

$$\frac{2\pi}{\kappa} = 2\pi\left(\frac{4}{\varepsilon^2} - \left(\frac{\xi_{11}}{a}\right)^2\right)^{-\frac{1}{2}}, \tag{4.6.12}$$

and in *dimensional* terms, these results are

$$\frac{\mathrm{U}}{\Omega \mathrm{r}_0} < \frac{2}{\xi_{11}}, \quad \frac{2\pi}{\kappa} = 2\pi\left(\frac{4\Omega^2}{\mathrm{U}^2} - \left(\frac{\xi_{11}}{\mathrm{r}_0}\right)^2\right)^{-\frac{1}{2}}.$$

Values appropriate to fig. 4.9 are $\mathrm{r}_0 = 14{\cdot}5$ cm, and $\mathrm{U}/\Omega\mathrm{r}_0 = 0{\cdot}276$; the theoretical wavelength is $14{\cdot}79$ cm. Far behind the obstacle, this wave should represent the major part of the disturbance; the measured value of the wavelength for the wave shown agrees with the theoretical prediction.

To actually solve the problem of flow about an obstacle, requires the superposition of all possible modes and the complete satisfaction of boundary conditions including that of no upstream wave motion. This is not a simple task because fundamental solutions fulfilling the prescribed wave requirements are difficult to construct. A notable success in this direction was achieved by Stewartson[190], who obtained the solution of (4.6.11) subject to the stated conditions in the case of flow past a sphere in an unbounded container.

The first step of his procedure is to determine the class of solutions that satisfies all the imposed conditions except those at the surface of the unit sphere. These functions, each of which implies no upstream motion, are

$$\psi_{2m}(\mathrm{r}, \Theta) = \mathrm{r}^{\frac{1}{2}}(1-\mu^2)\left\{\mathrm{J}_{-\frac{1}{2}-2m}\left(\frac{2}{\varepsilon}\mathrm{r}\right)\frac{\mathrm{d}}{\mathrm{d}\mu}\mathrm{P}_{2m}(\mu)\right.$$
$$\left.+\sum_{j=0}^{\infty}\frac{(4j+3)(2j)!(2m+1)!\mathrm{J}_{2j+\frac{3}{2}}\left(\frac{2\mathrm{r}}{\varepsilon}\right)\frac{\mathrm{d}}{\mathrm{d}\mu}\mathrm{P}_{2j+1}(\mu)}{j!(j+1)!m!(m-1)!(m+j+1)(2j-2m+1)2^{2j+2m+1}}\right\}, \tag{4.6.13}$$

and

$$\psi_{2m-1}(r,\Theta) = r^{\frac{1}{2}}(1-\mu^2)\left\{J_{\frac{1}{2}-2m}\left(\frac{2r}{\varepsilon}\right)\frac{d}{d\mu}P_{2m-1}(\mu)\right.$$

$$\left. -\sum_{j=0}^{\infty}\frac{(4j+5)(2j+1)!(2m)!\,J_{2j+\frac{5}{2}}\left(\frac{2r}{\varepsilon}\right)\frac{d}{d\mu}P_{2j}(\mu)}{j!(j+1)!\,m!(m-1)!(m+j+1)(2j-2m+3)\,2^{2j+2m+1}}\right\}, \tag{4.6.14}$$

where r and Θ are spherical co-ordinates, $\mu = \cos\Theta$ and $P_m(\mu)$ is a Legendre polynomial. Both odd and even functions may be written as

$$\psi_i = (1-\mu^2)\,r^{\frac{1}{2}}\left\{J_{-i-\frac{1}{2}}\left(\frac{2r}{\varepsilon}\right)\frac{d}{d\mu}P_i(\mu) + \sum_{j=0}^{\infty}\alpha_{ij}\,J_{j+\frac{1}{2}}\left(\frac{2r}{\varepsilon}\right)\frac{d}{d\mu}P_j(\mu)\right\}, \tag{4.6.15}$$

where the coefficients can be identified by comparison with the preceding equations. The general solution is represented by

$$\psi = \sum_{i=1}^{\infty} A_i\,\psi_i(r,\Theta), \tag{4.6.16}$$

and only the determination of the unknown Fourier constants remains. The coefficients are found by satisfying the boundary condition on the surface of the sphere $r = 1$,

$$\psi = -r^2 = -(1-\mu^2).$$

Therefore
$$\sum_{i=1}^{\infty} A_i\,\psi_i(1,\Theta) = -(1-\mu^2),$$

but according to (4.6.15) this is equivalent to

$$\sum_{i=1}^{\infty} A_i \sum_{j=1}^{\infty}\left[\delta_{ij}\,J_{-\frac{1}{2}-j}\left(\frac{2}{\varepsilon}\right) + \alpha_{ij}\,J_{j+\frac{1}{2}}\left(\frac{2}{\varepsilon}\right)\right]\frac{d}{d\mu}P_j(\mu) = -1,$$

where
$$\delta_{ij} = \begin{cases} 0 & i \neq j, \\ 1 & i = j. \end{cases}$$

The functions, $(1-\mu^2)(d/d\mu)P_j(\mu)$, $j = 1, 2, 3, \ldots$ constitute a complete set in the range $|\mu| < 1$ and the orthogonality integral for Legendre polynomials is used to establish that

$$\sum_{i=1}^{\infty} A_i\left[\delta_{ij}\,J_{-\frac{1}{2}-j}\left(\frac{2}{\varepsilon}\right) + \alpha_{ij}\,J_{\frac{1}{2}+j}\left(\frac{2}{\varepsilon}\right)\right] = -\delta_{1j} \quad j = 1, 2, \ldots. \tag{4.6.17}$$

This infinite set of linear equations must be solved numerically for the coefficients and Stewartson provides the following table of results:

ε	A_1	A_2	A_3	A_4	A_5
2	0·91	0·01	—	—	—
1	2·05	0·34	0·01	—	—
$\frac{2}{3}$	3·80	2·45	0·31	0·06	—
$\frac{1}{2}$	3·32	10·10	3·82	0·95	0·11

Stewartson concludes, upon investigating the properties of the numerical solution and the coefficient determinant, that the solution certainly breaks down below $\varepsilon = 0{\cdot}347$. At this point, the upstream boundary conditions definitely become untenable and a cylindrical component of flow develops. This critical Rossby number is very close to Taylor's experimental value, but the agreement may be accidental because the drag increases by two orders of magnitude between $\varepsilon = 1$ to $\varepsilon = \frac{1}{2}$, and the behaviour is also rather erratic in this region. This is an indication that slug flow occurs at these values of ε too; indeed, there is every indication that some two dimensionality exists at *all* values of ε. Therefore, columnar motion seems certain if $\varepsilon < 0{\cdot}347$, most probable for $\varepsilon < 0{\cdot}5$ and extremely likely at all larger values.

The modifications due to a closed geometry, such as flow within a cylinder, probably change the quantitative rather than the qualitative aspects of the solution. If a two-dimensional component is always present, then little else can be added by considering different configurations. On the other hand, if this is not true, then a definite relationship should exist between container shape (tube radius) and Rossby number that separates the wave motion and columnar flow regimes. This would be of significant interest but as a matter of pure speculation, the former possibility seems more likely.

4.7. Time-dependent considerations

The complete equation governing the axially symmetric stream function is

$$\left(\frac{\partial}{\partial t} - \varepsilon\frac{\partial}{\partial z} - E\mathscr{D}\right)^2 \mathscr{D}\psi + 4\frac{\partial^2\psi}{\partial z^2} = 0, \qquad (4.7.1)$$

which is comparable to (2.2.4) for the pressure. This equation in conjunction with the usual viscous constraints and initial conditions constitute a well-posed boundary value problem. The special theory for slow inviscid motion, $\varepsilon = E = 0$, is also well-set if the spatial boundary conditions are diminished in number to one (for the normal velocity component), because (4.7.1) is then only of second order.

The subject for discussion here is streaming rotational flow corresponding to the values $E = 0$, $\varepsilon \neq 0$. Equation (4.7.1) which then reduces to a fourth order partial differential equation stands, seemingly, somewhere between the two preceding cases. Some care is required in setting proper boundary conditions. (Recall that wave motion in streaming rotational inviscid flow—studied earlier in §4.6—is a composite of three different wave systems. Only one represents propagation in the upstream direction whereas all consist of waves moving downstream.)

Stewartson[190] and Trustrum[211] have applied transform methods to obtain solutions of (4.7.1). This procedure leads to complicated algebraic relationships among the transform parameters which must be sorted out before the inversion integrals can be successfully performed. As yet, no initial value problem has been solved that concerns flow about a body with the parameter setting $\varepsilon \neq 0$, $E = 0$.

Solutions have been obtained by Fraenkel[54] and Trustrum for flow in a cylindrical pipe girdled by a surface ring source in a plane perpendicular to the rotation axis. No obstruction exists in the flow and the vertical co-ordinate, z, varies from plus infinity to minus infinity along any axial line in the fluid domain. The question of type and number of boundary conditions is circumvented by requiring all variables, and their derivatives to all orders, to be zero at infinity. Though somewhat artificial, the solutions have many interesting features. Fraenkel analysed steady motion utilizing the exact equation of motion (4.6.11) and the assumption of no upstream disturbance; Trustrum solved the initial value problem based on (4.7.1). The relationship of the two theories has already been commented upon. The most significant difference is that the steady state developed from the transient flow has an extra two-dimensional or geostrophic component. Thus, columnar motion, specifically excluded by hypothesis in steady theories of §4.4, is a natural result of an initial value problem. Furthermore, the structure of the geostrophic flow is dependent on the initial conditions and the manner in which the final state is approached.

Though the last word has yet to be spoken, it seems that a columnar formation is an intrinsic feature of rotating flows and cannot be dismissed. The assumption of no upstream disturbance must be interpreted instead as an approximation appropriate in certain situations.

CHAPTER 5

DEPTH-AVERAGED EQUATIONS: MODELS FOR OCEANIC CIRCULATION

5.1. Introduction

This chapter deals with rotating fluid motions that are relevant to the understanding of the large scale oceanic circulations. For the most part, emphasis is placed on the study of flows produced *in the laboratory* which exhibit phenomena also observed in the ocean. A complete exposition of planetary fluid motions is not intended nor attempted. However, the theory of previous chapters is formulated in a manner that invites direct comparison with a fundamental and much analysed oceanic model. The uses of the former and the limitations of the latter then become apparent.

5.2. Depth-averaged equations

Consider the steady, forced circulation of a contained rotating fluid in a configuration of very slight depth variation and without closed geostrophic contours. Fig. 5.1 illustrates a basin of this type which is a section of a spherical annulus. Although the basin shape can be irregular, the sidewalls are assumed vertical for simplicity. Motion is produced by subjecting the top surface of the fluid to an applied stress (the wind) or by prescribing the velocity there as in most earlier work. The motion within a disk-driven sliced cylinder, § 2.20, which is a rather typical problem, has already demonstrated some striking parallels between laboratory and planetary circulations. In fact, the information obtained from that analysis will be valuable in motivating and substantiating aspects of the procedure to follow.

The *dimensionless* equations of motion are

$$\left.\begin{aligned} \nabla\cdot\mathbf{q} &= 0, \\ \varepsilon\mathbf{q}\cdot\nabla\mathbf{q}+2\hat{\mathbf{k}}\times\mathbf{q} &= -\nabla p+E\nabla^2\mathbf{q}, \end{aligned}\right\} \tag{5.2.1}$$

and the boundary conditions state that the velocity, or the surface stress, is known on the top surface Σ_T, and $\mathbf{q}\equiv 0$ on all others. There are at least three distinct regions in which the solution can exhibit markedly different behaviour. These are the Ekman layers at the top and bottom boundaries, the inviscid interior and the boundary layer adjacent to the sidewall. To account for this, the velocity and pressure are represented as

$$\left.\begin{aligned} \mathbf{q} &= \tilde{\mathbf{q}} + \mathbf{q}_I + \tilde{\mathbf{q}}_L, \\ p &= \tilde{p} + p_I + \tilde{p}_L, \end{aligned}\right\} \tag{5.2.2}$$

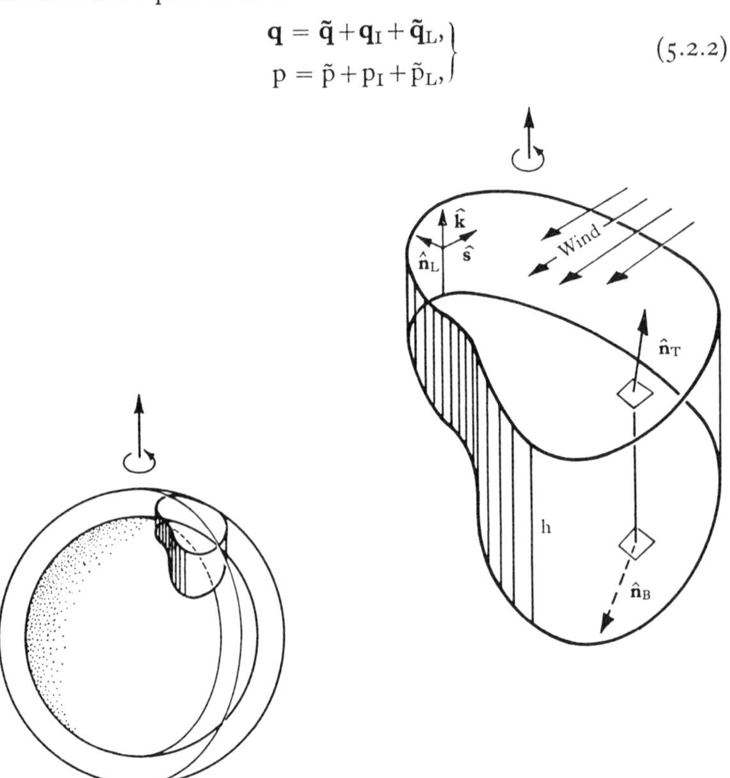

Fig. 5.1. An ocean basin having no closed geostrophic contours. The sidewalls are vertical and the top surface is subjected to a 'wind-stress'. The basin may be part of a spherical annulus as shown in the inset.

but it is not necessarily implied by this that the relative magnitudes of all terms are of the same order. If we were to proceed as before, a formal perturbation series would be constructed and a problem sequence resolved. However, the general approach is too cumbersome, if not impractical (there are three small parameters: ε, E, $\alpha \doteq |\nabla h|$) and a more flexible procedure is adopted.

An approximate theory is formulated which might correspond

to the first step of a consistent asymptotic development. The variables appearing therein could be identified with the lowest order terms of the appropriate expansions in the different regions and they are referred to as such. But the adoption of certain terminology (lowest order term, primary unknown, etc.) is really the limit of our involvement with the larger formalism and there should be no doubt on this point.

The principal objective is a theory governing the motion *outside* the Ekman layers. These layers may be excluded from the discussion because their net effect on the interior motion, the induced mass flux at the boundaries, is known. Equation (2.17.3) gives the boundary condition on $\mathbf{q}_I$, at a surface moving with a relative velocity, $\mathscr{U}$, of unit magnitude. The non-linear correction to this formula is neglected from the outset because ε is assumed small. The same boundary condition applies to $\mathbf{q}_I + \tilde{\mathbf{q}}_L$ in any sidewall layer whose thickness is much larger than $E^{\frac{1}{2}}$.

The prescription of the surface stress on Σ_T (scaled by $\varepsilon E^{\frac{1}{2}} \rho\, \Omega^2 L$) is equivalent to making

$$\mathbf{J} = \frac{\partial}{\partial \zeta} \tilde{\mathbf{q}}, \tag{5.2.3}$$

at $\zeta = 0$, a known function in problem $\mathscr{A}_2$, p. 42; the magnitude of this stress is the same as that produced at a moving plate. Subsequent discussion is restricted to motions generated in this way. It follows from the calculation of the Ekman boundary layer at Σ_T that the normal flux at this surface is

$$\hat{\mathbf{n}} \cdot \mathbf{q}_I = -\tfrac{1}{2} E^{\frac{1}{2}} \hat{\mathbf{n}} \cdot \nabla \times \left\{ \frac{\mathbf{J}}{\hat{\mathbf{n}} \cdot \hat{\mathbf{k}}} \right\}. \tag{5.2.4}$$

The particular problem solved in § 2.20 showed that topography imposes a severe restriction on the internal motion; the steady interior velocity is $O(E^{\frac{1}{2}})$ when the container has no closed geostrophic contours. As a direct consequence, the normal flux into the bottom Ekman layer can only be $O(E)$ because $\mathscr{U}_B = 0$; the velocity constraint at the bottom wall is mainly geometrical, that is

$$\hat{\mathbf{n}}_B \cdot \mathbf{q}_I = 0 \quad \text{on } \Sigma_B.$$

We conclude that *the bottom Ekman layer has a negligible effect on the primary circulation* in all such problems of *steady, forced motion.*

To lowest order in ε and E, the interior velocity is geostrophic:

$$2\hat{\mathbf{k}}\times\mathbf{q}_I = -\nabla p_I,$$

with

$$\nabla\cdot\mathbf{q}_I = 0.$$

Both $\mathbf{q}_I$ and p_I are independent of the vertical co-ordinate, to this degree of accuracy, but z reappears when higher order corrections are made.

The boundary conditions on the interior flow arising from the Ekman layer calculation can be absorbed into the equations of motion by introducing the depth-averaged internal velocity. Let

$$\mathfrak{q}_I + \mathfrak{w}_I\hat{\mathbf{k}} = \frac{1}{h}\langle\mathbf{q}_I\rangle = \frac{1}{h}\int_{-g}^{f}\mathbf{q}_I\,dz; \tag{5.2.5}$$

it is important to observe that $\mathfrak{q}_I(x,y)$ is just the horizontal component of the interior velocity vector. The vertical integration of the continuity equation across the interior, from the bottom to the top Ekman layers, provides the relationship

$$\nabla\cdot h\mathfrak{q}_I = -\mathbf{n}_T\cdot\mathbf{q}_I - \mathbf{n}_B\cdot\mathbf{q}_I.$$

However, the normal flux on Σ_B is effectively zero to the order of approximation used here and this becomes simply

$$\nabla\!\!\!\!/\cdot h\mathfrak{q}_I = -\mathbf{n}_T\cdot\mathbf{q}_I. \tag{5.2.6}$$

Here $\nabla\!\!\!\!/$ is the horizontal gradient,

$$\nabla\!\!\!\!/ = \nabla - \hat{\mathbf{k}}\frac{\partial}{\partial z},$$

and

$$\mathbf{n}_T\cdot\mathbf{q}_I = -\tfrac{1}{2}E^{\frac{1}{2}}\mathbf{n}_T\cdot\mathfrak{T} = -\tfrac{1}{2}E^{\frac{1}{2}}\mathbf{n}_T\cdot\nabla\times\left\{\frac{\mathfrak{I}_T}{\hat{\mathbf{n}}_T\cdot\hat{\mathbf{k}}}\right\}. \tag{5.2.7}$$

Since, in effect,

$$\mathfrak{q}_I = \mathbf{q}_I - \hat{\mathbf{k}}\cdot\mathbf{q}_I,$$

the momentum equation can be written as

$$2\hat{\mathbf{k}}\times\mathfrak{q}_I = -\nabla\!\!\!\!/ p_I, \tag{5.2.8}$$

and then solved for the horizontal velocity in terms of pressure,

$$\mathfrak{q}_I = \tfrac{1}{2}\hat{\mathbf{k}}\times\nabla\!\!\!\!/ p_I.$$

Hence

$$\nabla\!\!\!\!/\cdot\mathfrak{q}_I = 0, \tag{5.2.9}$$

and this allows the reduction of (5.2.6) to

$$\mathfrak{q}_I\cdot\nabla\!\!\!\!/ h = \tfrac{1}{2}E^{\frac{1}{2}}\mathbf{n}_T\cdot\mathfrak{T}. \tag{5.2.10}$$

Any two of the three equations, (5.2.6), (5.2.9) and (5.2.10), may be taken as an independent set. Actually, the velocity component that is perpendicular to constant-depth contours is evaluated from (5.2.10), which already implies that its magnitude is $O(E^{\frac{1}{2}}/|\nabla h|)$; (5.2.9) is used to find the tangential component in conjunction with the calculation of the sidewall boundary layer. This is the procedure employed in §2.20.

The equation for the pressure, p_I,

$$\hat{\mathbf{k}}\cdot\nabla p_I \times \nabla h = -E^{\frac{1}{2}}\mathbf{n}_T\cdot\mathfrak{T}, \tag{5.2.11}$$

completes the specification of the interior flow field.

We proceed next to develop the equations governing the sidewall boundary layer. This may be viscous or inertial in character but in either case, $\tilde{\mathbf{q}}_L$ is clearly larger there than $\mathbf{q}_I$. The exact linear analysis of the disk-driven flow, culminating in (2.20.21), showed that $\tilde{p}_L$ and the horizontal components of $\tilde{\mathbf{q}}_L$ are essentially independent of height but that $\hat{\mathbf{k}}\cdot\tilde{\mathbf{q}}_L$ varies linearly with z. In fact, the vertical component of velocity *must* vary with height in order to allow for the stretching of vortex lines as the fluid element is transported across depth contours.

The assumption is now made on the basis of earlier work, and subject to confirmation *a posteriori*, that the *horizontal components of velocity in the sidewall boundary layer do not depend on the height*, to lowest order in all parameters. The pressure is then nearly hydrostatic and suffers a change in magnitude across the layer which is only a second order effect. The vertical component of velocity is a linear function of z to the same order of approximation. It will be shown that this boundary layer structure is consistent with the form of the linear, inviscid, interior flow.

The applied surface stress is assumed not to vary appreciably across the sidewall boundary layer and the same condition is imposed on the total depth, h, although its variation along the layer is important.

Let

$$\tilde{\mathfrak{q}}_L + \tilde{\mathfrak{w}}_L\hat{\mathbf{k}} = \frac{1}{h}\int_{-g}^{f}\tilde{\mathbf{q}}_L\,dz.$$

The relationships,

$$\mathbf{q} = \mathfrak{q}_I + \tilde{\mathfrak{q}}_L + w\hat{\mathbf{k}}, \quad p = p_I + \tilde{p}_L,$$

are now substituted for velocity and pressure in the momentum equations. If only the important terms are retained—those which are not negligible in direct comparison with others in a boundary layer scaling—then the momentum boundary layer equations for the horizontal components of velocity are, approximately,

$$\varepsilon(\mathbf{q}_I + \tilde{\mathbf{q}}_L)\cdot\nabla\tilde{\mathbf{q}}_L + 2\hat{\mathbf{k}}\times\tilde{\mathbf{q}}_L = -\nabla\tilde{p}_L + E\nabla^2\tilde{\mathbf{q}}_L. \qquad (5.2.12)$$

The equations governing the interior motion have been used in this reduction.

The correct form of the continuity equation within the vertical boundary layer is

$$\nabla\cdot h\tilde{\mathbf{q}}_L = 0. \qquad (5.2.13)$$

In the problems considered, circulations which close via the Ekman *and* sidewall boundary layers are small compared to the main interior motion if $E^{\frac{1}{2}} \ll |\nabla h| \ll 1$, and these conditions are assumed.

Explicit mention of the pressure function (and its variation) can be avoided, if this is thought desirable, by considering instead the vorticity equation. If

$$\tilde{\mathfrak{B}}_L = \hat{\mathbf{k}}\cdot\nabla\times\tilde{\mathbf{q}}_L,$$

then it follows directly that the equation for this component is

$$(\mathbf{q}_I + \tilde{\mathbf{q}}_L)\cdot\nabla\left(\frac{2+\varepsilon\tilde{\mathfrak{B}}_L}{h}\right) = \frac{E}{h}\nabla^2\tilde{\mathfrak{B}}_L, \qquad (5.2.14)$$

which may be identified as a version of Ertel's theorem proved in §1.5. Hence, if $E = 0$, the potential vorticity

$$\frac{2+\varepsilon\tilde{\mathfrak{B}}_L}{h},$$

is conserved as a particle moves within the boundary layer.

The introduction into (5.2.14) of a boundary layer stream function,

$$h\tilde{\mathbf{q}}_L = \hat{\mathbf{k}}\times\nabla\tilde{\psi}_L = -\nabla\times\tilde{\psi}_L\,\hat{\mathbf{k}},$$

leads to a single boundary layer equation:

$$-\hat{\mathbf{k}}\cdot\left(\frac{h}{2}\nabla p_I + \nabla\tilde{\psi}_L\right)\times\nabla\left(\frac{2+\varepsilon\nabla\cdot\frac{1}{h}\nabla\tilde{\psi}_L}{h}\right) = \frac{E}{h}\nabla^2\nabla^2\tilde{\psi}_L. \qquad (5.2.15)$$

The relationship of $\mathbf{q}_I$ to ∇p_I has been employed and only the important part of the viscous term is written. (Here, $\nabla = \nabla$, because z does not appear explicitly.) The problem is then to find solutions of (5.2.11) and (5.2.15) which at the sidewall, satisfy the boundary condition,

$$\tilde{\mathfrak{q}}_L + \mathfrak{q}_I = 0. \qquad (5.2.16)$$

Thus far, the equations governing the largest and most important flow components in each domain have been identified and a well-set boundary value problem posed. Instead of separating the problem in this manner, a theoretical *model* can be formulated in terms of a combined depth-averaged velocity given by

$$\mathfrak{Q} + \mathfrak{W}\hat{\mathbf{k}} = \frac{1}{h}\int_{-g}^{f} (\mathfrak{q}_I + \tilde{\mathfrak{q}}_L)\,dz = \mathfrak{q}_I + \tilde{\mathfrak{q}}_L + (\mathfrak{w}_I + \tilde{\mathfrak{w}}_L)\,\hat{\mathbf{k}}. \qquad (5.2.17)$$

It is only necessary to retain in the development all terms which assume major importance *anywhere* in the container. For example, the appropriate form of the continuity equation is

$$\nabla \cdot h\mathfrak{Q} = -\mathbf{n}_T \cdot \mathfrak{q}_I = \tfrac{1}{2}E^{\frac{1}{2}}\mathbf{n}_T \cdot \mathfrak{T}. \qquad (5.2.18)$$

The right-hand side of this expression is crucial in the interior, where $\mathfrak{Q} \cong \mathfrak{q}_I$, but it is entirely negligible in the boundary layer when $\mathfrak{Q} \cong \tilde{\mathfrak{q}}_L + \mathfrak{q}_I$. The momentum equation can be replaced by

$$\varepsilon\mathfrak{Q} \cdot \nabla\mathfrak{Q} + 2\hat{\mathbf{k}} \times \mathfrak{Q} = -\nabla p + E\nabla^2\mathfrak{Q}, \qquad (5.2.19)$$

and the equation for the vertical component of the vorticity becomes

$$\mathfrak{Q} \cdot \nabla\left(\frac{2+\varepsilon\mathfrak{V}}{h}\right) = -\frac{2+\varepsilon\mathfrak{V}}{2h^2}E^{\frac{1}{2}}\mathbf{n}_T \cdot \mathfrak{T} + \frac{E}{h}\nabla^2\mathfrak{V}. \qquad (5.2.20)$$

It is seen that when terms multiplied by ε or E are neglected, (5.2.20) reduces to (5.2.10). On the other hand, all terms in which the curl of the 'stress', $\mathfrak{T}$, appears are of no consquence in the lateral boundary layer and there, the equations reduce correctly to (5.2.12) and (5.2.14). Note that the applied stress enters these equations in the form of a distributed *fluid source*. Not all of the terms in the preceding equations are essential, but all essential terms are present. In specific studies, a proper boundary layer analysis in terms of a scaled variable will reduce the formulation to the bare minimum.

The internal fluid velocity, $\mathfrak{Q}$, is $O(E^{\frac{1}{2}}\alpha^{-1})$ because the motion is steady and no closed geostrophic contours exist. The violation of either of these conditions generally leads to an interior flow of unit magnitude. A depth-averaged theory can still be developed in this circumstance if the motion remains quasi-geostrophic, i.e., the time variation is very slow.

An unsteady motion of this type was considered in § 2.16, in connection with spin-up in the sliced cylinder. The major conclusion there was that low frequency Rossby waves are generated to compensate for the loss of a purely geostrophic mode. This result turns out to be of general validity.

The *slow* change of a quasi-geostrophic flow should really be characterized by a long time scale $\tau\Omega^{-1}$, and the scaling rule $[\![L, \tau\Omega^{-1}, \varepsilon\Omega L]\!]$, in which case the dimensionless term

$$\frac{1}{\tau}\frac{\partial}{\partial t}\mathbf{q},$$

must be added to the momentum equation, (5.2.1). The large dimensionless parameter τ is often related to the depth variation, $|\nabla h|$, and this connection was clearly illustrated in § 2.16 where $\tau = 1/\alpha$, α being the small inclination angle.

The construction of a quasi-steady, depth-averaged model proceeds as before, with the same basic assumptions in force. Although the applied stress may now have a slow variation in time, the induced mass flux from the Ekman layer is calculated using steady conditions. However, a zeroth order interior velocity means that the contribution from the bottom Ekman layer is of the same magnitude as that from the top and must be included in the analysis. The depth-averaged continuity equation is then

$$\nabla \cdot h\mathfrak{Q} = -\mathbf{n}_T \cdot \mathbf{q}_I - \mathbf{n}_B \cdot \mathbf{q}_I,$$

and, in particular,

$$\nabla \cdot h\mathfrak{Q} = \frac{E^{\frac{1}{2}}}{2}\{\mathbf{n}_T \cdot \mathfrak{T} + \mathbf{n}_B \cdot \nabla \times [(\hat{\mathbf{n}}_B \times \mathfrak{Q} - \mathfrak{Q})\,|\hat{\mathbf{n}}_B \cdot \hat{\mathbf{k}}|^{-\frac{1}{2}}]\}. \tag{5.2.21}$$

The inhomogeneity, once again, is negligible in the sidewall boundary layer. The revised form of (5.2.20) is found to be

$$\frac{1}{\tau h}\frac{\partial \mathfrak{B}}{\partial t} + \mathfrak{Q} \cdot \nabla \frac{2 + \varepsilon\mathfrak{B}}{h} = -\frac{(2 + \varepsilon\mathfrak{B})}{2h^2}E^{\frac{1}{2}}\{\ \} + \frac{E}{h}\nabla^2\mathfrak{B}, \tag{5.2.22}$$

where the bracket is that defined by the previous equation. Ertel's theorem is recovered from this by setting $E = 0$; the potential vorticity of a particle is conserved in the absence of dissipation.

Consider the important class of problems dealing with the unforced relaxation of an initial disturbance in the configuration that is described at the beginning of this section. To be definite, let

$$h = h_0 + \frac{1}{\tau}\mathfrak{d}(x, y), \qquad (5.2.23)$$

where h_0 is a constant, and take

$$\varepsilon, E^{\frac{1}{2}} \ll \frac{1}{\tau} \ll 1.$$

The ordering of the parameters, ε, E, and τ, implies that in the interior (5.2.22) is approximately

$$\frac{\partial \mathfrak{B}}{\partial t} - \frac{2}{h_0}\mathfrak{Q}\cdot\nabla\mathfrak{d} = 0. \qquad (5.2.24)$$

Although the depth law allows no closed geostrophic circuits, the interior velocity is still in a state of near geostrophic balance so that

$$\mathfrak{q}_I = \mathfrak{Q} = \tfrac{1}{2}\hat{\mathbf{k}} \times \nabla p_I.$$

The substitution of this expression into (5.2.24) yields the fundamental relationship governing p_I:

$$\frac{\partial}{\partial t}\nabla^2 p_I - \frac{2}{h_0}\nabla\mathfrak{d}\cdot(\hat{\mathbf{k}} \times \nabla p_I) = 0. \qquad (5.2.25)$$

The proper boundary condition at the lateral wall is that the normal velocity component be zero there. This is the same problem for Rossby waves that was encountered in §2.16.

If convective terms are retained in the equations of motion, then the basis is laid for an analytical study of non-linear interactions of Rossby waves. A numerical solution of the complete depth-averaged boundary value problem, consisting of (5.2.18) and (5.2.20) and requisite boundary conditions, is feasible and results of recent investigations are reported later.

If, for any purpose, the motion is scaled against the usual characteristic time Ω^{-1}, then the factor, $1/\tau$, must appear multiplying $\nabla\mathfrak{d}$ in (5.2.24) and (5.2.25). This form of the equations is employed in §5.5.

5.3. Oceanic models

The ocean is for our purposes a very thin layer of a radially stratified fluid that covers part of the surface of a large globe. Movement within the fluid sheet is induced by an applied surface stress or by differential heating.

Fluid stratification, though slight, is an essential property because it implies that the direction of the gravitational force (and not just the direction of the rotation vector) is important over the entire spherical surface. As a result, the component of $\mathbf{\Omega}$ along the surface normal, $\hat{\mathbf{n}}$, enters the theory as a key factor. A severe geometrical constraint on the motion of a homogeneous, constant density fluid, like that posed by a spherical annulus of small gap-width, also gives special significance to the normal direction.

Another consequence of either stratification or a restrictive geometry is the emphasis placed on the component of absolute vorticity along the unit normal to the sphere:

$$\hat{\mathfrak{r}}\cdot(2\mathbf{\Omega}+\mathfrak{B}).$$

The role played by stratification in this connection can be established directly from Ertel's theorem (1.5.11), wherein $\mathfrak{S}$ is identified as the equilibrium density field $\rho_e(\mathfrak{r})$. In this case, $\nabla\rho_e$ is almost parallel to $\mathfrak{r}$, and it follows at once that in an incompressible, non-dissipative medium,

$$\frac{1}{\rho_e}\nabla\rho_e\cdot(2\mathbf{\Omega}+\mathfrak{B})$$

is conserved in the motion of a fluid particle.

The geometrical constraint described above also leads to a similar conclusion. This can be shown by constructing a theory for motion in the spherical annulus that is based upon approximations involving the ratio of gap-width to radius, and the proximity of the fluid element to the equator. Restrictions on separation distance and position, of this type, necessarily arise for although geometry can exercise tremendous influence on the motion of a homogeneous fluid, it cannot completely subvert the basic fact that the rotation axis is the only true direction of note. In view of this, it is anticipated that discrepancies between a constant density model, utilizing $(\mathbf{\Omega}\cdot\hat{\mathfrak{r}})\hat{\mathfrak{r}}$ instead of $\mathbf{\Omega}$, and an exact theory will depend to some degree on the extent to which the rotation axis and surface normal are aligned. The geometrical constraint is more effective as the gap-width decreases, but a shallow water theory, constructed from integrated averages in the radial direction, cannot be valid uniformly over the entire global surface. However, it may sometimes happen that in regions where this type of theory ceases to be valid on its own merits, its use and relevance can still be rationalized by arguments invoking the stratification of the real ocean.

Attention will continue to be focused primarily on the precise analyses of laboratory-produced rotating flows and the applicability of results so obtained to the study of oceanography. Comparison is made between the depth-averaged theory developed in the last section and similarly constructed oceanic models in current vogue, but not much consideration is given to a critical overall evaluation of the oceanic models per se (accuracy, relevance, etc.).

In the most widely studied model for the large scale circulation of a homogeneous ocean, only the velocity components tangent to the sphere and averaged over the ocean depth (the gap-width of a spherical annulus) appear in the formulation. Since variations with depth are averaged out, the curl of the surface wind-stress enters the equations as a body force, $\mathfrak{F}$. A coefficient of eddy viscosity, ν^*, usually replaces ν. The model equations governing the depth-averaged, tangential velocity vector, $\mathfrak{Q}_{\mathfrak{o}}$,[1] are

$$\nabla \cdot \mathrm{H}\mathfrak{Q}_{\mathfrak{o}} = 0, \tag{5.3.1}$$

$$\frac{\partial}{\partial t}\mathfrak{Q}_{\mathfrak{o}} + \mathfrak{Q}_{\mathfrak{o}} \cdot \nabla \mathfrak{Q}_{\mathfrak{o}} + 2\Omega \cos\Theta\, \hat{\mathbf{r}} \times \mathfrak{Q}_{\mathfrak{o}} = -\frac{1}{\rho}\nabla p + \nu^* \nabla^2 \mathfrak{Q}_{\mathfrak{o}} + \mathfrak{F}, \tag{5.3.2}$$

where H is the true ocean depth, $\Omega \cos\Theta\, \hat{\mathbf{r}}$ is the component of $\mathbf{\Omega}$ in the direction of the spherical normal, $\hat{\mathbf{r}}$, ∇ denotes the surface gradient, and all variables are functions of position on the surface of the sphere. A fairly complete discussion of the derivation of these equations can be assembled from the works of Morgan[131], Carrier and Robinson[28], Fofonoff[51], and Stommel[194]. A formal development would involve a perturbation analysis with suitable restrictions imposed on ocean depth, location, etc., but this is not

[1] The depth-averaged velocity here differs from that in § 5.2 because the flow within the Ekman layers is included in the definition.

our intention. However, it should be noted that the shallowness of the ocean, with respect to both the lateral scale of typical planetary motions and the radius of the earth, is a vital requirement, lying at the very heart of the development. The formulation in § 5.2, which also makes use of depth-averaging techniques, is *not* limited to thin strata of fluid but requires instead that the total depth variation be small, i.e., $E^{\frac{1}{2}} \ll |\nabla h| \ll 1$. This is an important distinction especially for the planning and execution of laboratory experiments designed to simulate oceanic phenomena.

In a *dimensionless* formulation, using the scales $[\![H_0, \Omega^{-1}, U]\!]$, (5.3.1) and (5.3.2) are

$$\left.\begin{aligned} &\nabla\cdot H\mathfrak{Q}_0 = 0,\\ &\frac{\partial}{\partial t}\mathfrak{Q}_0 + \varepsilon\mathfrak{Q}_0\cdot\nabla\mathfrak{Q}_0 + \mathfrak{f}\hat{\mathfrak{r}}\times\mathfrak{Q}_0 = -\nabla p + E\nabla^2\mathfrak{Q}_0 + \mathfrak{F}, \end{aligned}\right\} \qquad (5.3.3)$$

with $\qquad \mathfrak{f} = 2\cos\Theta, \quad \varepsilon = U/\Omega H_0, \quad E = \nu^*/\Omega H_0^2.$

A characteristic velocity, U, is chosen to make the motion due to the applied stress of unit dimensionless magnitude; H_0 is the typical ocean depth. The solution of these equations is required that satisfies prescribed initial conditions, and boundary conditions at the coastlines surrounding the ocean basin.

Certain features of the relationship between this system of equations and those of the last section, can easily be uncovered by considering steady, unforced, inviscid motion. For example, it follows that the form of Ertel's vorticity theorem for the steady model equations is

$$\mathfrak{Q}_0\cdot\nabla\left(\frac{\mathfrak{f}+\varepsilon\mathfrak{B}_0}{H}\right) = 0. \qquad (5.3.4)$$

(The quantity, $\nabla\cdot\mathfrak{f}\hat{\mathfrak{r}}$, is negligible because the radius of the earth is very much larger than the ocean depth.) Equation (5.3.4) should be compared with the inviscid version of (5.2.20),

$$\mathfrak{Q}\cdot\nabla\left(\frac{2+\varepsilon\mathfrak{B}}{h}\right) = 0. \qquad (5.3.5)$$

The extent to which these two equations are substantially the same is a measure of the accuracy and validity of the oceanic model. A comparison of this sort is made by Veronis[214, 215].

Both of the preceding equations are expressions of the conservation of vorticity, in some sense. Since the separation distance, H, between the concentric spheres (see fig. 5.9, p. 264) is small compared to the spherical radius, a, the following approximate formulas hold as long as $\mathrm{h}/a \ll \cos\Theta$, a condition which precludes the equatorial latitudes:

$$\frac{\mathrm{H}}{a} \cong \frac{\mathrm{h}}{a}\cos\Theta + \frac{1}{2}\left(\frac{\mathrm{h}}{a}\right)^2 \cong \frac{\mathrm{h}}{a}\cos\Theta.$$

But then, the vorticity component that appears in (5.3.5) can be written as

$$\frac{2+\varepsilon\mathfrak{B}}{\mathrm{h}} \cong \frac{2\cos\Theta + \varepsilon\mathfrak{B}\cos\Theta}{\mathrm{H}} \cong \frac{\mathfrak{f}+\varepsilon\mathfrak{B}\cos\Theta}{\mathrm{H}},$$

and this is very nearly the form required in (5.3.4). Hence, the two equations are approximately equal if $\mathfrak{B}\cos\Theta$ and $\mathfrak{Q}\cdot\nabla$ are nearly $\mathfrak{B}_0$ and $\mathfrak{Q}_0\cdot\nabla$, respectively. This is *exactly* true only at the poles; elsewhere, the approximation loses accuracy as the surface normal, $\hat{\mathbf{r}}$, and $\mathbf{\Omega}$ become less parallel. The error involved is $O(\mathrm{H}/a)$ over most of the spherical surface excluding a band about the equator (see [214]).

It is important to note that $1/\mathrm{h}$ *plays the same role in the general depth-averaged theory governing laboratory flows as the Coriolis parameter* $\mathfrak{f}$ *does in the oceanic model.* Therefore, the analogy of the depth-averaged equations of § 5.2 with the oceanic model equations really has a *dynamic* rather than a *geometric* basis. Oceanic phenomena can be studied in the laboratory by suitably arranging the depth topography of a tank so that the variation of the Coriolis force with latitude is correctly simulated. It is not necessary for many purposes to construct a container that is geometrically similar to the earth, as for example a spherical annulus with a small gap-width to radius ratio. Of course, the flows in such a container are interesting in themselves, but at present that is beside the point.

Assuming that the model equations are relevant and useful, it now becomes clear why many contained flows studied in chapter 2 can have an extremely close relationship with large scale phenomena in the oceans. The sliced cylinder configuration, which has a total depth that varies linearly with position, corresponds to a Coriolis parameter of the form

$$\mathfrak{f} = \mathfrak{f}_0 + \beta \mathrm{y}, \tag{5.3.6}$$

where y is a Cartesian co-ordinate of latitude. This formula for $\mathfrak{f}$, the normal component of planetary vorticity, is a central assumption underlying the β-plane approximation in oceanography. Spin-up in the cylinder, which was effected through the propagation of vorticity by Rossby waves, most probably has its oceanographic analogue in the transition from one steady oceanic circulation to another. A change of seasons could bring this about in certain parts of the world.

In the sliced cylinder problem, § 2.20, the bottom slope simulates the variation of Coriolis force with latitude on the surface of the earth and the shear produced by the top disk acts like an applied 'wind-stress'. The sidewall boundary layer that forms on half the perimeter of the basin is dynamically analogous with the Gulf Stream, and there is some indication that the separation of this current from the Atlantic coastline may also be reproduced. Many of the gross features of oceanic circulation are present in this simple experiment. Indeed, von Arx[220, 221], Faller and von Arx[48], Stommel, Arons and Faller[196], and Fultz[61] have utilized this important principle to construct a number of imaginative experiments, some of which have already been described (see p. 117). It seems conclusive that much can be learned about the basic dynamic processes in the controlled environment of the laboratory. The essential point is to arrange a correct dynamic analogy using the total depth variation to simulate $\mathfrak{f}$.

5.4. Steady circulations and inertial boundary layers

The analysis in § 2.20 is rather typical of all investigations of steady flows in enclosed basins where constant-depth contours are open curves. No other viscous flow problems of this type will be solved for this reason.

Solutions of the linearized, 'viscous' model equations for oceanic circulation, which are inherently of the same class, were obtained by Stommel[193], Munk[133], and Munk and Carrier[134], among others. Discrepancies between these theoretical predictions and oceanic observations led to Stommel's suggestion that non-linear, rather than viscous, processes constitute the dominant mechanism controlling the great ocean currents.

The theory of non-linear or inertial boundary layers, developed by Charney[31] and Morgan[131], provides a reasonably accurate description of the Gulf Stream in regions where it borders a coastline. Non-linear processes also seem to dominate when the stream leaves the boundary to become a free inertial jet, Warren[222]. Comprehensive accounts of the entire subject are given by Stommel [195] and Fofonoff[51].

The extreme importance of non-linear processes in oceanography motivates the study of inertial boundary layers in contained rotating fluid motions. The correspondence between the theory pertinent to laboratory motions and that concerned with geophysical circulations has been demonstrated. Consider then, a rather general configuration of the type shown in fig. 5.1, and the depth-averaged equations of motion, (5.2.18), (5.2.19) and (5.2.20).

Boundary layers are a feature of singular perturbation problems in which a very small parameter multiplies the most highly differentiated terms. Viscous boundary layer theory stems from the fact that E is small. However, even if the Ekman number were identically zero, in (5.2.15) for example, the most highly differentiated terms that would remain, would still be multiplied by the small Rossby number, ε. Hence, the inviscid, non-linear theory is also singular in this sense, and the existence of a boundary layer can be inferred wherein the *convective* terms become as important as any others. If E and ε are both small but non-zero, then the structure of any boundary layer will combine features of both the viscous and inertial layers, depending on the relative magnitudes of the two parameters.

The interior circulation is assumed to be known; the problem is to find the *inertial* boundary layer that corresponds to a prescribed interior state. In actuality, both flow regimes exert a strong influence on each other and the precise determination of the interior flow (there is usually an unspecified function to be found) depends on the properties of the boundary layer and vice versa. The non-linear problem is intrinsically more complicated to solve than the corresponding linear, viscous problem, although the singular perturbation techniques used in both are comparable. Aside from questions of the greater skill and ingenuity required, difficulties of a fundamental nature can be involved. For example, steady, inviscid, non-linear problems may not even be well-posed. The reason for this is that an arbitrary applied wind-stress will usually create vorticity in the fluid which cannot be removed in the absence of viscosity. A steady state cannot exist in such a situation unless dissipation occurs somewhere in the ocean basin (or the applied stress is of a very special form). Evidently, inertial effects do not predominate over the entire ocean; dissipative processes are not only important in certain regions, but are essential to the over-all motion and vorticity balance. These conclusions are confirmed by the numerical work of Bryan[20] and Veronis[217]. There is ample reason then to study only the local structure of an inertial current without attempting a complete determination of the entire flow field.

A difficulty of much less seriousness is the inability of a strictly inertial theory to satisfy the no-slip boundary condition. A very thin viscous sublayer within the inertial layer fulfils this requirement.

In terms of the perturbation boundary layer stream function, $\tilde{\psi}_L$, the inviscid, inertial boundary layer equation is, from (5.2.15),

$$-\hat{\mathbf{k}}\cdot\left(\frac{h}{2}\nabla p_I+\nabla\tilde{\psi}_L\right)\times\nabla\left(\frac{2+\varepsilon\nabla\cdot\frac{1}{h}\nabla\tilde{\psi}_L}{h}\right)=0. \qquad (5.4.1)$$

By assumption, p_I *is a known function* and the total depth, h, has no boundary layer character.

The inertial boundary layer exists because rapid variations of the relative vorticity must occur somewhere, if a fluid element is to overcome the strong topographical constraint and complete a closed circuit. Thus, the relative vorticity, $\varepsilon\nabla\cdot(h^{-1}\nabla\tilde{\psi}_L)$, is of unit magnitude in the boundary layer and this implies at once that the dimensionless thickness of the layer is $\varepsilon^{\frac{1}{2}}$, if $\tilde{\psi}_L$ is itself $O(1)$.

Let $\hat{\mathbf{n}}_L$ be the outwardly directed unit normal at the sidewall and $\hat{\mathbf{s}}$ be a unit tangential vector to this surface, so that $\hat{\mathbf{s}}$, $\hat{\mathbf{n}}_L$, and $\hat{\mathbf{k}}$ form a right-handed, orthogonal triad, shown in fig. 5.1. (The arc length, s, increases in the clockwise direction about the perimeter of the basin.) Furthermore, let ζ be the boundary layer co-ordinate, defined by

$$\hat{\mathbf{n}}_L\cdot\nabla\simeq-\varepsilon^{-\frac{1}{2}}\frac{\partial}{\partial\zeta}.$$

By hypothesis, only $\tilde{\psi}_L$ varies with ζ, and (5.4.1) can be reduced to

$$\left(\frac{\partial\tilde{\psi}_L}{\partial s}+\frac{h}{2}\frac{\partial p_I}{\partial s}\right)\frac{\partial}{\partial\zeta}\left(\frac{1}{h}\left\{2+\frac{1}{h}\frac{\partial^2\tilde{\psi}_L}{\partial\zeta^2}\right\}\right)$$
$$-\frac{\partial\tilde{\psi}_L}{\partial\zeta}\frac{\partial}{\partial s}\left(\frac{1}{h}\left\{2+\frac{1}{h}\frac{\partial^2\tilde{\psi}_L}{\partial\zeta^2}\right\}\right)=0. \qquad (5.4.2)$$

The interior pressure and the depth are prescribed functions of s at the lateral boundary, $\zeta = 0$, and we may write

$$\frac{d\phi}{ds} = \frac{h}{2}\frac{\partial p_I}{\partial s}.$$

With this transformation, (5.4.2) can be recast as a Jacobian,

$$\partial\left(\frac{1}{h}\left\{2+\frac{1}{h}\frac{\partial^2\tilde{\psi}_L}{\partial\zeta^2}\right\}, \tilde{\psi}_L+\phi\right)\Big/\partial(\zeta, s) = 0,$$

and integrated once:

$$\frac{1}{h^2}\frac{\partial^2\tilde{\psi}_L}{\partial\zeta^2}+\frac{2}{h} = F(\tilde{\psi}_L+\phi). \tag{5.4.3}$$

Here, F is an arbitrary function which can be evaluated at $\zeta = \infty$, because $\tilde{\psi}_L$ is a boundary layer function and $\lim_{\zeta\to\infty} \tilde{\psi}_L = 0$. It follows that

$$F(\phi) = \frac{2}{h}, \tag{5.4.4}$$

where both ϕ and h are functions of s only. Necessary conditions for the existence of inertial boundary layers can now be ascertained. If (5.4.3) is rewritten as

$$\frac{1}{h^2}\frac{\partial^2\tilde{\psi}_L}{\partial\zeta^2} = F(\tilde{\psi}_L+\phi)-F(\phi),$$

then for large ζ,

$$\frac{1}{h^2}\frac{\partial^2\tilde{\psi}_L}{\partial\zeta^2} \cong \left(\frac{d}{d\phi}F(\phi)\right)\tilde{\psi}_L. \tag{5.4.5}$$

The function $\tilde{\psi}_L$ has the proper boundary layer character—an exponential decay for large ζ—only if

$$\frac{d}{d\phi}F(\phi) > 0, \tag{5.4.6}$$

and this is a criterion for the existence of inertial layers. It is readily expressed in terms of the total depth and the value of the interior horizontal velocity at the wall. The equivalent statement is

$$(\hat{\mathbf{n}}_L\cdot\mathbf{q}_I)\frac{dh}{ds} < 0, \tag{5.4.7}$$

which has a simple interpretation. For example, when h decreases with increasing s, a mass influx from the interior is required to sustain an inertial layer. Moreover, a sign reversal in the derivative of h necessitates a change in direction of the velocity vector or the layer cannot be continued along the wall. In that case, the inertial current may become time-dependent and/or break away from the boundary to become a free inertial jet. Of course, any one of the processes omitted in this formulation may possibly exert a strong influence in the neighbourhood of predicted separation and allow the jet to proceed along the boundary, unaware of any difficulty. The numerical studies reported in § 5.5 indicate that something like this can and does happen. Viscous processes offer sufficient control to abort separation in certain situations.

The form of (5.4.5) shows that the width, δ_i, of the inertial boundary layer is

$$\frac{\delta_i}{L} \doteq \frac{1}{h}\left(\frac{\varepsilon}{\frac{dF}{d\phi}}\right)^{\frac{1}{2}} \doteq \left|\frac{\varepsilon \mathbf{q}_I \cdot \hat{\mathbf{n}}}{\frac{2}{h}\frac{dh}{ds}}\right|^{\frac{1}{2}}; \tag{5.4.8}$$

$\mathbf{q}_I \cdot \hat{\mathbf{n}}$ is the magnitude of the normal component of the interior velocity at the edge of the layer, scaled with respect to $\varepsilon\Omega L$.

An analysis of the oceanic model, Greenspan[72], leads to the same criterion given above except that $1/h$ is replaced by the Coriolis parameter $\mathfrak{f}$ (or $\mathfrak{f}/H$, if H varies across the boundary layer). Since $\mathfrak{f}$ increases with latitude, and is an increasing function of s on the *western* side of an ocean basin, the interior velocity must be directed towards the western coastline to support an inertial jet there. Similarly, the velocity must point away from the eastern boundary. It is interesting to note that the Gulf Stream separates from the coastline at a position where the current suddenly enters deep water off Cape Hatteras. At this location, the ocean depth increases more rapidly along the stream path than does the Coriolis parameter and the derivative of $\mathfrak{f}/H$ suffers a sign reversal. Thus, separation occurs at the same position where the simple criterion is violated by changes in the local topography. The Japanese current also leaves its coastline under the same conditions, but whether this agreement is purely fortuitous is not absolutely clear. It seems likely that topographic changes are responsible for the separation of the currents from the coastlines (or vice versa), but the reasons behind this may be more complex than those provided by a theory based on a single, homogeneous fluid layer.

Laboratory experiments designed to take full advantage of the close theoretical analogy of $\mathfrak{f}$ and $1/h$ could be invaluable in the resolution of these questions. It should not be too difficult to construct a tank whose vertical depth varies in a manner that would lead to a separated inertial current. The factors controlling the process could then be examined closely, hopefully to provide new ideas and impetus for further theoretical investigations. It was in this spirit that Stommel, Arons and Faller[196] undertook the study of boundary layers in a closed basin. Extremely interesting results were obtained and these were described briefly in § 2.19.

The experiment with the sliced cylinder, discussed in § 2.20, had the same motivation. At very small values of the Rossby number, ε, the flow is essentially that predicted by the linear theory. As the Rossby number is increased, the centre of the interior gyre shifts sharply away from the wall, in the direction of the boundary layer current. The position of the vortex centre varies little at still larger values of ε, but the boundary layer becomes strongly non-linear and separates intact from the wall to penetrate the interior as a broad, steady, meandering stream. In the final flow regime observed, the meanders begin to fluctuate and vortices (Rossby waves) are seen propagating across that part of the basin while a steady motion persists elsewhere.

As non-linear effects become predominant over viscous processes, the structure of the western boundary layer must approach that of an inertial current although the limiting form may not be attainable. According to (5.4.7), an inertial layer can develop along this border only if the interior flow is directed towards the boundary and this is made possible by the movement of the main gyre in the manner described.

The numerical solutions of related theoretical problems, presented in § 5.6, reproduce all the important qualitative features of this experiment.

Inertial and viscous boundary layers seem always to arise on the same preferred side of the basin although they do not necessarily coincide. (The latter usually overlaps the former.) The reasons for this are connected with the properties of wave motion in the basin, specifically, the different propagation speeds and dissipation rates of long and short waves. For example, the 'western' boundary is a source of short Rossby waves which decay rapidly, and the prominent viscous boundary layer is the region wherein these waves are entirely dissipated. A complete discussion of the wave interpretations of viscous and inertial layers is given in the next section.

The eastern boundary layer exudes mass into the interior and, as a consequence, must be fed continually by a strong fluid source. This can occur if the western current empties directly into the eastern jet or if a strictly meridional current (along a constant-depth contour) transports mass across the basin from one side to the other. In either case, the flow circuit is completed by the slow, westward oceanic circulation.

An intense current flowing along a boundary that is a constant-depth contour is also governed by (5.4.3), or (5.4.5). The function $dF(\phi)/d\phi$ is then constant along this bounding curve, and necessarily positive for a boundary layer to exist. However, this condition cannot be converted directly into a criterion like (5.4.7) without more information about the interior solution.

Fofonoff[50] studied a class of *free*, non-linear, steady circulations for which the dimensionless equations of motion are the homogeneous versions of (5.2.18) and (5.2.20):

$$\nabla \cdot h\mathfrak{Q} = 0, \tag{5.4.9}$$

$$\mathfrak{Q} \cdot \nabla \left(\frac{2 + \varepsilon \mathfrak{B}}{h} \right) = 0. \tag{5.4.10}$$

The absence of an applied wind-stress removes the difficulties mentioned earlier about the generation and dissipation of vorticity. Introduction of a stream function,

$$h\mathfrak{Q} = \hat{\mathbf{k}} \times \nabla \psi,$$

permits the integration of the vorticity equation and the result is

$$\frac{\varepsilon}{h} \nabla \cdot \frac{1}{h} \nabla \psi + \frac{2}{h} = F(\psi). \tag{5.4.11}$$

(The constancy of the potential vorticity along streamlines was used in chapter 4 in the same way to obtain exact, non-linear solutions.) The boundary layer equation (5.4.3) is a special case of this general relationship. The function $F(\psi)$ may be determined from the nature of the unforced interior flow, assuming, of course, that the solution exhibits a boundary layer behaviour. The interior flow is directed along depth contours because

$$F(\psi_I) = \frac{2}{h},$$

and this formula is identical to (5.4.4) at the boundary. The existence criterion for an inertial layer along a boundary of constant depth can then be written as

$$\frac{d}{d\psi_I} F(\psi_I) = \frac{2}{h^3} \frac{\hat{\mathbf{n}} \cdot \nabla h}{\hat{\mathbf{s}} \cdot \mathbf{q}_I} > 0.$$

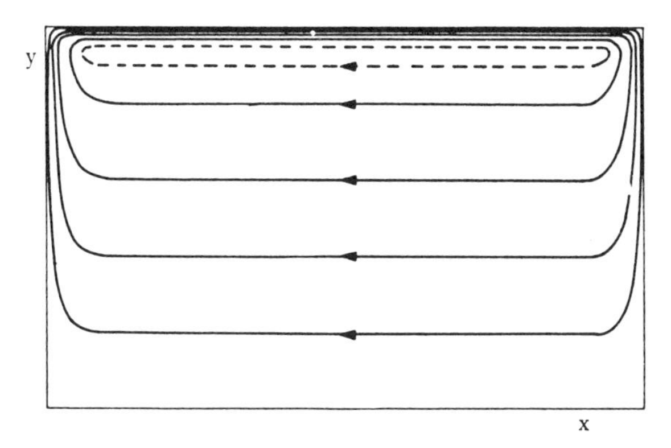

Fig. 5.2. Streamlines of a free flow in a homogeneous ocean computed by Fofonoff [50]. The circulation pattern is closed in this case with a meridional jet along the northern boundary.

Hence, the direction of the tangential flow along the outer edge of a meridional inertial layer is intimately related to the gradient of the depth. An interior flow that supports such a meridional jet is also one that is consistent with inertial layers on other boundaries at the same latitude (the same value of h). Therefore, a *free*, non-linear mode can exist in which mass from the slow, westward movement of the ocean is returned completely via inertial layers. Fofonoff has calculated some of these 'natural' solutions for the rectangular basin in the special case for which F is a linear function of ψ. One such pattern is shown in fig. 5.2. Closure of the current system can be made with a northern or southern jet, or both, depending on the particular form of F. The non-linear, numerical solutions for the wind-driven circulation, § 5.6, produce flow patterns which resemble this strongly.

5.5. Rossby waves: part two

Consider a container whose total depth variation is a linear function of one spatial co-ordinate:

$$h = 1 - \alpha y. \tag{5.5.1}$$

Here, α is a small number and the positive y axis is chosen as the northward direction to make $1/h$ an increasing function of y, just as the Coriolis parameter $\mathfrak{f}$ increases with latitude. The sidewalls form a vertical cylinder whose intersection with the plane, $z =$ constant, is a simple closed contour line of arbitrary shape.

Analysis of wave motion in the sliced cylinder shows that, in addition to ordinary inertial waves, a new class of very low frequency waves is generated to replace the loss of a purely geostrophic mode. The frequencies of these Rossby waves are proportional to 'angle' α; each modal function is independent of the depth, z, and represents a wave that propagates from east to west.

Rossby waves are discussed again in this section, but from a more general standpoint that is less dependent on a particular container shape. Plane waves are examined first, and the discussion parallels that given in § 4.2 for ordinary inertial waves.

Equation (5.2.25) for the interior pressure governs the slow time variations of a fluid in a container such as that described above. If, according to (5.2.23), we choose

$$\tau = 1, \quad h_0 = 1, \quad \mathfrak{d} = -\alpha y,$$

then

$$\frac{\partial}{\partial t} \nabla^2 p + 2\alpha \hat{\mathbf{j}} \cdot (\hat{\mathbf{k}} \times \nabla p) = 0. \tag{5.5.2}$$

(Note that the scaling rule here is $[\![L, \Omega^{-1}, \varepsilon\Omega L]\!]$. The time scale does not incorporate the factor α which then appears multiplying the eigen-frequency directly, as it did in § 2.16.)

Plane waves of the form

$$p = \Phi \exp i(\boldsymbol{\kappa} \cdot \mathbf{r} - \lambda t),$$

with

$$\boldsymbol{\kappa} \cdot \mathbf{r} = \kappa_1 x + \kappa_2 y,$$

are solutions of (5.5.2) provided

$$\lambda = \frac{-2\kappa_1 \alpha}{\kappa_1^2 + \kappa_2^2}. \tag{5.5.3}$$

If κ_1 is taken as positive for convenience, then λ is negative. For a given frequency, the loci of possible horizontal wave numbers is a circle in the wave number space, of radius $|\alpha/\lambda|$, centred at $(-\alpha/\lambda, 0)$, fig. 5.3 (*a*). Moreover, the fact that $-\kappa_1$ and λ have the same sign means that the wave pattern, as visualized by lines of constant phase, always propagates westward in the direction of decreasing values of x. The calculation of the phase speed yields the result

$$\mathbf{c}_p = -\frac{2\kappa_1 \alpha}{(\kappa_1^2 + \kappa_2^2)^{\frac{3}{2}}} \hat{\boldsymbol{\kappa}}. \tag{5.5.4}$$

Likewise, the group velocity

$$\mathbf{c}_g = \nabla_\kappa \lambda,$$

is

$$\mathbf{c}_g = 2\alpha \frac{\kappa_1^2 - \kappa_2^2}{(\kappa_1^2 + \kappa_2^2)^2} \hat{\mathbf{i}} + 4\alpha \frac{\kappa_1 \kappa_2}{(\kappa_1^2 + \kappa_2^2)^2} \hat{\mathbf{j}}, \tag{5.5.5}$$

which indicates a *westward propagation of energy for 'long' waves*, $\kappa_2^2 > \kappa_1^2$, but *an eastward flux for 'short' waves*, $\kappa_1^2 > \kappa_2^2$. Although the phase speed points to the west, the propagation of energy is not so restricted and the direction of the energy flux depends on the wavelength. This has an important bearing on the preferred formation of boundary layers on the western side of an ocean basin and we shall come to this shortly.

The phase and the group velocities are proportional, respectively, to the directed line segments $\overline{\text{BO}}$ and $\overline{\text{AB}}$ in the wave polar diagram. In particular, $\mathbf{c}_g$ is always directed away from the centre of the circle, A. For wave vectors lying on the arc OD, energy moves to the west as do the lines of constant phase.

The reflexion of ordinary inertial waves off a rigid wall exhibits rather unusual features (see § 4.2) and Rossby waves prove to be no exception to this rule. A Rossby wave propagating energy towards the plane boundary, $x = 0$, reflects as a Rossby wave with a different wave vector, whose group velocity is directed away from the boundary. If the pressure functions of the incoming and reflected waves are

$$A \exp i(\boldsymbol{\kappa} \cdot \mathbf{r} - \lambda t), \quad A' \exp i(\boldsymbol{\kappa}' \cdot \mathbf{r} - \lambda' t),$$

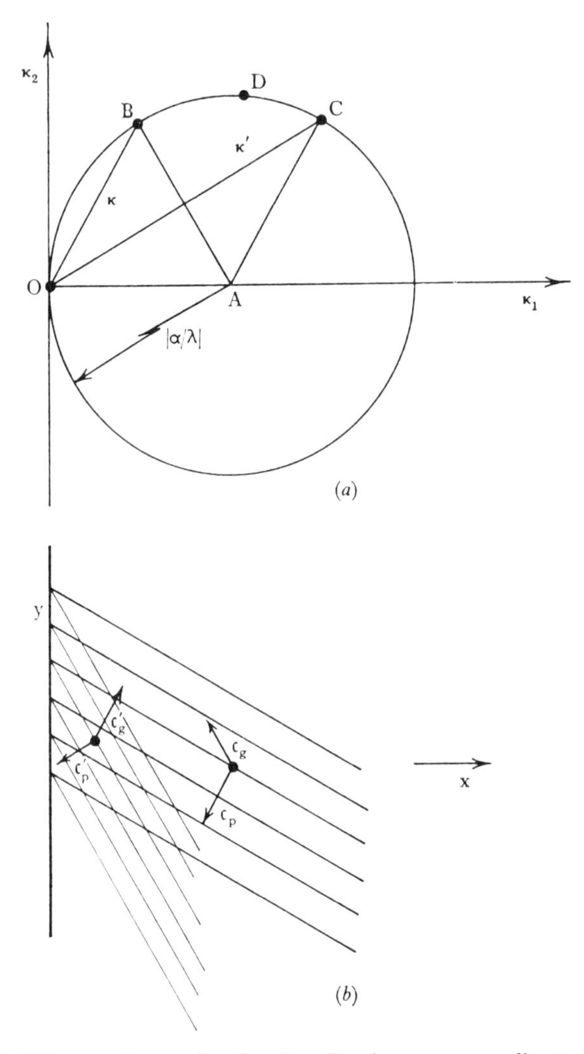

Fig. 5.3. (*a*) Wave number surface for plane Rossby waves according to (5.5.3). (*b*) The reflexion of a plane wave off the wall $x = 0$ with the phase speeds and group velocities shown.

then the condition of zero normal velocity at $x = 0$ implies

$$\kappa_2 A \exp i(\kappa_2 y - \lambda t) + \kappa_2' A' \exp i(\kappa_2' y - \lambda' t) = 0. \qquad (5.5.6)$$

It follows that $\quad \lambda = \lambda', \quad \kappa_2 = \kappa_2', \quad A = -A'. \qquad (5.5.7)$

In other words, the frequencies of both waves are identical,

$$\frac{\kappa_1}{\kappa_1^2+\kappa_2^2} = \frac{\kappa_1'}{\kappa_1'^2+\kappa_2'^2},$$

as are the components of the phase speeds parallel to the wall.

A complete graphical construction of the vectors $\boldsymbol{\kappa}'$, $\mathbf{c}_g$ and $\mathbf{c}_p$ is not as straightforward as that in §4.2, but it is enlightening to determine the directions of the various vectors in this way, (Longuet-Higgins[116]). If the incoming wave vector is $\overline{\mathrm{OB}}$ in fig. 5.3 (*a*), so that $\mathbf{c}_p$ and $\mathbf{c}_g$ are parallel to $\overline{\mathrm{BO}}$ and $\overline{\mathrm{AB}}$, then the reflected wave vector is $\overline{\mathrm{OC}}$ and $\mathbf{c}_p'$ and $\mathbf{c}_g'$ are proportional to $\overline{\mathrm{CO}}$ and $\overline{\mathrm{AC}}$. The reflected energy flux is outwards and the group velocity vectors make equal angles with the normal vector to the wall. Thus, the direction of energy flux, but not its phase, is reflected in the usual manner, i.e., the angles of incidence and reflexion are equal.

A wave propagating energy westward, to the left in the figure, reflects as a wave of shorter wave length, in order for the reflected energy flux to be properly directed away from the wall. Similarly, a short wave must reflect off an eastern boundary as a long wave. Hence, the western boundary acts as a source of short waves whereas long waves are created at the eastern boundary. But short waves are highly susceptible to viscous action and dissipate rapidly. As a consequence, manifestations of viscous phenomena should be most apparent in regions where these waves originate and this is the mechanism underlying the formation of the viscous, western boundary layer. Pedlosky[146] has shown that the characteristic widths of both viscous and inertial boundary layers can be calculated from the known properties of Rossby waves.

The group velocity of short waves, $\kappa_1^2 \gg \kappa_2^2$, is approximately

$$\mathbf{c}_g \simeq \frac{2\alpha}{\kappa_1^2}\,\hat{\mathbf{i}};$$

their decay rate is exponential, $e^{-\delta t}$, and δ can be estimated from the basic momentum equation by balancing the viscous term, $E\nabla^2$, against the time rate of change, $\partial/\partial t$. This implies that

$$\delta \doteq E\kappa_1^2,$$

i.e., short waves decay in the dimensionless time scale $E^{-1}\kappa_1^{-2}$ (see p. 187). Let the wave whose amplitude decreases by the factor

e^{-1} in the distance $\frac{\delta_v}{L} \doteq \frac{1}{\kappa_1}$, about one wavelength from the wall, define the region of dissipation, the boundary layer. This distance is traversed in the decay time by an energy packet travelling with the group velocity. Thus,

$$\frac{1}{\kappa_1} \doteq \frac{\delta_v}{L} \doteq \frac{|\mathfrak{c}_g|}{\mathfrak{s}},$$

and it follows that

$$\frac{\delta_v}{L} \doteq \left(\frac{E}{\alpha}\right)^{\frac{1}{3}},$$

which is the same thickness for the viscous boundary layer as determined by analysis of steady motion in the sliced cylinder.

In order to have an inertial layer on the western boundary, the interior flow must be directed towards that boundary. A short wave generated at that wall can penetrate back into the interior domain only if its group velocity is larger than the speed of the incoming current. In other words, the complete group velocity augmented by convection must have a positive horizontal component,

$$(\mathfrak{c}_g + \varepsilon\mathbf{q}_I)\cdot\hat{\mathbf{i}} > 0,$$

for a wave to avoid entrapment at the boundary. Waves that are trapped contribute to the formation of the *inertial* boundary layer and the longest of these is determined upon replacing the inequality by equality in the preceding expression:

$$\frac{2\alpha}{\kappa_1^2} = -\varepsilon\mathbf{q}_I\cdot\hat{\mathbf{i}}.$$

This yields the boundary layer thickness

$$\frac{\delta_i}{L} \doteq \frac{1}{\kappa_1} \doteq \left(\frac{-\varepsilon\mathbf{q}_I\cdot\hat{\mathbf{i}}}{\alpha}\right)^{\frac{1}{2}},$$

which as an estimate agrees with (5.4.8).

Rossby waves in an enclosed basin can sometimes be determined rather easily by superposing plane wave solutions. However, it is more instructive to relate the general problem with the problem of membrane oscillations, Longuet-Higgins[116]. To this end, let

$$p = \mathscr{P}(x, y)\exp\left[-i\left(\frac{\alpha}{\lambda}x + \lambda t\right)\right]; \qquad (5.5.8)$$

upon substituting this in (5.5.2), we find that

$$\nabla^2\mathscr{P}+\frac{\alpha^2}{\lambda^2}\mathscr{P}=0. \tag{5.5.9}$$

The boundary condition on the lateral surface is simply

$$\mathscr{P}=0.$$

This shows that a Rossby wave can be interpreted as an exponential carrier wave (moving to the west), modulated by an amplitude function, $\mathscr{P}$, which is itself a solution of a classical boundary value problem. Obviously, all that is known about membrane oscillations becomes applicable to the oceanographic problem. Any solution can be immediately associated with a Rossby wave and, perhaps of equal importance, is the acquired ability to make full use of all established approximate methods and techniques of analysis.

Rossby waves in the rectangular container, $0 \leqslant x \leqslant a$, $0 \leqslant y \leqslant b$, are related to the membrane eigenmodes

$$\mathscr{P}_{nm}(x,y)=\sin\frac{n\pi x}{a}\sin\frac{m\pi y}{b},$$

with

$$\lambda_{nm}=\pm\frac{\alpha}{\pi}\left(\frac{n^2}{a^2}+\frac{m^2}{b^2}\right)^{-\frac{1}{2}}.$$

Moreover, if (x', y') is a co-ordinate system obtained from system (x, y) by a rotation of axes through a fixed angle, then

$$\mathscr{P}_{nm}(x',y')\exp\left[-i\left(\frac{\alpha}{\lambda_{nm}}x+\lambda_{nm}t\right)\right],$$

are the Rossby waves in the new frame, as long as the depth is given by (5.5.1).

It is a relatively direct procedure to solve the general initial value problem in the rectangular domain. This has already been done for the sliced cylinder geometry, and since the results are much the same, discussion of this particular problem is omitted.

The mixing of two waves by non-linear convective terms generates additional waves at the sum and difference of the primary frequencies. In particular, a zero frequency response develops from a single wave interacting with itself, and this can be a substantial part of the total steady circulation pattern. The discussion of steady inertial currents requires consideration of the entire spectrum of the applied wind-stress and not just its average. The non-linear response to an oscillatory forcing function in a rectangular basin is considered by Pedlosky[145], Veronis[218], Veronis and Morgan[219], and Longuet-Higgins[117].

A linear variation of the total depth, h, corresponds to the approximation

$$\mathfrak{f} = \mathfrak{f}_0 + \beta y$$

for the dependence of the Coriolis parameter on latitude. This is often used in the oceanic model equations, (5.3.4), and it is an essential part of the β-plane approximation which also requires that βy be neglected whenever it appears in conjunction with $\mathfrak{f}_0$. The validity of results obtained on this basis and their relevance in oceanography has been the subject of many investigations, Longuet-Higgins[116, 118] and Veronis[214, 215], but this is not of immediate concern here. However, certain mathematical aspects are too interesting to be ignored. For example, Longuet-Higgins found a general class of unforced, linear, inviscid solutions to (5.3.3), with $\mathfrak{f} = 2\cos\Theta$, for an ocean covering the entire surface of a sphere. These can be compared directly with solutions of the β-plane equations and, in this way, the effects of spherical curvature can be gauged and the relative accuracy of β-plane analyses assessed. (Equations (5.3.3) are already depth-averaged approximations and strictly speaking do not apply over the entire spherical surface, being of questionable validity near the equator.)

Let H = 1, $\mathfrak{f} = 2\cos\Theta$ and the scaled radius of the sphere be 'a'. It follows from the continuity equation that

$$\mathfrak{Q}_0 = a\nabla \times [\psi(\theta, \Theta, t)\hat{\mathfrak{r}}] = -\frac{\partial\psi}{\partial\Theta}\hat{\boldsymbol{\theta}} + \frac{1}{\sin\Theta}\frac{\partial\psi}{\partial\theta}\hat{\boldsymbol{\Theta}};$$

the linear, inviscid equation for the radial component of vorticity is

$$\frac{\partial}{\partial t}\left(\frac{1}{\sin\Theta}\frac{\partial}{\partial\Theta}\sin\Theta\frac{\partial\psi}{\partial\Theta} + \frac{1}{\sin^2\Theta}\frac{\partial^2\psi}{\partial\theta^2}\right) + 2\frac{\partial\psi}{\partial\theta} = 0. \quad (5.5.10)$$

Bounded solutions of this equation are

$$\psi_{nm} = P_n^m(\cos\Theta)\exp\left[im\left(\theta + \frac{2}{n(n+1)}t\right)\right], \quad (5.5.11)$$

and represent waves which all move westward relative to the

rotating frame. (It is not really necessary for the polar axis of the co-ordinate system (θ, Θ) to coincide with the axis of rotation. The solution can be referred to an arbitrary co-ordinate system (θ', Θ'), as long as the new polar axis revolves about the rotation axis with the relative angular velocity $-2/[n(n+1)]$. However, the discussion here is restricted to the natural co-ordinate system (θ, Θ).)

A very interesting feature of Longuet-Higgins' analysis is the trapping of Rossby waves in a band about the equator. If attention is directed to waves of very large order n, for which m/n neither approaches zero nor one as $n \to \infty$, then the asymptotic forms of the

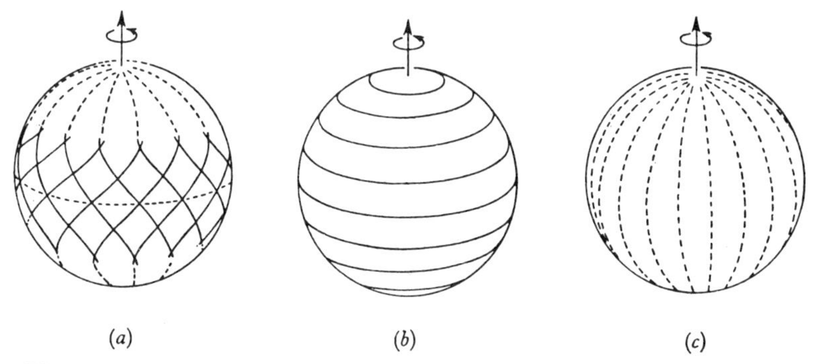

Fig. 5.4. Wave patterns on the surface of a sphere given by (5.5.11): (*a*) A trapped Rossby wave for the general case $m \neq n$, [116]. (*b*) $m = 0$ and the wave is sinusoidal over most of the surface. (*c*) $m = n$ and the 'wave' is confined to the equator.

spherical harmonics reveal markedly different behaviours depending on the condition

$$\sin \Theta \gtrless \frac{m}{n}.$$

In the case $\sin \Theta > m/n$, the formulas indicate an oscillatory variation with colatitude Θ, but only an exponential decay is possible when $\sin \Theta < m/n$. Hence, the wave pattern is effectively trapped within an equatorial belt; the limiting latitudes are ray caustics. Fig. 5.4 illustrates the wave patterns in three cases, (*a*) $m \neq n$; (*b*) $m = 0$; (*c*) $m = n$.

Although the preceding analysis is concerned with the model equations, it seems likely that entrapment of inertial modes near the equator is a general feature of motion in a spherical annulus. Indeed, Stern[184] and Bretherton[15] have shown this to be true and some rather interesting conclusions, drawn from these investigations, are discussed in § 5.8.

Some of the many references dealing with Rossby waves in a geophysical context are Margules[126], Haurwitz[78, 79], Rossby[174], Rattray and Charnell[157], and Phillips[149] (who provides an extensive bibliography).

An account of the experiments of Fultz and Frenzen[63], and Frenzen[55] concerning inertial waves in a spherical annulus appears in the last section of this chapter.

A very simple demonstration of Rossby waves in a rotating cylindrical annulus with a free surface is given by Ibbetson and Phillips[91]; its theoretical analysis is to be found in Phillips[150]. Among the various classes of surface waves that can be excited in this geometry, is one which corresponds exactly to the low frequency waves studied in this section. Waves are generated by oscillating a paddle at a fixed azimuthal position and the resultant disturbance is examined as the excitation frequency changes. Although the phase propagation is always to the west, the direction of group velocity depends on the frequency in the manner detailed earlier. Experimental results correlate well with theory and the photographs of the waves on one or the other side of the paddle, depending on frequency, are striking. Further discussion of this interesting experiment is omitted because its theoretical formulation stands somewhat apart from the general development pursued here.

5.6. Numerical studies

Most numerical investigations to date have been based on the oceanographic model equations, further simplified by the β-plane approximation (5.3.6). The depth of the ocean is usually taken as constant, $\hat{\mathbf{k}}$ is in the vertical direction and all dependent variables are functions of the cartesian co-ordinates (x, y), which measure distance to the east and north, respectively. In this case, the gradient, ∇, and the horizontal gradient, ∇, may be used interchangeably. The equations governing mass conservation and the vertical component of vorticity are derived from (5.3.3):

$$\nabla \cdot \mathfrak{Q}_0 = 0, \tag{5.6.1}$$

$$\frac{\partial \mathfrak{B}_0}{\partial t} + \varepsilon \mathfrak{Q}_0 \cdot \nabla \mathfrak{B}_0 + \beta \hat{\mathbf{j}} \cdot \mathfrak{Q}_0 = E \nabla^2 \mathfrak{B}_0 + \hat{\mathbf{k}} \cdot \nabla \times \mathfrak{F}. \tag{5.6.2}$$

Velocity and vorticity are expressible in terms of a stream function, $\psi_0(x, y)$, as

$$\left.\begin{aligned} \mathfrak{Q}_0 &= \hat{\mathbf{k}} \times \nabla\psi_0, \\ \mathfrak{B}_0 &= \nabla^2\psi_0; \end{aligned}\right\} \qquad (5.6.3)$$

the three numbered equations constitute the system to be solved subject to proper boundary conditions, which are usually homogeneous. The forcing function may be chosen freely.

The boundary of the ocean basin is most often a rectangle or a figure made up of perpendicular straight line segments. This makes the selection of a computational grid as simple as possible. The memory capacity of available computing machines severely limits the mesh size in either direction. Techniques for altering the grid dimensions wherever greater precision is required (for example in the boundary layers) have not been fully developed.

The ability to solve the non-linear boundary value problem rapidly and accurately provides the means of conducting informative numerical experiments. The relative importance of various processes can be quantitatively assessed in this manner, over a wide range of parameter settings which fall beyond the reach of analytical methods. The information from such studies often leads to the improvement of the model and motivates better approximations of both a physical and a mathematical character.

One major conclusion resulting from the treatment of the complete equation is that viscous dissipation is an essential mechanism somewhere in the basin, *in all problems* of 'wind-driven' circulation.

Bryan [20] considered a rectangular basin, $0 \leqslant x \leqslant 1$, $0 \leqslant y \leqslant 2$, subjected to an applied force field

$$\hat{\mathbf{k}} \cdot \nabla \times \mathfrak{F} = -\beta \sin \frac{\pi}{2} y. \qquad (5.6.4)$$

(The parameter β can be scaled out of the problem by redefining the characteristic time and setting $\varepsilon' = \varepsilon/\beta$, $E' = E/\beta$. Since this is so, β may simply be set equal to unity and equations (5.6.1) and (5.6.2) solved in their present form.)

One objective, difficult to achieve, is to make E really small enough so that the inertial boundary layer is much thicker than the viscous layer. Fig. 5.5 (*a*) shows the time-averaged results of four calculations as E decreases but ε remains fixed. As non-linear terms become more important, the centre of the main gyre shifts northward (see the discussion on p. 243). A second gyre forms in the top corner and a counter-current is produced on the eastern side of the main stream. Fluid motions in the boundary layer and the interior become

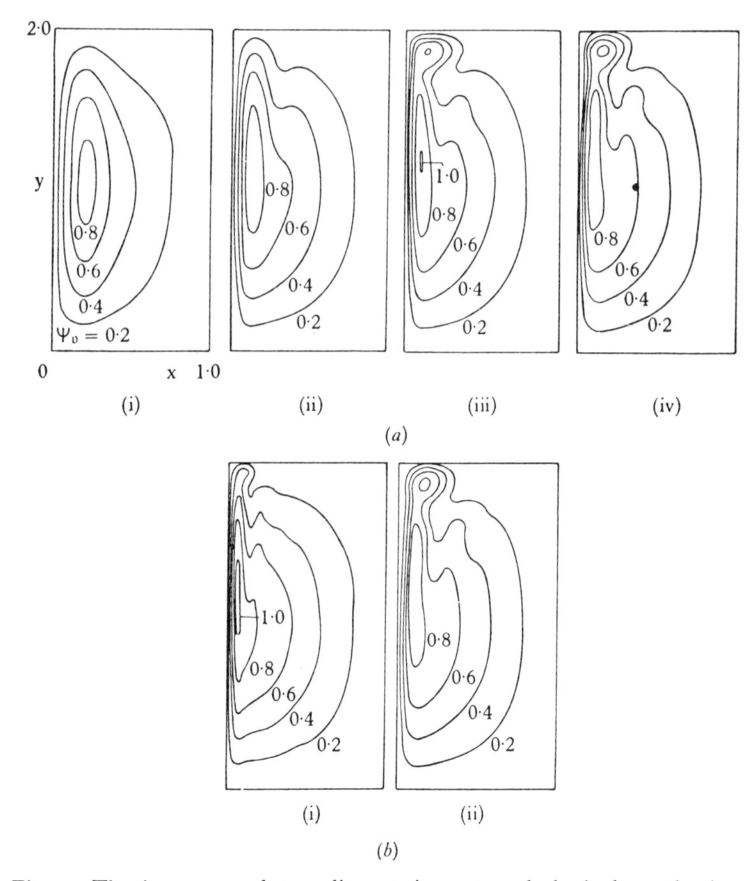

Fig. 5.5. The time-averaged streamlines, ψ_0, in a rectangular basin due to the sinusoidal wind-stress in (5.6.4), [20]. In (*a*) $\varepsilon = 1{\cdot}28\ 10^{-3}$ and the Ekman number is made progressively smaller; (i) $E = \varepsilon/5$, (ii) $E = \varepsilon/20$, (iii) $E = \varepsilon/40$, (iv) $E = \varepsilon/60$. In (*b*), the Rossby number is increased from $\varepsilon = 0{\cdot}32\ 10^{-3}$, $E = \varepsilon/60$ in (i) to $\varepsilon = 1{\cdot}28\ 10^{-3}$, $E = \varepsilon/60$ in (ii).

more inseparable from each other even though the boundary layer approximation remains valid over a large part of the enclosure. Fig. 5.5 (*b*) shows that increasing the Rossby number intensifies the secondary gyre and the inertial counter-current. The main stream does not truly separate from the wall, but propagating vortices do appear in the current north of the critical latitude. The effect of viscous processes, though reduced, is critical and every streamline passes through a region of intense dissipation.

Veronis[216, 217] simplified the numerical problem by using a velocity drag law so that $-E\mathfrak{Q}_0$ replaces $E\nabla^2\mathfrak{Q}_0$ in the momentum equation (i.e., $-E\mathfrak{V}_0$ instead of $E\nabla^2\mathfrak{V}_0$ in (5.6.2)). The number of boundary conditions must be diminished accordingly and, with these modifications, it is possible to study a strongly non-linear regime.

Veronis considers a square ocean basin, $0 \leqslant x, y \leqslant \pi$, subjected to a wind-stress

$$\hat{\mathbf{k}}\cdot\nabla\times\mathfrak{F} = -\beta \sin x \sin y, \tag{5.6.5}$$

which corresponds to a centred, high pressure area. Parameter β may again be absorbed into a redefinition of ε and E, or equivalently, we can simply make $\beta = 1$.

The associated, steady, *linear* problem with the simpler drag law is

$$E\left(\frac{\partial^2}{\partial x^2}+\frac{\partial^2}{\partial y^2}\right)\psi_0+\frac{\partial\psi_0}{\partial x} = -\sin x \sin y, \tag{5.6.6}$$

and $\psi_0 = 0$ on the boundary. The solution, given by Stommel[193], is

$$\psi_0 = \frac{\sin y}{1+4E^2}\left[2E\sin x+\cos x + \frac{1}{e^{\pi s_1}-e^{\pi s_2}}\{(1+e^{\pi s_2})\,e^{s_1 x}-(1+e^{\pi s_1})\,e^{s_2 x}\}\right], \tag{5.6.7}$$

with

$$s_{1,2} = \frac{-1\pm(1+4E^2)^{\frac{1}{2}}}{2E}. \tag{5.6.8}$$

A boundary layer analysis provides the approximate formula

$$\psi_0 = (1+\cos x-2e^{-x/E})\sin y; \tag{5.6.9}$$

fig. 5.6 (*a*) shows constant streamlines for $E = 0{\cdot}05$.

The ultimate steady solutions of the non-linear, initial value problem are obtained numerically. The details of the numerical analysis cannot be presented here, but the results are shown in figs. 5.6 and 5.7. The graphs are arranged in the order of increasing importance of the non-linearity and in fig. 5.7 (*b*), the thickness of the inertial boundary layer is eight times that of the viscous layer.

The centre of the main gyre, a high pressure area, shifts northward in response to non-linear processes. The increased pressure gradient to the north tends to press the current onto the boundary and an intensification is noted in that region. The southern part of the current relaxes into a wider stream because the pressure is diminished there. As non-linear effects increase in magnitude, a jet forms along the northern boundary, and ultimately on the eastern side as well. In that stage, the solution resembles the free mode found by Fofonoff[50], see p. 245.

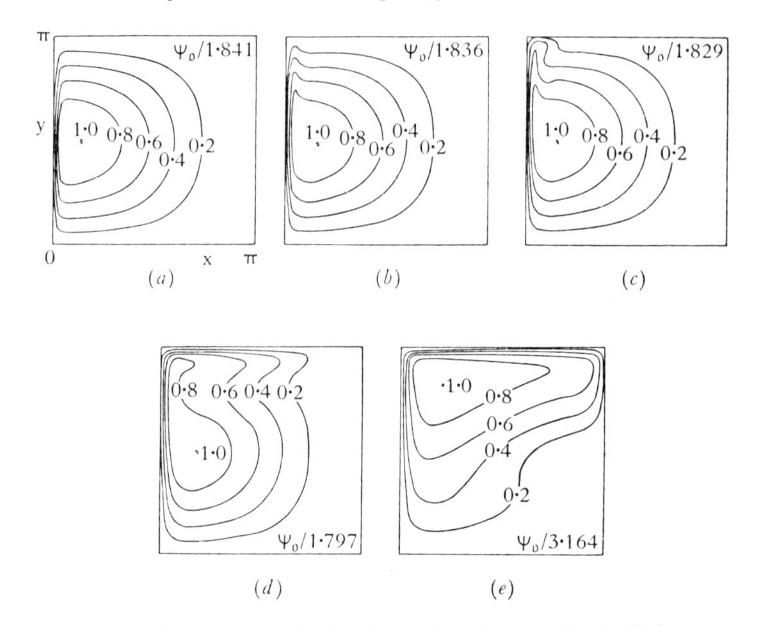

Fig. 5.6. Streamline patterns as the Rossby number is increased[217]. (*a*) $\sqrt{\varepsilon} = 0{\cdot}001$, $E = 0{\cdot}05$; (*b*) $\sqrt{\varepsilon} = 0{\cdot}003$, $E = 0{\cdot}05$; (*c*) $\sqrt{\varepsilon} = 0{\cdot}05$, $E = 0{\cdot}05$; (*d*) $\sqrt{\varepsilon} = 0{\cdot}1$, $E = 0{\cdot}05$; (*e*) $\sqrt{\varepsilon} = 0{\cdot}1$, $E = 0{\cdot}025$.

As a fluid particle moves northward in the western boundary layer, its relative vorticity decreases to compensate for an increase in planetary vorticity. The oscillation in the northwest corner develops because inertia makes the particle overshoot the latitude at which it should re-enter the interior of the ocean with negligible relative vorticity. The particle returns to this equilibrium position by passing through a region of intense dissipation where the appropriate adjustment is made in the level of relative vorticity. A typical display of the regions of positive and negative relative vorticity is shown in fig. 5.7 (*b*), which corresponds to fig. 5.6 (*c*) and (*d*).

Since there is a mass efflux from the northern boundary layer, the oscillations seen there can be ascribed to the escape, propagation, and interior penetration of Rossby waves formed at the western wall (see p. 250 and Moore [128]).

Separation from the coastline is not indicated by any of these results.[1] In fact, quite the contrary occurs; the limiting circulation

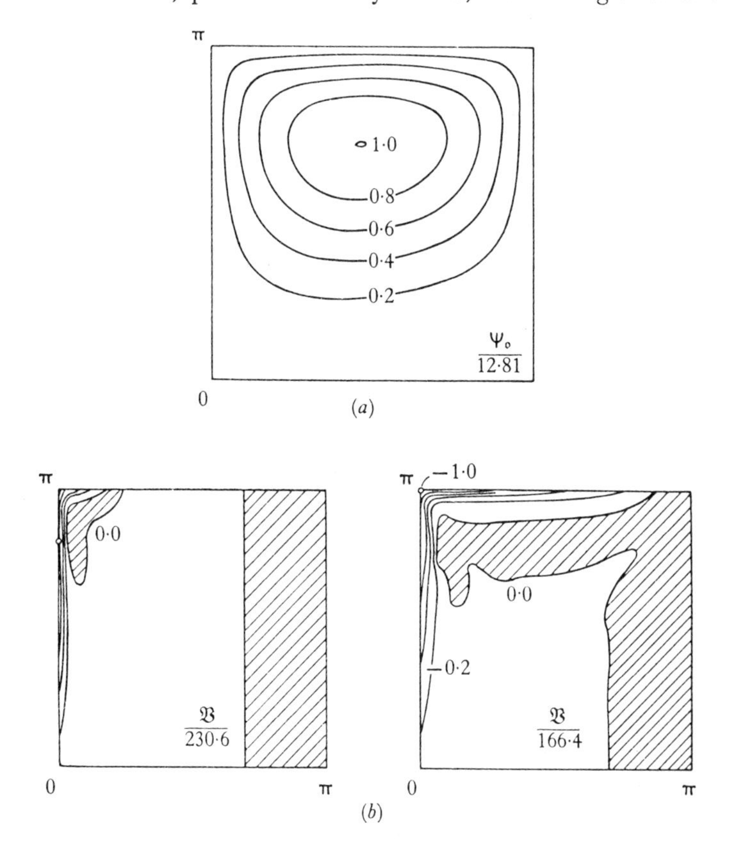

Fig. 5.7. (*a*) Streamlines for a highly non-linear motion $\varepsilon = 0{\cdot}2$, $E = 0{\cdot}025$. (*b*) Regions of positive vorticity (shaded) and negative vorticity corresponding to the circulations shown in fig. 5.6 (*c*) and (*d*).

[1] In this connection, it should be noted that Bryan's computations do not penetrate the non-linear regime sufficiently to allow much inference about purely inertial jets. On the other hand, the simple model analysed by Veronis may, by its very structure (the reduced equation with undisturbed initial conditions) preclude any final steady state that is not of the Fofonoff type. The question of inertial jets and separation is still open.

pattern, fig. 5.7(*a*), contains an intense inertial jet on three sides of the boundary and is very similar to a free mode. It would be interesting to include topographic changes in the ocean basin for this might force separation to occur. Work in this direction is in progress.

Niiler [138] examined the problem of forced circulation from an analytical standpoint. His procedure involves an appropriate scaling of the equations of motion and expansions in powers of a parameter related to the Ekman number. Fofonoff's free mode arises from this development as a sort of natural resonance produced by the applied stress. Moreover, analysis of the higher order field yields a circulation condition which removes the indeterminacy of the purely inertial theory (the arbitrary function $F(\psi)$).

The effect of topography on a *free* inertial jet has been studied by Warren [222] in connection with the meanders of the Gulf Stream. If the path of the current, and not its structure, is of prime concern, a rather simple theory can be constructed that yields surprisingly good results.

The assumptions are made that the narrow current carries a constant mass flux and that its curvature is very large compared to its width. With the neglect of viscous effects, the equation governing the motion is the conservation of potential vorticity,

$$\frac{\mathfrak{V}_0 + \mathfrak{f}}{\mathrm{H}} = F(\psi_0). \tag{5.6.10}$$

The vorticity in a meandering jet whose radius of curvature is large can be expressed in terms of the streamline curvature, $\mathscr{C}$. If (ξ_1, ξ_2, z) is a 'natural', orthogonal co-ordinate system with ξ_2 a constant on streamlines, then this relationship is

$$\mathfrak{V}_0 \cong \mathscr{C}\mathfrak{Q}_0 + \frac{\partial \mathfrak{Q}_0}{\partial \xi_2}, \tag{5.6.11}$$

where $\mathfrak{Q}_0$ is the speed of the current. Equation (5.6.10) can then be written as

$$\mathscr{C}\mathfrak{Q}_0 + \frac{\partial \mathfrak{Q}_0}{\partial \xi_2} + \mathfrak{f} = \mathrm{H}\, F(\psi_0). \tag{5.6.12}$$

The simplest approximation results by applying the preceding equation to only the central streamline $\overline{\overline{\psi}}_0$, defined by $(\partial/\partial \xi_2)\, \mathfrak{Q}_0 = 0$.

This is equivalent to the assumption that conditions are uniform across the width of the stream. If $(\bar{\bar{x}}, \bar{\bar{y}}(\bar{\bar{x}}))$ are the co-ordinates of $\overline{\overline{\Psi}}_0$, and $\mathfrak{Q}_0 H = \mathscr{M}$ on each streamline (mass conservation), then (5.6.12) is very nearly the same as

$$\overline{\overline{\mathscr{C}}}\mathscr{M} + \mathfrak{f}\overline{\overline{H}} = \overline{\overline{H}}^2 F(\overline{\overline{\Psi}}_0) \tag{5.6.13}$$

with
$$\overline{\overline{H}} = \overline{\overline{H}}(\bar{\bar{x}}, \bar{\bar{y}}),$$

$$\overline{\overline{\mathscr{C}}} = \left(\frac{d^2\bar{\bar{y}}}{d\bar{\bar{x}}^2}\right)\left(1 + \left(\frac{d\bar{\bar{y}}}{d\bar{\bar{x}}}\right)^2\right)^{-\frac{3}{2}}.$$

The function $F(\overline{\overline{\Psi}}_0)$ is evaluated at a position of zero curvature on the jet, i.e., a point of inflexion. More sophisticated equations, of the same type, can be developed by taking appropriate averages across the stream width. The known properties of the stream can then be introduced and utilized in a manner similar to that employed in momentum-integral methods. This is essentially Warren's procedure.

The ocean depth is a prescribed function of position, and the numerical integration of the ordinary differential equation (5.6.13), a prototype for the entire class, is a relatively simple task. Warren's results apply to conditions encountered by the Gulf Stream, and the calculations are in good agreement with many of the meander patterns actually observed, fig. 5.8. This is fairly strong evidence that the free stream is topographically controlled.

If the increased pressure gradient tends to intensify the northern part of the current due to non-linear interactions, but an adverse depth gradient or some other effect separates it from the coastline, then the free jet must find a favourable topography to ride northward. A free jet, flowing northward with deep water to the east, experiences an increase in either mean velocity or breadth as it moves into shallow water, in order to preserve a given mass transport. Either change results in an imbalance in which the Coriolis force exceeds the local pressure gradient and the stream is forced to return to deeper water. A current moving into deeper water is forced back for similar reasons; the pressure gradient overcomes the Coriolis force. The topographic gradient tends to stabilize the motion of the current proceeding northward, but makes it meander

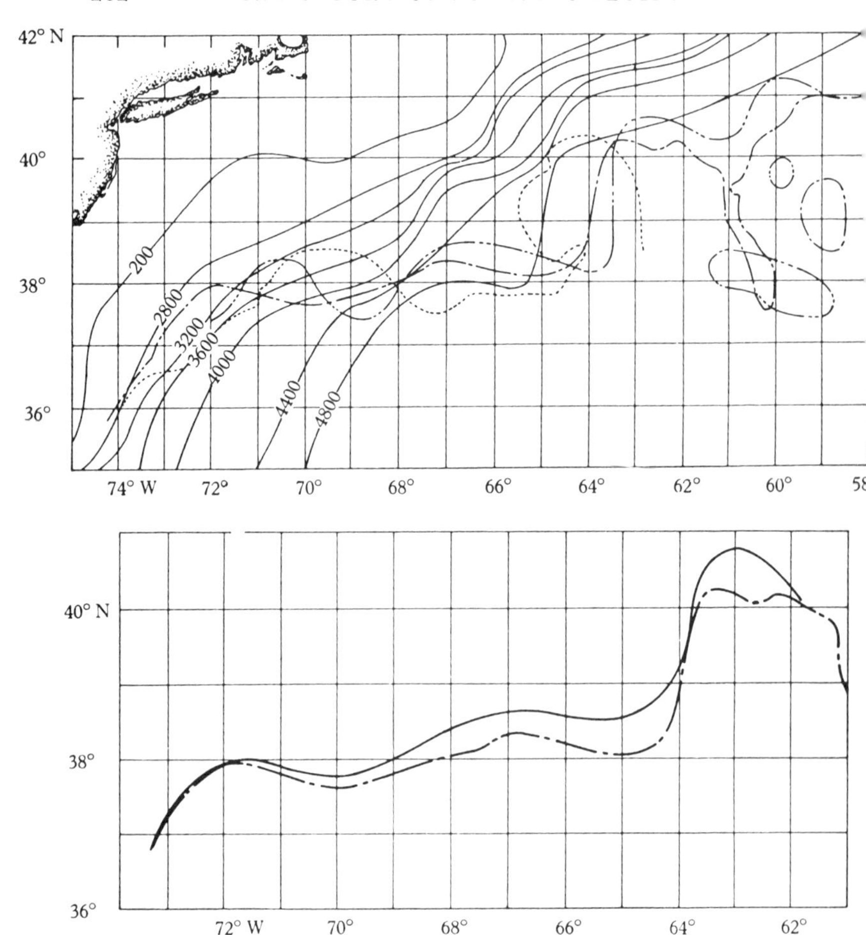

Fig. 5.8. (*a*) Three observed paths of the Gulf Stream overlying smoothed bottom contours (solid lines). (*b*) Comparison of one stream path with Warren's[222] numerical calculation (heavy solid curve).

about an equilibrium depth. A slow drift eastward into deeper water is dictated by vorticity conservation.

Calculations of a free, time-dependent jet and the analysis of the meanders observed in the laboratory are incomplete at this time.

5.7. Flow between concentric spheres

Fluid motions in a rotating spherical annulus have been the object of a great deal of research in the expectation that they resemble atmospheric and oceanic circulations. As long as the annular gap-width is small compared to the radius, the flows are indeed analogous in many problems with homogeneous fluids. Examples have already been presented which confirm this; more are given in this section. (It should be recognized that the concentric sphere configuration is challenging and interesting in its own right.) Experience with this particular geometry seems to indicate that a higher *a priori* probability of success is attached to experimental efforts than to theoretical investigations. However, neither approach is easy. Purely dynamical analogies of geophysical motions are somewhat simpler because analytical difficulties attributable solely to geometry can be avoided.

Motion between concentric spheres differs appreciably from that inside a single sphere. The total vertical depth is a discontinuous function of cylindrical radius in the former case, but not in the latter, and this has a pronounced effect on the nature of the internal modes. For example, the inviscid geostrophic mode is discontinuous across the surface of a vertical cylinder circumscribing the inner sphere. It would not be too surprising to find that some of the inertial waves also exhibit singular behaviour. Explicit formulas for all the eigenmodes in the spherical annulus do not exist. Bryan's transformation, p. 64, the essential step in the solution procedure for the spherical container, does not apply and for this reason very little is known about the inertial waves.[1] However, the narrow-gap annulus is a somewhat simpler geometry to analyse and it has been shown by Stern [184] and Bretherton [15] that in this case there is a class of inertial modes trapped in the vicinity of the equator. These correspond to the trapped Rossby waves found by Longuet-Higgins which were discussed in section 5.3.

The entrapment of waves was described qualitatively by

[1] Certain of the spherical modes given by (2.12.7) are also proper modes for any spherical annulus. This subclass corresponds to the index setting $n = k+1$; the modes and eigenvalues are

$$\Phi = zr^k \, e^{ik\theta}, \quad \lambda = \frac{2}{k+1} \quad \text{for } k \geqslant 2.$$

For large k, these modes are essentially confined to the equatorial region; their phase speeds are identical to those of the trapped inertial waves discussed on p. 253.

Bretherton in terms of a ray theory and his presentation is followed here. This theory is restricted to wavelengths, $2\pi/\kappa$, that are small compared to the separation distance H, i.e., $\kappa H \gg 1$. Because of this, the spherical curvature in the azimuthal direction is relatively unimportant, and locally, the analysis is comparable to that for the concentric cylinders geometry shown in fig. 5.9(*a*). We continue with a discussion of this simpler container shape.

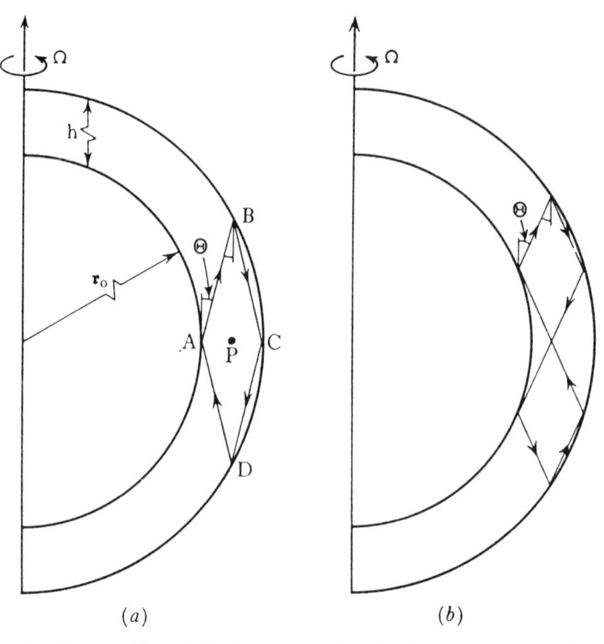

Fig. 5.9. (*a*) Ray paths of the lowest order Rossby wave trapped near the equator[15]. (*b*) A higher order trapped mode.

The properties of inertial wave propagation are set out in §4.2 and many of the results obtained there on energy flux, etc., are of use now. The following deductions are especially pertinent: the direction of the ray is always parallel to the local group velocity; the group velocity is perpendicular to the wave vector; the angle between the ray direction and the rotation axis is preserved in a reflexion off a boundary, see fig. 4.1.

The lowest order mode is described by the ray ABCD of fig. 5.9(*a*); a more complex mode is shown in fig. 5.9(*b*). Ray AB, which makes an angle Θ with the rotation axis, is the direction of energy propagation of a plane wave whose wave vector is perpendicular to the ray. Moreover, the frequency of this plane wave, according to (4.2.11), is then

$$\lambda = 2\mathbf{\Omega}\cdot\hat{\mathbf{\kappa}} = 2\Omega\sin\Theta \tag{5.7.1}$$

in *dimensional* units. This ray undergoes multiple reflexions at positions B, C and D on the outer boundary and returns to its point of origin, A. The angle Θ with the vertical axis remains the same for each straight segment.

The maximum northward penetration of the wave, distance BP, is approximately $(H\mathrm{r}_0)^{\frac{1}{2}}$ if $H \ll \mathrm{r}_0$. Therefore, the inertial wave frequency, as calculated from (5.7.1), is

$$\lambda = \Omega\left(\frac{H}{\mathrm{r}_0}\right)^{\frac{1}{2}}.$$

These results are all in agreement with those obtained by Stern[184] in a more analytical fashion.

Thus far, the only requirement imposed is that the ray makes a closed circuit. In addition, the wave must also return to point A with its original phase if it is part of a genuine mode. To prove that this is possible, the travel time about the circuit and the wave period are computed. The former is proportional to $\kappa H(H/\mathrm{r}_0)^{-\frac{1}{2}}\Omega^{-1}$, while the latter is the order of $(H/\mathrm{r}_0)^{-\frac{1}{2}}\Omega^{-1}$. Since $\kappa H \gg 1$, a very slight change in wave number can modify the travel time by several complete periods. Suitable values of κ can easily be found for which the wave will return with exactly the correct phase and this implies the existence of trapped inertial modes.

A series of experiments to produce Rossby waves in a spherical annulus were conducted by Fultz[56, 57], Fultz and Long[64], Long[114], Frenzen[55], and Fultz and Frenzen[63]. The basic apparatus, shown in fig. 5.10, consisted of two concentric, rigidly mounted glass hemispheres. Various obstacles, placed in the gap between the spheres, were connected to a separate driving shaft so that they could revolve at a rate different from that of the container. Relative motions produced in this manner were made visible by a suspension of white tracer particles. The obstacles were usually circular cylinders cut to fit the spherical curvature and their width ranged to 20° of latitude. The heights of the objects varied from half the gap-width to the full separation distance.

Turning the obstacle more slowly than the spheres, generates a relative westerly flow past it while a larger rotation rate means a relative easterly motion. Since Rossby waves drift to the west, an oncoming westerly flow can confine the waves downstream and freeze the disturbance into a steady pattern. A relative easterly flow accentuates upstream propagation.

The characteristic patterns observed during the westerly flow past a 20° diameter, full-depth obstacle at 30°, 45° and 60° latitude

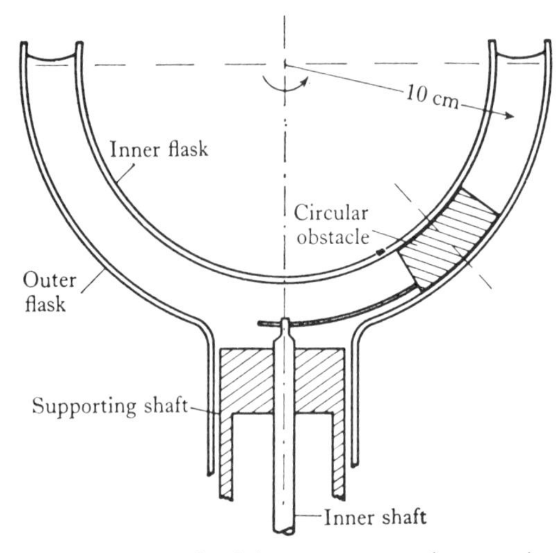

Fig. 5.10. A schematic of the apparatus used to examine flow past an obstacle in a spherical annulus.

were summarized by Frenzen [55]. At low latitudes, a strong steady wave pattern is observed in the downstream direction, fig. 5.11. The wave amplitude is seen to be larger than the body diameter and a small, closed anticylone occurs in the first ridge. At 45°, in the mid-latitudes, a similar wave pattern is observed but with a smaller amplitude and no counter-vortex. At high latitudes, 60°, *no steady wave pattern* is discernable. Instead, a periodic array of moving cyclonic vortices develops; each forms at the obstacle and eventually attains a diameter roughly equalling that of the body.

Relative easterly motion does not produce a wave pattern. In this case, the obstacle drags along all the fluid in its path and little flow occurs outside the latitudinal band occupied by the cylinder.

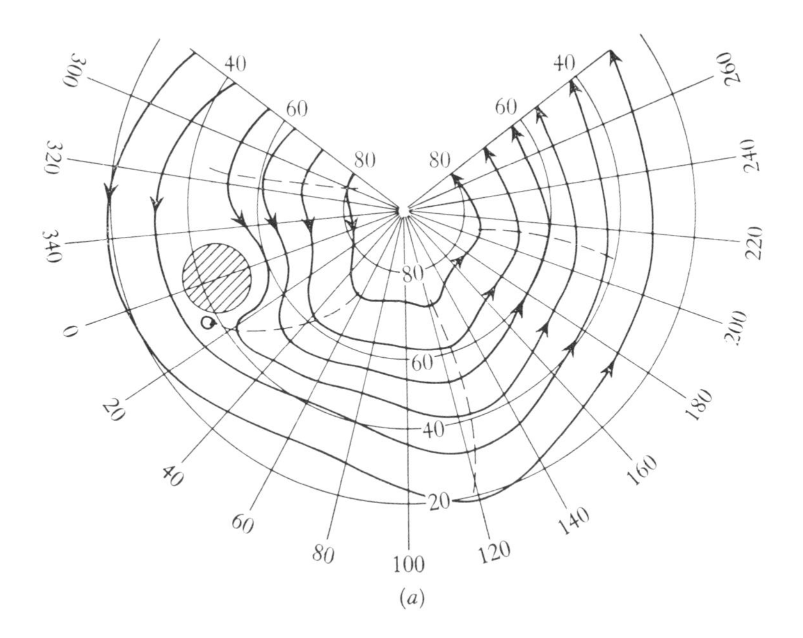

(*a*)

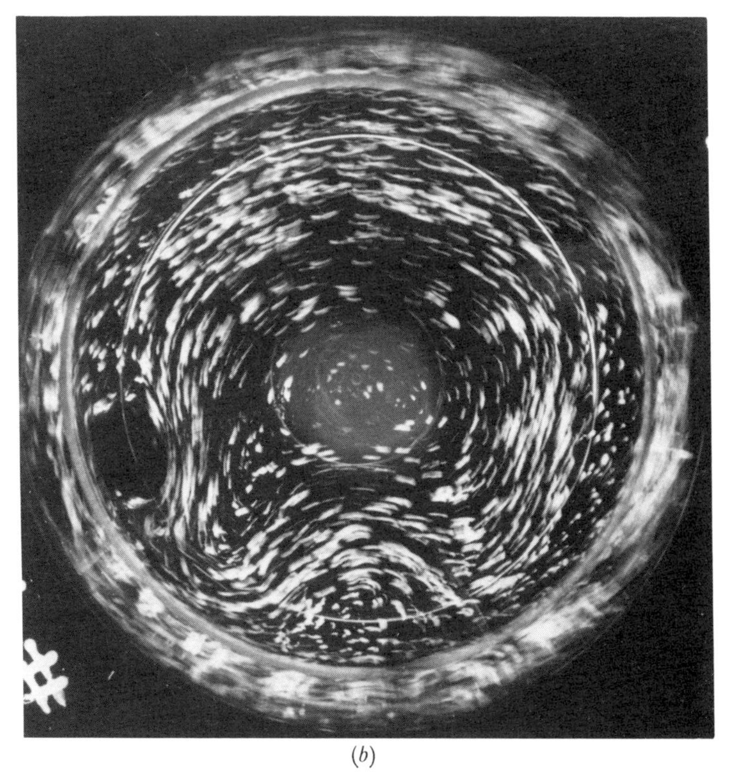

(*b*)

Fig. 5.11. (*a*) Streamlines traced from a streak photograph of a four-wave pattern generated by westerly flow past an obstacle at 30° latitude with $\varepsilon = 0{\cdot}078$. (*b*) A five-wave pattern at 30° latitude; $\varepsilon = 0{\cdot}063$.

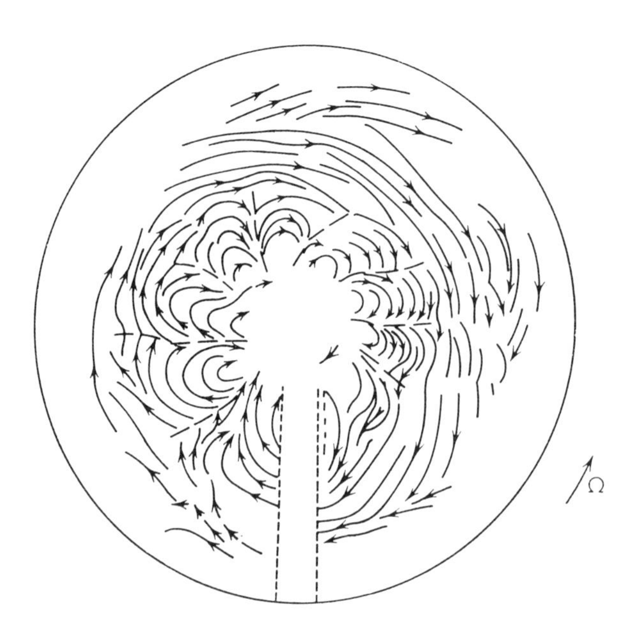

Fig. 5.12. A streak photograph of a slow easterly current relative to the 90° half-depth zonal barrier. The hemisphere is seen looking straight down the rotation axis [63].

Similar results are found when the body does not fill the entire gap between container walls.

Fultz and Frenzen repeated these experiments with a half-depth meridional barrier which extended from the pole to the equator. Results for westerly flow are qualitatively similar to those obtained with a cylindrical obstacle. However, an easterly current veers sharply northward after going over the barrier and then appears to cusp. The pattern repeats all the way about the sphere, fig. 5.12. Whether or not the cusps have any relationship to trapped Rossby waves is not known.

It is an extremely difficult and complex task to obtain a reasonably detailed solution to any of these problems. Wave motion, non-linear interactions, inertial boundary layers along the ridge and elsewhere, a fairly complicated geometry—in fact, the whole gamut of rotational phenomena and complications are involved. It should not be surprising then to find that progress on the theoretical front has been somewhat retarded. A start in the development of a theory was made by Long [114] who considered certain implications of vorticity conservation in a steady flow field.

Let the equations of motion be written with respect to the *co-ordinate system fixed in the obstacle* which rotates with frequency Ω_0. The rotation rate of the annulus is then $\Omega = \Omega_0(1+\varepsilon)$ and positive ε signifies a westerly wind relative to the body. The radius of the inner hemisphere, r_0, is taken as the characteristic length so that the scaling rule is $[\![r_0, \Omega_0^{-1}, \varepsilon\Omega_0 r_0]\!]$. For reasons of simplicity, and because this is only an exploratory study, the model equations of § 5.3 are used as the basis of discussion. For a constant gap-width H, the equations are

$$\left.\begin{aligned} \nabla\cdot\mathfrak{Q} &= 0, \\ \mathfrak{Q}\cdot\nabla(\varepsilon\mathfrak{B}+\mathfrak{f}) &= 0, \end{aligned}\right\} \tag{5.7.2}$$

where $\mathfrak{Q}$ is the tangential velocity vector averaged over H and ∇ is the surface gradient. With the introduction of a stream function,

$$\mathfrak{Q} = \hat{\mathfrak{r}}\times\nabla\psi, \tag{5.7.3}$$

the system can be integrated once to obtain the familiar conservation law

$$\varepsilon\nabla^2\psi+\mathfrak{f} = F(\psi). \tag{5.7.4}$$

There is *no* way at present to determine the arbitrary function $F(\psi)$ without analysing viscous effects. This has been a continuing source of difficulty and, as before in §§ 4.6 and 5.4, a 'reasonable' choice is made so that a particular class of solutions can be studied with maximum simplicity. An appeal to 'upstream' conditions is even less reliable than in the uniform motion problems of chapter 4 because the streamlines here form closed circuits. However, let

$$\psi = \psi_\infty + \psi,$$

where

$$\psi_\infty = -\cos\Theta$$

is the undisturbed stream function (when there is no obstacle). If it is assumed that *the flow is undisturbed along some meridian* where $\psi = 0$, then it follows that

$$F(\psi_\infty) = 2(1+\varepsilon)\cos\Theta, \tag{5.7.5}$$

or

$$F(y) = -2(1+\varepsilon)\,y.$$

The equation for the disturbance stream function is then

$$\frac{1}{\sin\Theta}\frac{\partial}{\partial\Theta}\sin\Theta\frac{\partial\psi}{\partial\Theta}+\frac{1}{\sin^2\Theta}\frac{\partial^2\psi}{\partial\theta^2}+2\frac{1+\varepsilon}{\varepsilon}\psi = 0. \tag{5.7.6}$$

Solutions of this equation are wave-like for a relative westerly flow, $\varepsilon > 0$, but are of a boundary layer character for $\varepsilon < 0$. The problem is entirely similar to that studied by Fofonoff [50] for oceanic circulations (see § 5.3).

The spherical harmonics $P^m_{n+1}(\cos\Theta)\,e^{im\theta}$ are solutions of (5.7.6) when ε is positive and

$$\frac{n(n+1)}{2} = \frac{1+\varepsilon}{\varepsilon}. \tag{5.7.7}$$

For the hemisphere, $\psi = 0$ is the condition at the equator so that $n-m = 2k+1$, in which case

$$\varepsilon = \frac{2}{(m+2k)(m+2k+3)}. \tag{5.7.8}$$

This relationship of the Rossby number to the longitudinal wave number of the inertial waves was verified experimentally by Fultz and Frenzen [63], fig. 5.13. Waves corresponding to $k = 0$ are the dominant ones excited.

The solution for an easterly wind, which must involve inertial boundary layers, has not been analysed successfully. This interesting problem should be tractable with the methods and information presently available. A second theoretical attack is warranted and there is reasonable prospect for success.

Pearson's numerical work on the spherical annulus problem was described briefly in § 3.5. There seems to be no reason why the same techniques would not work as well in the case of a thin spherical annulus. The behaviour of the fluid near the equator when the inner sphere rotates would be of considerable theoretical interest and might even shed some light on the dynamics of the Cromwell Equatorial Current, see Knauss [97, 98], Robinson [168].

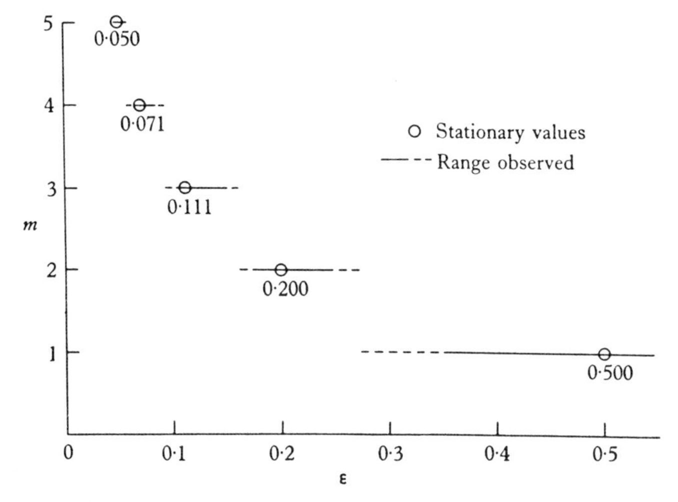

Fig. 5.13. Results of the experimental verification [63], of (5.7.8) which relates the Rossby number ε to the wave number m, for $k = 0$.

The Cromwell Current has stimulated much theoretical and experimental research. In the experiments of Baker and Robinson [2], the fluid within a closed basin that is a section of a spherical annulus is driven by rotating the bounding spheres at different speeds. The basin can occupy any position on the surface with respect to the rotation axis; in particular, it can be placed so that it straddles the equator. Several oceanic phenomena have been observed in this apparatus including the western boundary layer, a broad, geostrophic, mid-ocean circulation and an equatorial undercurrent.

Carrier [26] examined an equatorial flow controlled by viscous processes in an annulus of very small width, H. His analysis shows that viscous processes alone tend to distort the velocity profile in a manner that appears to lead to a surface current and subsurface counter-current.

Much will yet be discovered about flows in the equatorial regions from continuing research on all three fronts, numerical, experimental and analytical. Accordingly, this section must conclude in a rather incomplete state, its purpose served if the open problems and the current research trends have been at least partially illuminated.

CHAPTER 6

STABILITY

6.1. Introduction

A steady viscous laminar fluid motion is stable with respect to infinitesimal disturbances if all such disturbances ultimately decay to zero leaving the basic flow unchanged. Conversely, a given motion is unstable if the effects of any small disturbance lead to the development of either another laminar flow or a state of turbulence. (A more general concept of stability, one that applies to time-dependent motions or finite amplitude disturbances, will not be required.)

This chapter is limited, for the most part, to a discussion of the infinitesimal stability of fluid motions examined in previous sections. Accordingly, much peripheral material on the stability of rotating flows has either been mentioned briefly when necessary or omitted entirely. This is also true of those topics which, though relevant, receive extensive coverage elsewhere. The last category includes the stability of rotating fluids in the following circumstances: between concentric cylinders; heated from below; in a magnetic field; constituting a self-gravitating medium. Most of these topics may be found in the treatise of Chandrasekhar [30], (see also Lamb [103] and Lyttleton [121]). The works of Lin [112], Stuart [197] and Drazin and Howard [39], are cited as general references for the subject of hydrodynamic stability.

Though the scope of this investigation is severely restricted, there remains a wide variety of interesting and important instabilities that occur in quite ordinary circumstances. Some of these have been studied at great length and are thoroughly understood. With others, the very fact that the underlying mechanisms are not as evident, motivates their presentation here as part of a survey of common instabilities.

6.2. Rayleigh's criterion

A very simple type of inviscid rotational instability results when the fundamental balance of the centrifugal force and the radial pressure gradient is upset in an axisymmetric motion. This was first discussed by Rayleigh [160] but the best physical description, given next, is due to Kármán [94].

Suppose that an axisymmetric disturbance appears in an inviscid region of a rotating fluid where the principal balance of forces is that just mentioned. Suppose further that fluid elements at radii r_o and r_1, $r_o > r_1$, are caused to interchange by the disturbance. Then the flow is inherently stable if the force field tends to restore the particles to their original positions and it is unstable otherwise.

Since angular momentum is conserved, the particle initially at r_1 with absolute velocity v_1 attains the velocity $(r_1/r_o)\, v_1$ in moving out to radius r_o. The centrifugal force on this element, $(r_i^2 v_i^2)/r_o^3$, opposes the inward force of the equilibrium pressure field, whose magnitude is v_o^2/r_o. Thus, if

$$\frac{v_o^2}{r_o} > \frac{r_i^2 v_i^2}{r_o^3},$$

the net force on the particle is directed radially inwards and acts to restore the original undisturbed motion. On the other hand, the basic flow is unstable if

$$(r_o v_o)^2 < (r_1 v_1)^2, \tag{6.2.1}$$

in which case slight deviations are accentuated by the force imbalance. Hence, Rayleigh's criterion, (6.2.1), asserts that a steady laminar motion is unstable when the *square* of the absolute circulation about the rotation axis decreases with increasing cylindrical radius. The mathematical derivation of this condition was provided by Synge [198] and is discussed in [112].

Rayleigh's criterion is of central importance in the stability theory for fluid motion between concentric cylinders which revolve at different rates. (Although viscosity is necessary to establish the primary flow, it occupies a subsidiary position in stability considerations for a wide parametric range.) This problem was first studied by Taylor [204] from both an experimental and a theoretical standpoint. A great many papers have appeared since (see [30] and [39] for a partial list of references); the linear stability theory is fairly complete. The definitive experiments of Coles [35] seem to end that particular avenue of research while providing a continuing stimulus for theoretical efforts on the non-linear aspects of the problem.

Only a few of the results of these investigations will be described, and briefly at that. The present state of knowledge is summarized in fig. 6.1, which is taken from the work of Coles. The data is arranged according to the values of the Ekman numbers defined by

$$E_o = \frac{\nu}{\Omega_o r_o^2}, \quad E_1 = \frac{\nu}{\Omega_1 r_1^2},$$

where Ω_o and Ω_1 are the rotation rates of the two cylinders of radii r_o and r_1, $r_o > r_1$. According to Rayleigh's criterion, instability should occur when

$$E_o > E_1$$

(to the *left* of the dashed line in the figure), but viscous processes tend to stabilize the flow thereby modifying this condition, especially

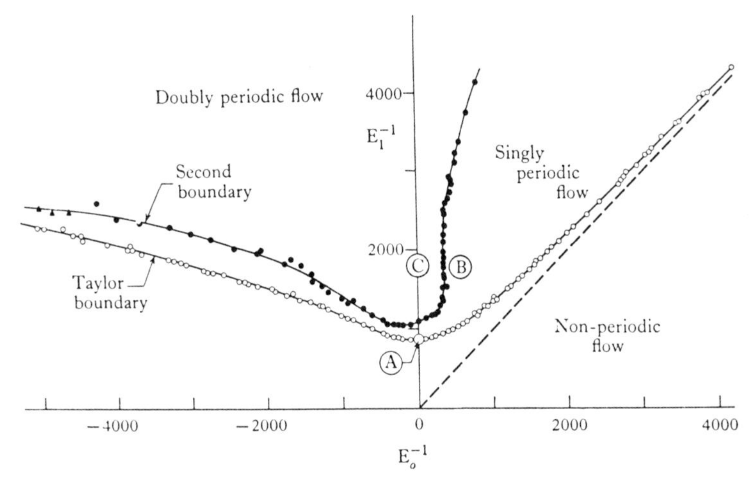

Fig. 6.1. Different regimes in circular Couette flow from visual observations (rough cylinders, $r_o/r_1 = 1{\cdot}135$; $\nu = 0{\cdot}11$ cm²/sec), [35]. Ekman numbers E_o, E_1 are defined in the text. Circled letters refer to photographs in fig. 6.2.

at large positive values of E_o. Actually, instability is first observed along the Taylor boundary and has the form of toroidal vortex cells which are regularly spaced along the longitudinal axis. The structure of this periodic wave shows no variation in the azimuthal direction. Experimental and theoretical agreement on the stability border line is exact for all practical purposes. A second boundary line in the unstable (and definitely non-linear) regime demarcates waves that are doubly periodic in the circumferential direction from those which are singly or non-periodic. Higher modes can occur and for E_1^{-1} large enough there is a transition to turbulence. Photographs of the flow at conditions corresponding to positions marked Ⓐ, Ⓑ and Ⓒ in the diagram are shown in fig. 6.2. As the Ekman number of the inner cylinder is decreased, the motion passes from complete stability through a number of stages of instability, corresponding to the excitation of higher order harmonics, and finally to a turbulent state. Energy feeds continually from the low to high frequencies by complicated non-linear interactions, a process typical of transition.

Most of these phenomena are discernible in spin-down experiments if the Rossby number is larger than a small but critical value. A simple qualitative demonstration to illustrate these effects and others of importance, can be arranged from the equipment described in § 1.1. Specifically, the fluid, rotating rigidly within a right circular cylinder, is illuminated by an open lamp, so that the suspended aluminium particles are easily observed on a broad surface section of the vertical wall. The rotation rate of the container is impulsively decreased and within a few revolutions, regular Taylor vortices appear as shown in fig. 6.3. These develop into a more complicated pattern that lacks axial symmetry and ultimately decays as spin-down is achieved.

The relationship of these wave forms to the instabilities that occur in the concentric cylinder configuration seems clear although a detailed and explicit identification has not been made. Fluid in the sidewall boundary layers is, in effect, contained between the rigid outer casing and an inner 'cylinder' consisting of the inviscid core which rotates almost steadily at the original frequency for a considerable time span. The width of this 'annulus' increases during the transient evolution and this marks a corresponding variation in the relative difference of the two Ekman numbers. However, the changing character of the instability with time, as indicated in fig. 6.3, appears due to other causes.

The development of an unstable and possibly turbulent flow in the boundary layer can be a means of exciting the inviscid inertial modes. If the frequencies of the unstable modes are less than 2Ω, the disturbance can be propagated away from the container walls. Fig. 6.4 corresponds to fig. 6.3 (*c*) but is made with slit-beam lighting to illuminate the interior. The unstable zone near the lateral wall is clearly visible and so are characteristic cones which emanate from this disturbance to permeate the inviscid core. These characteristics

(*a*) (*b*)

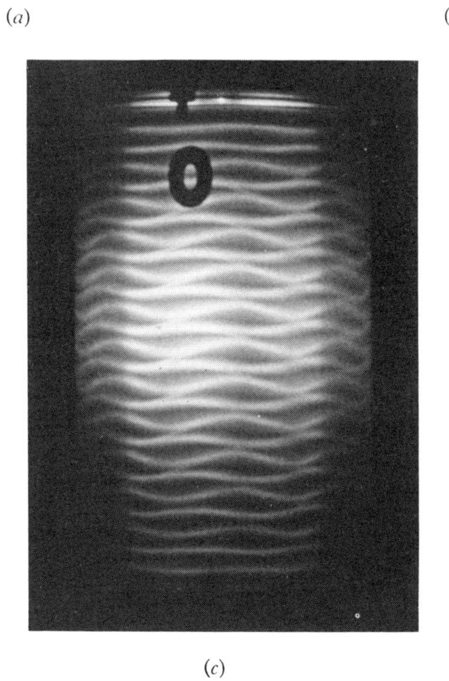

(*c*)

Fig. 6.2. Photographs of the flow, between cylinders, [35], at conditions marked Ⓐ, Ⓑ, Ⓒ in fig. 6.1.

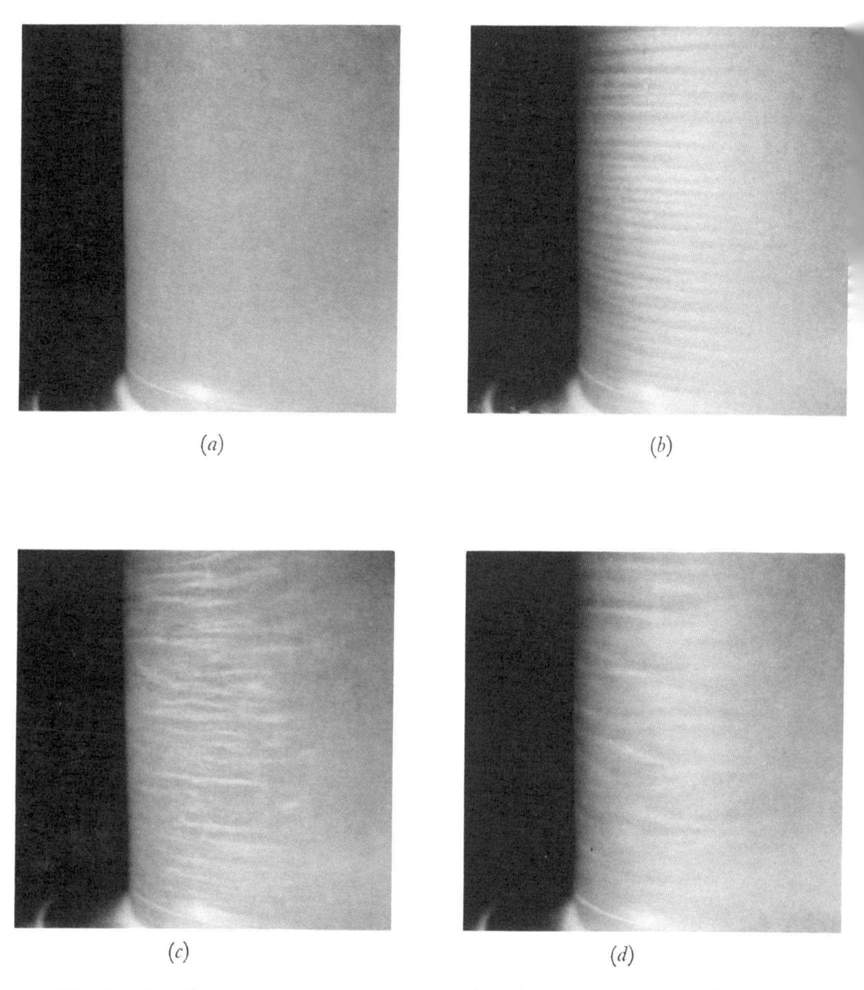

Fig. 6.3. Instabilities observed on the vertical surface of a cylinder during spin-down. Plate (*a*) corresponds to the initial state of rigid rotation; plates (*b*), (*c*), (*d*) show the transient evolution.

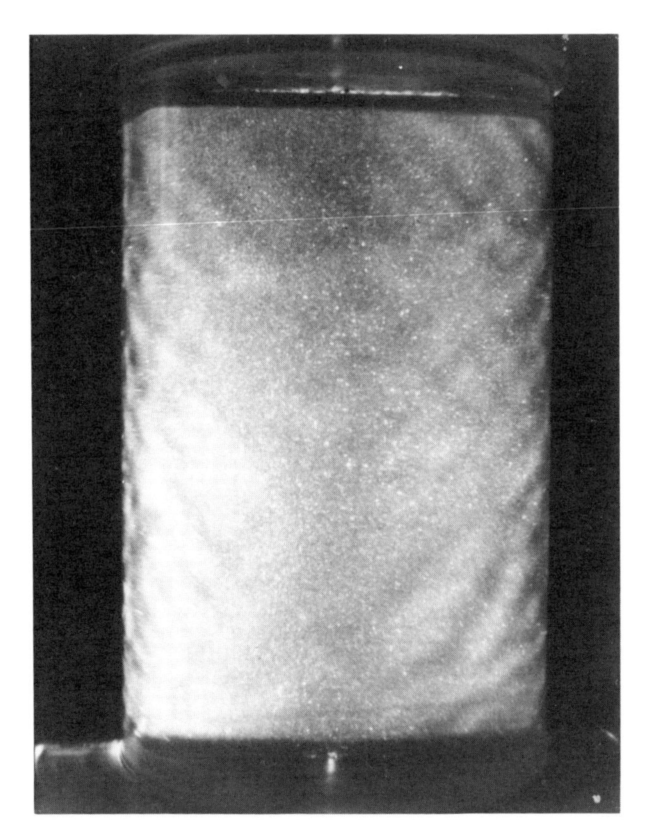

Fig. 6.4. The disturbance at the wall, created by impulsive spin-down, propagates throughout the fluid medium. The characteristic cones observed in the interior are apparently produced by an excitation at a single definite frequency. This photo corresponds to plate (*c*) of fig. 6.3.

seem to correlate with a single excitation frequency, most probably the dominant unstable mode in the boundary layer. This speculation requires confirmation but the main point has been made and should be emphasized. With the right conditions, the mechanism of wave propagation enables a locally unstable or turbulent flow to affect the entire contained fluid. This also seems to happen in connection with the unstable Ekman layer, the next order of business.

6.3. Stability of the Ekman layer: experiments

The motion in a cylindrical annulus, whose inner and outer vertical walls are, respectively, a sink and source of fluid, was analysed in § 2.19. The mass transport was found to take place entirely within the boundary layers. The relative interior flow is a geostrophic potential vortex, the strength of which is related to the net efflux according to (2.19.39). Furthermore, the horizontal Ekman boundary layer on $z = 0$, $(\hat{\mathbf{n}} = -\hat{\mathbf{k}})$, is divergence free and its structure is given by (2.6.12):

$$\tilde{\mathbf{q}} = [\hat{\mathbf{k}} \times \mathbf{q}_I \sin(E^{-\frac{1}{2}}z) - \mathbf{q}_I \cos(E^{-\frac{1}{2}}z)]\, e^{-E^{-\frac{1}{2}}z}. \qquad (6.3.1)$$

The stability of this Ekman layer, established in the manner described, was studied experimentally by Faller [46], Faller and Kaylor [47], and Tatro and Mollö-Christensen [199]. The investigations of the former authors involved a water tank with a free surface and measurements were made by visual observation of the motion after the insertion of permanganate dye crystals. The latter utilized air as the working medium and a completely closed cylindrical annulus; very accurate measurements were made with assemblies of hot wire anemometers that traversed the flow.

The presentation here is based primarily on the experiments of Tatro and Mollö-Christensen for two reasons. First, their apparatus is the exact counterpart of the theoretical model discussed earlier. (The physical dimensions of the annulus are as follows: height—3 inches, outer radius—36 inches, inner radius—3 inches.) Secondly, their measurements are undoubtedly the most accurate to date.

The basic experimental procedure of Tatro and Mollö-Christensen is to scan the flow with a hot wire anemometer whose signal is monitored on a recorder. The first objective is to examine the stable laminar flow established and to compare the results with theory. A slight discrepancy is noted in as much as the geostrophic flow is always somewhat weaker than predicted, and this is a consequence of non-linearity, a finite Rossby number. Since the structure of the boundary layer depends critically on the *local* external flow field, the stability problem is characterized in terms of a Rossby number and a Reynolds number based on the local measured values:

$$\varepsilon = \frac{V_I}{\Omega r}, \quad R_E = \frac{\delta V_I}{\nu}. \tag{6.3.2}$$

Here, V_I is the *relative* azimuthal component of the interior velocity; δ is the local value of the boundary layer thickness which is computed by measuring the vertical co-ordinate z of the maximum radial velocity and using the formula

$$\delta = \frac{4z}{\pi}.$$

Component V_I can be related to the mass flux at the outer wall and the linearized form of this relationship is (2.19.39). However, this is not really necessary, even though, in fact, the different flows are established by varying the source strength, $\mathscr{S}$.

In the actual experiments, $\mathscr{S}$ is increased by increments and all other conditions are held fixed. Records of the voltage are made at several points in the layer for each value of the flux and one of these is shown in fig. 6.5 (*a*). At relatively small values of $\mathscr{S}$, the recording line is essentially straight but at some point in the increase, an oscillation appears. The lowest value of the flux at which the oscillation is observed is taken as the critical point. The quantities V_I and δ are measured and the critical values of ε and R_E determined.

The results of these experiments, fig. 6.5 (*a*, *b*) for example, show conclusively that there are two different types of instability associated with this boundary layer. These are designated as Class A and Class B instabilities according to the order in which they appear.[1] The waves of both families form a series of horizontal roll vortices whose spacing is related to the depth of the boundary layer. Class A waves, as observed by Faller and Kaylor [47], are shown in

[1] The designation in the literature is based on the order in which the instabilities were discovered. Thus, Faller's type I is Class B and his type II is Class A.

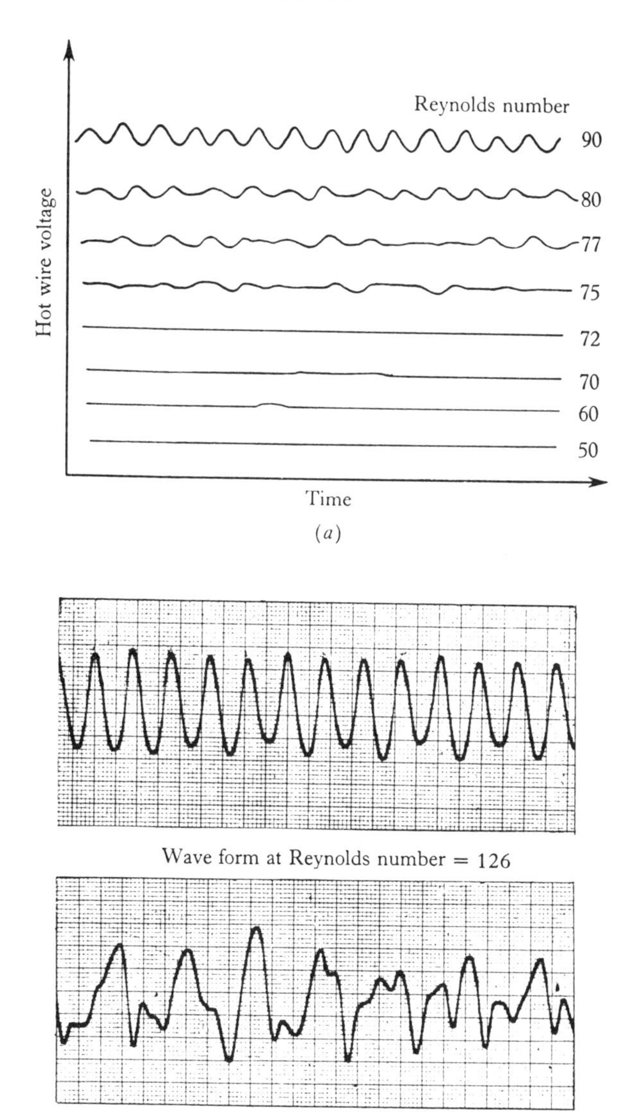

Fig. 6.5. (*a*) A record of hot wire voltage by Tatro and Mollö-Christensen[199] which shows the onset of instability and Class A waves. (*b*) The wave form undergoes a second change at $R_E \cong 126$ which corresponds to the development of Class B modes.

fig. 6.6. The local orientation of these waves is anywhere from 0° to −8° with respect to the geostrophic flow; the wave lengths vary between 25δ and 33δ and the phase speed is approximately $0{\cdot}16V_I$, directed radially inward. Fig. 6.7 shows the critical Reynolds number plotted versus the Rossby number; the relationship is approximately

$$R_E^{(A)} = 56{\cdot}3 + 58{\cdot}4\varepsilon^{(A)}. \tag{6.3.3}$$

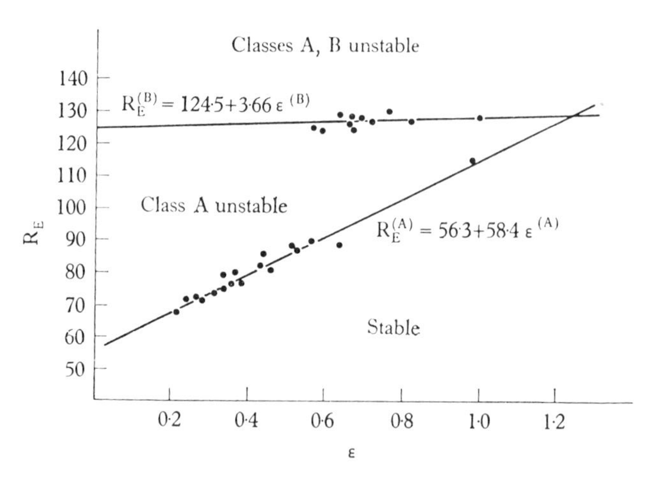

Fig. 6.7. The critical Reynolds number vs. Rossby number for Class A and Class B instabilities [199].

Waves of this family develop first and are very sensitive to the value of ε. As ε increases, the disturbance ceases to be confined to the boundary layer and the effects propagate throughout the interior, fig. 6.8, much like the process pictured in fig. 6.4 of the last section. Since the primary frequency of the disturbance is greater than 2Ω, a non-linear wave interaction within the boundary layer resulting in a lower frequency wave may be responsible for the interior excitation.

Class B waves form an angle of almost exactly 14·6° with respect to the geostrophic flow. The wavelength is 11·8δ and the phase velocity is $0{\cdot}034V_I$ directed radially inward. There is only a slight dependence of critical Reynolds number on the Rossby number given by

$$R_E^{(B)} = 124{\cdot}5 + 3{\cdot}66\varepsilon^{(B)}. \tag{6.3.4}$$

This family of waves is shown in fig. 6.9 (a), and both sets appear in

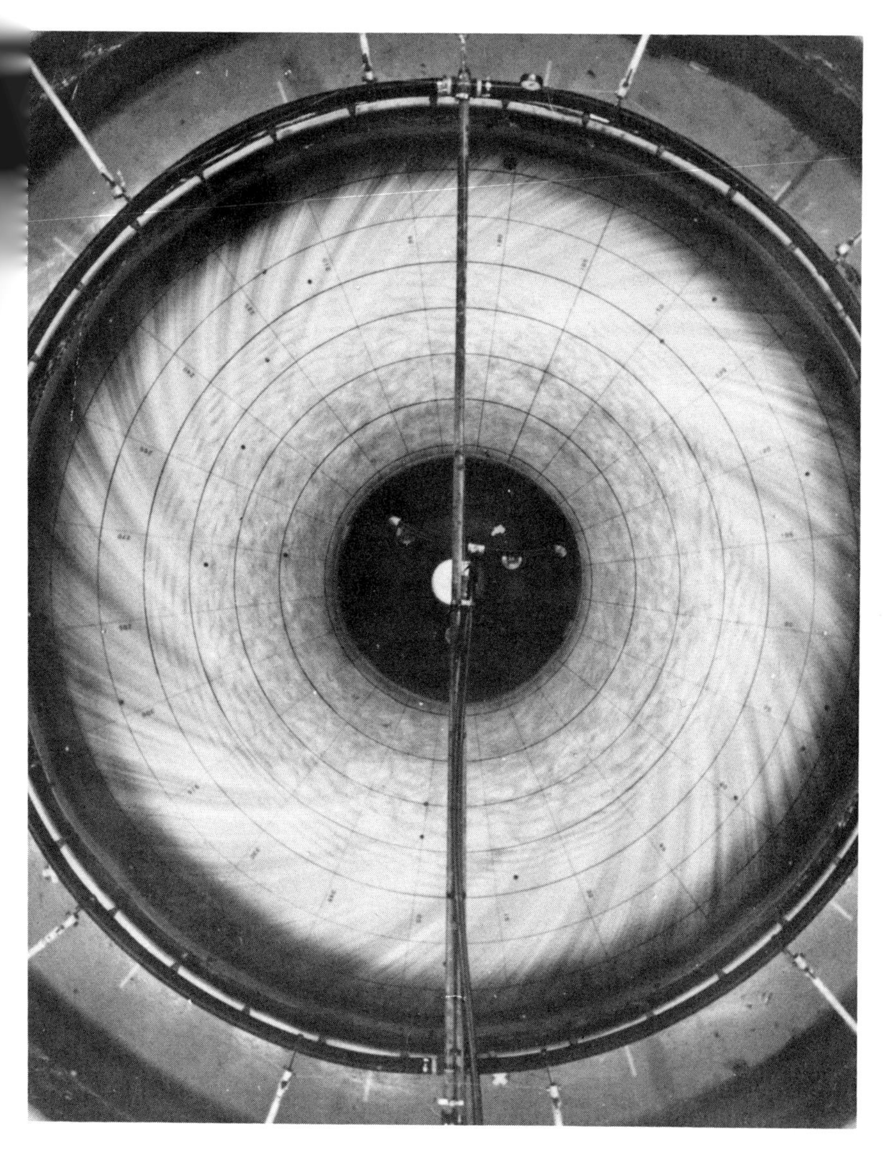

Fig. 6.6. A photograph by Faller and Kaylor [47] showing Class A waves and turbulence at smaller radii.

fig. 6.9(*b*). The spatial relationships are summarized in fig. 6.10, where the different types of waves are displayed schematically.

The analysis of turbulent transition is still incomplete. Preliminary observations indicate that turbulent bursts do appear and that transition may depend on the relative development of the different kinds of waves, [47].

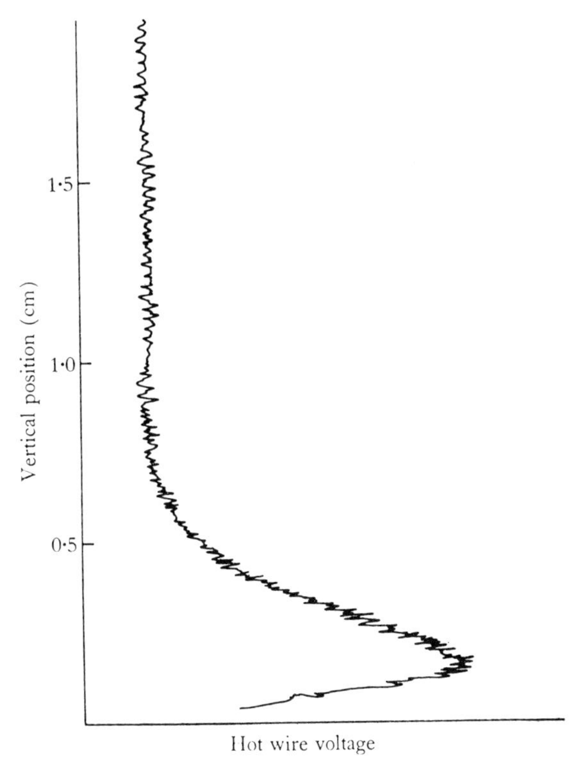

Fig. 6.8. Record of hot wire voltage with height showing the penetration or propagation of the disturbance into the interior [199].

The fact that the instabilities depend only on the local external flow field implies that they are a general feature of all rotational boundary layers and not just a singular manifestation in a particular configuration. This is confirmed by the investigations of the Kármán boundary layer on a disk that rotates in stationary air, Theodorsen and Regier [206], Smith [180], Gregory, Stuart and Walker [77].

Indeed, the last cited reference was really the first thorough experimental and theoretical examination of Class B waves.

It is relatively easy to demonstrate that the transient Ekman layer can go unstable in either spin-up or spin-down. Permanganate crystals dropped about the periphery of a uniformly rotating cylindrical container produce a thin annular layer of dyed fluid at the bottom plate, fig. 6.15 (*a*). During spin-down, this coloured

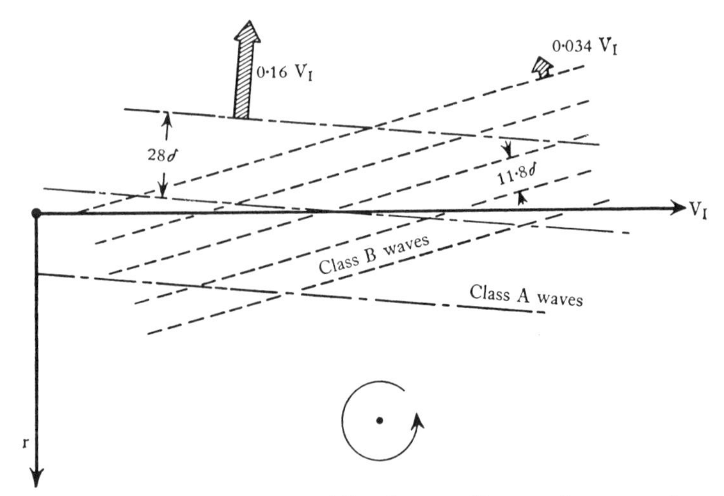

Fig. 6.10. Diagram of the two instability classes to illustrate the measured wave lengths, phase velocities and wave front orientations.

fluid is drawn radially inward by the efflux from the Ekman layer and within a few revolutions, a series of rolls develop, plate (*b*). (The classification of these waves, A or B, is uncertain.) The waves amplify considerably and by the time spin-down is achieved, the dyed fluid occupies an almost perfectly circular lens, of modest thickness, at the bottom of the tank. This 'disk' is separated from the outer wall by a ring of clear fluid drawn down from the vertical surface of the cylinder. (The remaining plates of fig. 6.15 concern a spin-down experiment with a stratified fluid to be discussed shortly.) Spin-up can exhibit the same type of instability but the techniques of visualization must be altered slightly.

If the Reynolds number defined in (6.3.2) is written in terms of the Ekman number, a gross but general criterion for the onset of

(*a*)

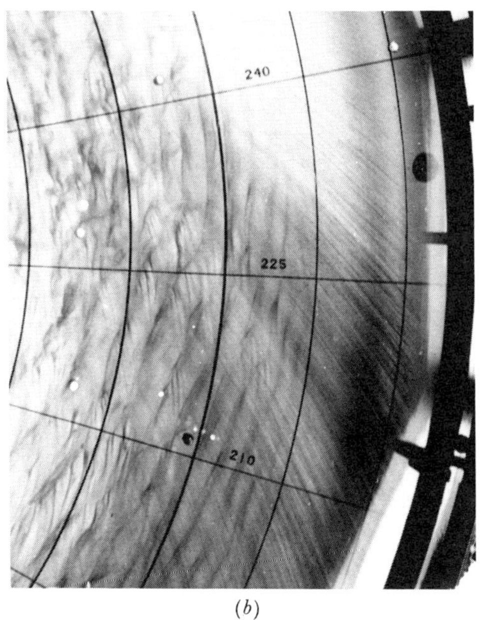

(*b*)

Fig. 6.9. (*a*) Class B waves observed by Faller and Kaylor [47]. (*b*) Class A waves with intermittent Class B bands.

instability can be derived. Let $\varepsilon = (V_I/\Omega L)$ and $\delta \doteq (\nu/\Omega)^{\frac{1}{2}}$, then

$$R_E \doteq \varepsilon E^{-\frac{1}{2}},$$

where $E = \nu/\Omega L^2$ as usual. According to (6.3.3) boundary layer instability may first be expected when

$$\varepsilon E^{-\frac{1}{2}} \doteq 56. \tag{6.3.5}$$

For $\varepsilon = 0{\cdot}1$ and 1, $E \doteq 4{\cdot}10^{-6}, 4{\cdot}10^{-4}$ and these estimates show that rotational flows remain laminar for exceedingly small values of the Ekman number. The extensive use of laminar boundary layer theory is thereby vindicated.

6.4. Stability of the Ekman layer: theory

We turn now to a theoretical discussion of the instabilities described in the last section. Since both the length and time scales of the phenomena in question are closely related to the thickness of the boundary layer, it is advantageous to make the equations of motion dimensionless using the scaling rule

$$\left[\!\left[\left(\frac{\nu}{\Omega}\right)^{\frac{1}{2}}, \left(\frac{\nu}{\Omega}\right)^{\frac{1}{2}}(\varepsilon\Omega L)^{-1}, \varepsilon\Omega L\right]\!\right].$$

The Reynolds number, $R_E = \varepsilon E^{-\frac{1}{2}}$, is then the only parameter that appears explicitly in the fundamental equations:

$$\left.\begin{aligned} \nabla\cdot\mathbf{q} &= 0, \\ R_E\left(\frac{\partial}{\partial t}\mathbf{q}+\mathbf{q}\cdot\nabla\mathbf{q}\right)+2\hat{\mathbf{k}}\times\mathbf{q} &= -\nabla p+\nabla^2\mathbf{q}. \end{aligned}\right\} \tag{6.4.1}$$

The stability of a steady laminar motion, $\mathbf{q}_\ell(\mathbf{r})$, which satisfies (6.4.1) and all prescribed boundary conditions, can be determined by considering the growth characteristics of any small disturbance, $\epsilon\mathbf{q}'(\mathbf{r}, t)$, superposed on the mean flow. If the velocity and pressure functions are written as

$$\begin{aligned} \mathbf{q} &= \mathbf{q}_\ell(\mathbf{r})+\epsilon\mathbf{q}'(\mathbf{r}, t), \\ p &= p_\ell(\mathbf{r})+\epsilon p'(\mathbf{r}, t), \end{aligned}$$

then the substitution of these expressions in (6.4.1) and the linearization of the equations with respect to the small amplitude parameter ϵ results in

$$\left.\begin{aligned} \nabla\cdot\mathbf{q}' &= 0, \\ R_E\left(\frac{\partial}{\partial t}\mathbf{q}'+\mathbf{q}_\ell\cdot\nabla\mathbf{q}'+\mathbf{q}'\cdot\nabla\mathbf{q}_\ell\right)+2\hat{\mathbf{k}}\times\mathbf{q}' &= -\nabla p'+\nabla^2\mathbf{q}'. \end{aligned}\right\} \tag{6.4.2}$$

The disturbance velocity satisfies homogeneous boundary conditions which, together with the preceding equations, constitute the linear stability theory. The object now is to determine whether this homogeneous system has natural modes which amplify with time. If this proves to be the case, then the basic flow is unstable because a small disturbance in background noise would receive energy from the mean motion sufficient to grow beyond the bounds of validity of a linear theory. A new, stable, laminar motion could result from the interaction, or a transition to turbulence might take place. (To prove instability, only one unstable mode need be found. A motion is stable if it can be shown that every possible type of small disturbance ultimately decays to zero.)

The manner in which energy is transferred to the disturbance from the steady flow is illuminated by establishing the following relationship from the invariant form of (6.4.2):

$$\tfrac{1}{2}R_E \frac{\partial}{\partial t}\int \mathbf{q}'\cdot\mathbf{q}'\,dV + R_E\int \mathbf{q}_\ell\cdot\mathbf{q}'\times\nabla\times\mathbf{q}'\,dV + \int(\nabla\times\mathbf{q}')\cdot(\nabla\times\mathbf{q}')\,dV = 0. \qquad (6.4.3)$$

The first term represents the time rate of change of the total kinetic energy in the perturbation; the last is the rate at which disturbance energy is dissipated by viscous processes. The middle expression is the rate of conversion of energy from the basic flow into the perturbed motion through the action of the Reynolds stress. Were it not for the last mentioned process, there could be no source of energy for secondary motions and all disturbances would eventually attenuate. However, the Reynolds stress presents a means of counteracting dissipation if R_E is large enough (larger than some critical value) and the velocity components are phased properly to make the sign of the stress integral positive.

To examine the stability of the Ekman layer, the extent of the fluid domain may be supposed infinite, bounded by a single plane wall. The experimental results indicate that the observed instabilities are dependent mainly on *local* flow conditions and this is incorporated into the analysis as a simplifying approximation. The local, interior velocity field is assumed to be constant and uni-directional, just $\hat{\boldsymbol{\theta}}$ in scaled units. The structure of the boundary layer instability consists of two-dimensional vortex rolls and it is advantageous to align the horizontal co-ordinates so that x measures distance along the bands and y is in the normal direction. The local x axis is obtained from the direction of the interior velocity $\hat{\boldsymbol{\theta}}$ (see fig. 6.10) by a rotation through a positive or counter-clockwise angle α:

$$\hat{\boldsymbol{\theta}} = \cos\alpha\,\hat{\boldsymbol{\imath}} - \sin\alpha\,\hat{\boldsymbol{\jmath}}.$$

With respect to cylindrical co-ordinates, the local x co-ordinate is directed circumferentially for small α whereas y is a radial co-ordinate (which increases as r decreases). The complete velocity profile in the laminar boundary layer is then

$$\begin{aligned}\mathbf{q}_\ell = \mathrm{u}_\ell\hat{\boldsymbol{\imath}} + \mathrm{v}_\ell\hat{\boldsymbol{\jmath}} &= [\cos\alpha - \mathrm{e}^{-\mathrm{z}}\cos(\alpha+\mathrm{z})]\,\hat{\boldsymbol{\imath}} \\ &\quad - [\sin\alpha - \mathrm{e}^{-\mathrm{z}}\sin(\alpha+\mathrm{z})]\,\hat{\boldsymbol{\jmath}} \qquad (6.4.4)\end{aligned}$$

and it is the instability of this motion that is of greatest interest. (Note that $\mathbf{q}_\ell = \tilde{\mathbf{q}} + \mathbf{q}_\mathrm{I}$ in (6.3.1).)

The nature of the observed disturbances motivates a search for two-dimensional waves that are independent of the x co-ordinate. In this case, the continuity equation permits the introduction of a stream function defined by

$$\mathrm{v}' = \frac{\partial\psi'}{\partial \mathrm{z}}, \quad \mathrm{w}' = -\frac{\partial\psi'}{\partial y}. \qquad (6.4.5)$$

Furthermore, a disturbance is assumed of the form

$$\left.\begin{aligned}\mathrm{u}'(y,\mathrm{z},\mathrm{t}) &= \mathrm{U}(\mathrm{z})\exp\mathrm{i}\kappa(y-\mathrm{ct}),\\ \psi'(y,\mathrm{z},\mathrm{t}) &= \Psi(\mathrm{z})\exp\mathrm{i}\kappa(y-\mathrm{ct}),\end{aligned}\right\} \qquad (6.4.6)$$

and the replacement of these expressions in the x components of the momentum and vorticity equations leads to the following ordinary differential equations:

$$\frac{\mathrm{d}^2\mathrm{U}}{\mathrm{dz}^2} - \kappa^2\mathrm{U} + 2\frac{\mathrm{d}\Psi}{\mathrm{dz}} - \mathrm{i}\kappa\mathrm{R_E}\left[(\mathrm{v}_\ell - \mathrm{c})\,\mathrm{U} - \Psi\frac{\mathrm{du}_\ell}{\mathrm{dz}}\right] = 0, \quad (6.4.7)$$

$$\left(\frac{\mathrm{d}^2}{\mathrm{dz}^2} - \kappa^2\right)^2\Psi - \mathrm{i}\kappa\mathrm{R_E}\left[(\mathrm{v}_\ell - \mathrm{c})\left(\frac{\mathrm{d}^2}{\mathrm{dz}^2} - \kappa^2\right)\Psi - \Psi\frac{\mathrm{d}^2\mathrm{v}_\ell}{\mathrm{dz}^2}\right] - 2\frac{\mathrm{dU}}{\mathrm{dz}} = 0. \qquad (6.4.8)$$

The boundary conditions at the plate are simply

$$U = 0, \quad \Psi = \frac{d\Psi}{dz} = 0; \tag{6.4.9}$$

those at infinity, the outer edge of the boundary layer, require the perturbation velocity components to decay to zero. Lilly[110] applies the outer constraints of zero stress

$$\frac{\partial u'}{\partial z} = \frac{\partial v'}{\partial z} = 0,$$

and zero vertical velocity, $w' = 0$. These are consistent with the previous conditions and are perhaps especially appropriate for a symmetrical internal state as in the flow between concentric disks.

Equations (6.4.7) and (6.4.8) with boundary conditions constitute an eigenvalue problem in which the phase speed c is the characteristic value, Ψ and U are the eigenfunctions and α, κ, R_E are specified parameters. For each natural mode, c is a single-valued complex function of the basic parameters. Should the imaginary part of c be positive for one or more of these modes, as expected, then a positive amplification is implied and the wave is unstable. The neutral stability surface,

$$\mathcal{Im}\, c(\alpha, \kappa, R_E) = 0 \tag{6.4.10}$$

forms the boundary between stable and unstable conditions and the critical Reynolds number at which instability is first encountered is the smallest absolute value of R_E on this surface. The determination of this number, and the associated values of α and κ is the prime objective of the stability analysis.

The occurrence of instabilities at rather large Reynolds number motivates an asymptotic analysis of (6.4.7) and (6.4.8). The lowest order theory, corresponding to $R_E = \infty$, consists of

$$U = \frac{\Psi}{v_\ell - c}\frac{du_\ell}{dz}, \tag{6.4.11}$$

$$(v_\ell - c)\left(\frac{d^2}{dz^2} - \kappa^2\right)\Psi - \Psi\frac{d^2 v_\ell}{dz^2} = 0. \tag{6.4.12}$$

The equations uncouple because rotational effects are completely suppressed in this limit; the formulation reduces to a classical inviscid stability problem whose long history is detailed in [112].

A necessary condition for inviscid instability was derived by Rayleigh [158]. If (6.4.12) and its complex conjugate are multiplied by $\Psi^\dagger/(v_\ell - c)$ and $\Psi/(v_\ell - c^\dagger)$ respectively, the two can be subtracted and integrated over the entire range of the vertical co-ordinate with the result

$$(\mathscr{I}m\, c) \int_{z_B}^{z_T} \frac{\Psi\Psi^\dagger}{(v_\ell - c)(v_\ell - c^\dagger)} \frac{d^2 v_\ell}{dz^2} dz = 0. \qquad (6.4.13)$$

(The boundary conditions allow the elimination of all other terms which arise from exact integration.) If $\mathscr{I}m\, c > 0$, then the integrand cannot be positive in the entire interval and this means that at some interior point

$$\frac{d^2 v_\ell}{dz^2} = 0. \qquad (6.4.14)$$

Hence, an unstable laminar flow *must* necessarily have an inflexion point. (This conclusion does not hold for stable flows, $\mathscr{I}m\, c < 0$, for reasons connected with a turning point analysis and the limit $R_E \to \infty$. The fluid motion is required to be the limit of a real viscous flow; but see Lin [112] for a full explanation.) Fjørtoft [49] extended this condition by showing that it is also necessary for $|(d/dz)\, v_\ell|$ to be a maximum at the point of inflexion.

The physical basis of inviscid instability was explained by Lin [111] (p. 58 in [112]) in terms of vortex filaments as follows. Suppose $(d/dz)\, v_\ell$ does not change sign—let it increase with increasing z—then a particular fluid element displaced downwards will have an excess of vorticity over that of the surrounding parallel shear flow. This is so because the motion is inviscid. Other fluid elements to the right of this displaced vortex in fig. 6.11 will be shifted upwards by it into regions of higher vorticity. Those to the left are displaced downwards so that the net effect of the redistribution is to induce an upward velocity on the original filament. (Analogous conclusions hold when the particular element has a relative vorticity deficit.) Thus, the filament is restored towards its equilibrium position and stability is implied. However, when $(d/dz)\, v_\ell$ has an extremum, the interchange of fluid elements across the extremum does not involve either an excess or defect of vorticity. The restorative action is then severely restricted, so much so that the motion is not necessarily stable, in which case the distortion may amplify. This is the meaning of the necessary condition for instability expressed in (6.4.14).

The Ekman layer velocity profiles shown in fig. 2.1 have numerous inflexion points each of which may give rise to instabilities. Note that for small α, v_ℓ is identified with the *radial* velocity component on a circular plate and it is the structure of this component which is the incipient cause of the unstable waves. The first inflexion point lying nearest the plate is the most prominent and Stuart's analysis (in [77]) of the inviscid instabilities generated in this region gives results in good agreement with the observations of Class B waves. The

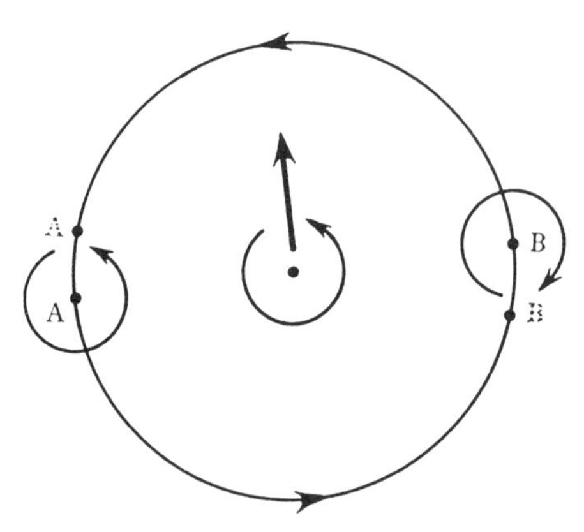

Fig. 6.11. A vortex filament, centre, which is displaced downwards in a uniform shear, acquires vorticity and induces the elements at A and B to move as shown. The modified vorticity distribution tends to restore the centre filament to its original position.

theoretical value of orientation angle is 13·5°, corresponding to the observed value of 14·6°, but the wave number of the most unstable mode is too large. Viscosity evidently has a minor influence in the determination of α, but seems to be quite essential in the selection of κ. The omission of the last term in (6.4.8) reduces it to the Orr–Sommerfeld equation and the numerical work of Lilly [110] confirms that Class B waves are governed mainly by this classical equation.

Class A waves are missed entirely by an analysis of inviscid instabilities because they arise from a more complicated interaction, an overturning that involves the Coriolis and shear forces. Stern [183] established the possibility of such an instability which draws its energy from the non-geostrophic component of the primary flow. Lilly, by means of a highly simplified analysis, explicitly exhibited unstable waves associated with the Coriolis force and designated them as 'parallel instabilities'. These vanish at high Reynolds numbers when the Coriolis and shear forces tend to uncouple. Another very simple theoretical demonstration of Class A instability, given by Stuart in a private communication, consists of setting $v_\ell = 0$ and $(d/dz)u_\ell =$ constant, in (6.4.7) and (6.4.8) and then determining the parametric conditions which sustain bounded oscillatory solutions in the infinite domain, $|z| < \infty$. If $\Psi = e^{-i\mu z}$ then it follows directly from (6.4.7) and (6.4.8) that

$$-ic\kappa = \frac{1}{R_E}\left[-(\mu^2+\kappa^2) \pm \left(\frac{2\mu\kappa R_E \dfrac{du_\ell}{dz} - 4\mu^2}{\mu^2+\kappa^2}\right)^{\frac{1}{2}}\right];$$

a positive growth rate is indicated for

$$R_E > \frac{4\mu^2 + (\mu^2+\kappa^2)^3}{2\mu\kappa \dfrac{du_\ell}{dz}}.$$

This model quickly illustrates that there is a close connection between the circumferential velocity component and Class A waves, and reveals that the growth factor decreases to zero as R_E increases.

The numerical solutions of the complete boundary value problem by Faller and Kaylor [47] and Lilly [110] confirm the presence of two distinct classes of unstable waves. The location of the Class B vortices coincides with the first inflexion point of the radial velocity profile; Class A disturbances arise mainly from the constant shear of the azimuthal velocity component and are centred much nearer to the plate. These and other results are in good quantitative agreement with observations.

Lilly made extensive calculations of the complete system consisting of (6.4.7) and (6.4.8), the boundary conditions (6.4.9), and

$$\Psi = \frac{d^2\Psi}{dz^2} = \frac{dU}{dz} = 0$$

at $z = \infty$. The results of these computations are exhibited in fig. 6.12. The first diagram, a plane cut of the surfaces of constant $\mathcal{I}m\, c(\alpha, \kappa, R_E)$ at $R_E = 65$, shows the regions of stability and instability (shaded) and the point of maximum growth rate. These particular values correspond to Class A disturbances. The slices at $R_E = 110, 150, 500$ show a continual shift of the unstable zone to angles of positive orientation and to small phase speeds, both symptomatic of the emergence of Class B modes. Waves of the second kind prevail completely at high Reynolds number. Fig. 6.13 illustrates the maximum growth rates with corresponding phase velocities of the most unstable solutions as a function of R_E. Class A

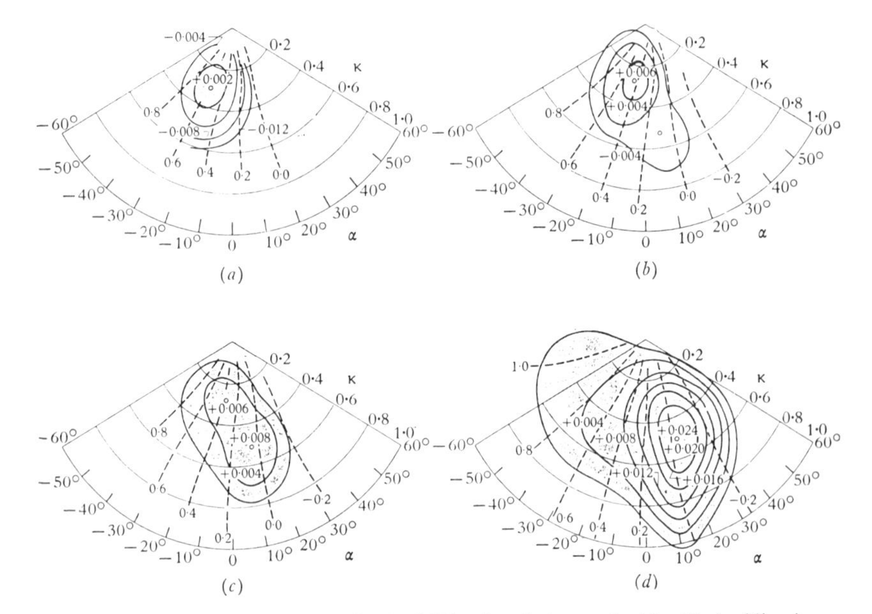

Fig. 6.12. The growth rates, $\kappa \mathscr{I}m\, c$ (solid lines) and phase velocities (dashed lines) of the most unstable modes as functions of wave number, κ, and orientation angle α: (*a*) $R_E = 65$; (*b*) $R_E = 110$; (*c*) $R_E = 150$; (*d*) $R_E = 500$. Shaded regions indicate instability and o locates the position of maximum growth rate. Results by Lilly[110].

modes appear at the critical value $R_E = 55$ and those of Class B develop at a later stage, $R_E = 110$. Obviously, the theory is in substantial agreement with experiment.

6.5. Vertical shear layers

The stability of the transient sidewall boundary layer was discussed in § 6.2 as an illustration of Rayleigh's criterion. There it was shown that the form of the observed disturbance is much the same as the regular horizontal rolls which develop between rotating concentric cylinders.

Free, vertical shear layers are also susceptible to instabilities, whose source may be the inflexion points in the local structure of the tangential velocity component. Hide [82] studied the stability of free shear layers surrounding an internal disk that revolves at a slightly different rate from that of the external casing. As the rate differential increases, the layer becomes unstable. When the frequency of the

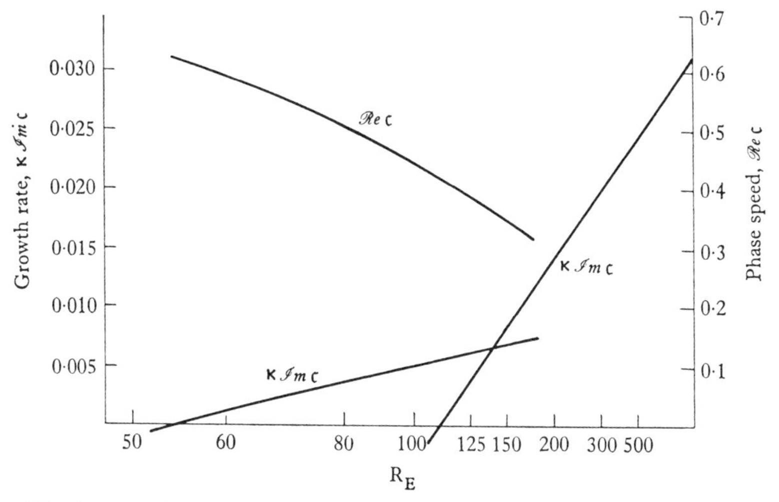

Fig. 6.13. Maximum growth rates and the corresponding phase velocities of the most unstable modes versus Reynolds number [110]. The two straight lines are associated with the two classes of Ekman layer instability.

disk is greater than that of the container, a series of symmetrically placed, vertical (two-dimensional) vortex rolls develop on the shear layer. Fig. 6.14, a photograph taken from the top of the tank looking down, shows four such waves. The layer is also unstable when the disk rotates more slowly than the container but the form of the disturbance is different, being more irregular and not as sharply defined.

This investigation indicates that instabilities arise when the Rossby number exceeds a definite critical value (which may depend on the location of the disk). Busse [23] has recently studied instabilities of this sort and the essential arguments in this, as yet, unpublished analysis can be restated in a manner that invites generalization.

If all the variables are made dimensionless according to the scaling rule $[\![L, \Omega^{-1}, \varepsilon\Omega L]\!]$, then the equations governing the perturbation field (see (6.4.2)) are

$$\left.\begin{aligned} \nabla\cdot\mathbf{q}' &= 0, \\ \frac{\partial}{\partial t}\mathbf{q}' + \varepsilon[\nabla(\mathbf{q}_\ell\cdot\mathbf{q}') + (\nabla\times\mathbf{q}_\ell)\times\mathbf{q}' + (\nabla\times\mathbf{q}')\times\mathbf{q}_\ell]& \\ +2\hat{\mathbf{k}}\times\mathbf{q}' &= -\nabla p' + E\nabla^2\mathbf{q}', \end{aligned}\right\} \quad (6.5.1)$$

with the boundary condition, $\mathbf{q}' = 0$ on Σ.

Consider first, motion within a right circular cylinder and let the basic laminar state be a purely zonal flow,

$$\mathbf{q}_\ell = V(r)\,\hat{\boldsymbol{\theta}}. \quad (6.5.2)$$

Since the total depth is a constant, geostrophic motion exists in a multiplicity of forms other than that given above and the interactions of these can produce instabilities. To examine this possibility, set $\varepsilon = R_E E^{\frac{1}{2}}$, $t = E^{-\frac{1}{2}}\tau$, and expand all variables in powers of $E^{\frac{1}{2}}$, i.e.,

$$\mathbf{q}' = \mathbf{q}_0'(\mathbf{r}, \tau) + E^{\frac{1}{2}}\mathbf{q}_1' + \ldots + \tilde{\mathbf{q}}_0' + \ldots. \quad (6.5.3)$$

From here on, the development parallels that in § 2.6; for example, the expression that corresponds to (2.6.17) is

$$\mathbf{q}_1' = \frac{z}{2}\nabla\times\left\{\frac{\partial}{\partial\tau}\mathbf{q}_0' + R_E[(\nabla\times\mathbf{q}_\ell)\times\mathbf{q}_0' + (\nabla\times\mathbf{q}_0')\times\mathbf{q}_\ell]\right\} + \mathbf{A}(x, y), \quad (6.5.4)$$

where $\mathbf{A}$ is arbitrary. Satisfaction of the boundary conditions (see (2.6.13)) at the end plates of the cylinder, $z = 0$, $z = 1$, leads to a compatability equation for $\mathbf{q}_0'$ which is

$$\hat{\mathbf{k}}\cdot\nabla\times\frac{\partial}{\partial\tau}\mathbf{q}_0' + R_E\,\hat{\mathbf{k}}\cdot\nabla\times[(\nabla\times\mathbf{q}_\ell)\times\mathbf{q}_0' + (\nabla\times\mathbf{q}_0')\times\mathbf{q}_\ell] + 2\hat{\mathbf{k}}\cdot\nabla\times\mathbf{q}_0' = 0; \quad (6.5.5)$$

at the vertical sidewall, $\mathbf{q}_0'\cdot\hat{\mathbf{n}} = 0$. If the disturbance is assumed to be of the following form

$$\mathbf{q}_0' = \tfrac{1}{2}\hat{\mathbf{k}}\times\nabla[\phi\, e^{i(m\theta - c\tau)}], \quad (6.5.6)$$

the substitution of this in the foregoing yields a stability equation for ϕ having c as an eigenvalue. This is the problem studied by

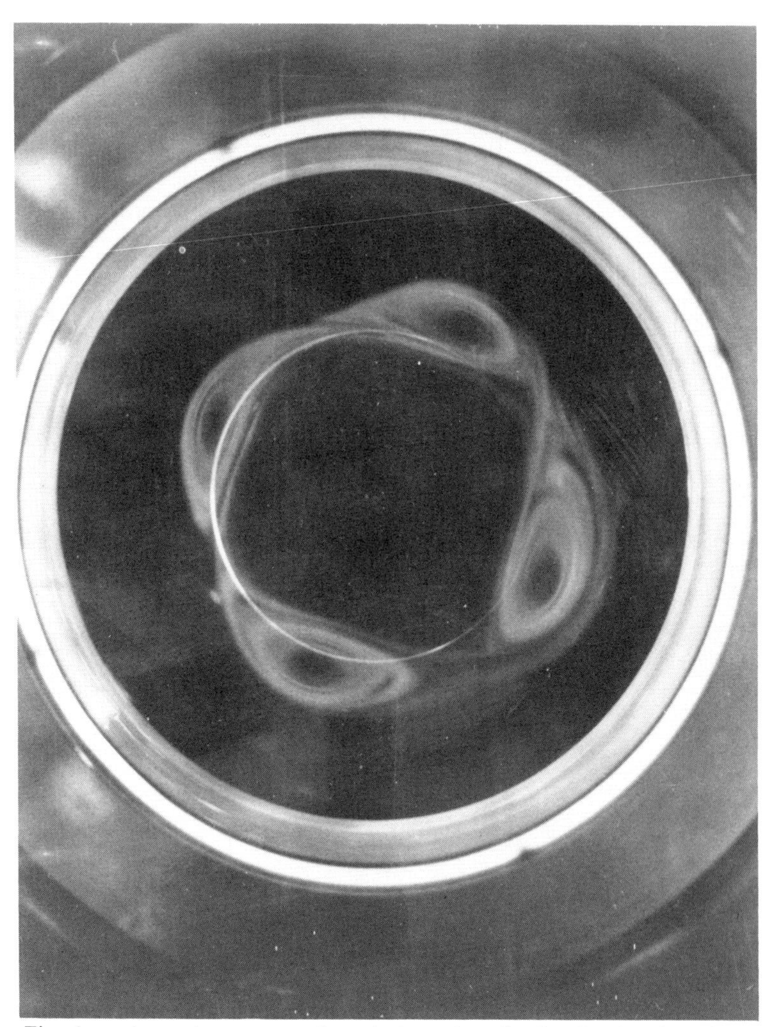

Fig. 6.14. A regular pattern of vertical vortices develop in the shear layer surrounding a disk which rotates a little faster than the ambient fluid. This photograph by Hide [82], was taken from the top looking down.

Busse who reports good agreement with Hide's experiments when $\mathbf{q}_\ell$ models the structure of the free $E^{\frac{1}{4}}$ shear layer.

Consider next, the situation in which the total depth has a variation that is small, but much larger than the thickness of the Ekman layer. Let the end plates of the cylinder be $z = 1$ and $z = -\alpha\mathfrak{d}(x, y)$.

The stretching and tilting of vortex lines is the main agent in the instability process (see also [224]) and such action can be produced either by Ekman layer suction, as in the preceding case, or by purely topographical and inertial effects, as in the present example. To examine this type of inviscid instability of a geostrophic flow, we return to (6.5.1) and set $E = 0$, $\varepsilon = R_E\,\alpha$. Furthermore, let

$$\mathbf{q}' = (\mathbf{Q}_0' + \alpha\mathbf{Q}_1' + \ldots)\exp it(\alpha\lambda^{(1)} + \alpha^2\lambda^{(2)} + \ldots).$$

The analysis now parallels that in § 2.16; $\mathbf{Q}_0'$ is one of the geostrophic modes in the right circular cylinder whose specific form is dictated by the boundary value problem for $\mathbf{Q}_1'$. The expression that corresponds to (2.16.8) is

$$\mathbf{Q}_1' = \tfrac{1}{2}i\lambda^{(1)}(z-1)\nabla\times\mathbf{Q}_0'$$
$$+\tfrac{1}{2}R_E(z-1)\nabla\times[(\nabla\times\mathbf{q}_\ell)\times\mathbf{Q}_0' + (\nabla\times\mathbf{Q}_0')\times\mathbf{q}_\ell] + \mathbf{A}(x, y). \quad (6.5.7)$$

To satisfy the boundary conditions at the end plates, the velocity, $\mathbf{Q}_0' = \tfrac{1}{2}\hat{\mathbf{k}}\times\nabla\Phi_0'$, must be a solution of

$$i\lambda^{(1)}\hat{\mathbf{k}}\cdot\nabla\times\mathbf{Q}_0' + R_E\,\hat{\mathbf{k}}\cdot\nabla\times[(\nabla\times\mathbf{q}_\ell)\times\mathbf{Q}_0' + (\nabla\times\mathbf{Q}_0')\times\mathbf{q}_\ell]$$
$$-2\nabla\mathfrak{d}\cdot\mathbf{Q}_0' = 0, \quad (6.5.8)$$

subject to $\mathbf{Q}_0'\cdot\hat{\mathbf{n}} = 0$ on the lateral wall. This is the direct analogue of (6.5.5) and only the mechanisms of vortex line stretching differ (the last terms on the left-hand sides of the respective equations).

It was shown in § 2.16 that the variety of geostrophic motions that can exist in a cylinder is severely restricted when the bottom surface is sloped slightly. For example, the disturbance represented in (6.5.6) would become a Rossby wave in the new configuration. It is anticipated, therefore, that the instability manifested in the viscous theory is associated with a Rossby wave in the distorted geometry. In particular, let $\mathfrak{d} = \mathfrak{d}(r)$ and $\Phi_0' = \phi(r)\,e^{im\theta}$, (the Rossby wave that corresponds exactly to (6.5.6)). The substitution of this into (6.5.8) results in an eigenvalue problem for ϕ and $\lambda^{(1)}$ which determines the properties of the instability. Similar equations receive much attention in meteorology; a recent discussion of the problem with an extensive bibliography is given by Drazin and Howard [39].

The conclusions are these: For a constant-depth container, the critical Reynolds number is some multiple of the ratio $\epsilon/E^{\frac{1}{2}}$ and the disturbance phase speed is proportional to $E^{\frac{1}{2}}$. In a container of slightly variable depth, the critical Reynolds number depends on ϵ/α, and the phase speed of the disturbance is proportional to α. When the depth variation is not small, either the critical number is a multiple of ϵ, or the wavelength of the disturbance becomes short enough so that viscous shear terms are no longer negligible.

Taylor–Proudman columns are extremely sensitive to flow conditions, and instabilities of the type just described may occur on the surface of the pillar. Moreover, when the external Rossby number exceeds a very small critical value, the column cannot hold rigidly to the protuberance that produced it. In this event, the column, after tilting, may be torn off the obstacle, thenceforth, to appear as a system of shed, longitudinal vortices migrating about the tank.

As a last example, it is observed that precession of a spheroid produces free internal shear surfaces, the stability of which may have some bearing on the problem of the geo-magnetic field created by motion in the core of the earth. It seems clear that future investigations of the stability of this particular flow, and indeed of all the linear motions discussed earlier which exhibit internal shears, may well have important and unexpected ramifications.

6.6. Stratified fluids

The transient motion of a stratified fluid seems to be notoriously prone to instabilities of all sorts. A simple qualitative experiment, that of spin-down, demonstrates this clearly.

A stratified fluid (salt in water) is in a near equilibrium state within a uniformly rotating, cylindrical container. Permanganate crystals are dropped about the periphery of the container to form a dyed band of fluid at the bottom, fig. 6.15 (*a*). The rotation rate of the container is decreased a modest amount so that the coloured water is convected inwards to replace the fluid emitted into the interior from the Ekman layer, fig. 6.15 (*b*, *c*). Ordinary Ekman layer instability is readily observed in this phase of motion which

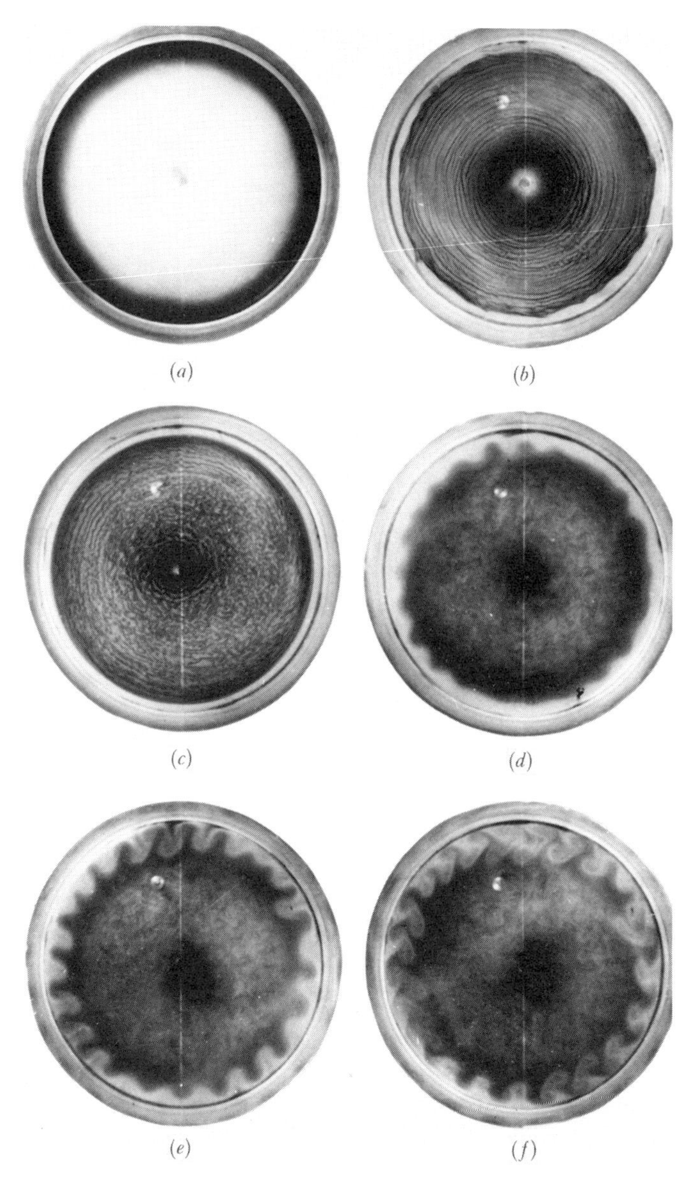

Fig. 6.15. The coloured rim of fluid, plate (*a*), is a thin layer on the base plate of a rotating cylindrical container that is filled with a stably stratified solution of water and salt. During spin-down, (*b*), waves are observed in the Ekman layer. The dyed heavy fluid is drawn into a thin circular disk, (*c*). This is separated from the wall by clear, lighter fluid that has been convected downwards along the boundary. Within a short time, waves develop on the disk, (*d*), (*e*) and (*f*), and complete mixing is effected.

lasts several revolutions following the impulsive change. In fig. 6.15 (*d*), the dyed fluid occupies a perfectly cylindrical disk at the bottom of the cylinder which is separated from the outer wall by a ring of clear water brought down from above. If there were no density differences, this pattern would persist for a considerable time, affected only by diffusive processes. However, the dyed fluid is heavier than the colourless liquid in the outer ring, a situation analogous to having heavy fluid over light fluid in a gravitational field. (The outwards centrifugal force plays the role of gravity.) The configuration is unstable and within a few minutes what seems to be rotationally modified Rayleigh–Taylor waves (Rayleigh [159], Taylor [205]; see also Chandrasekhar [30]) develop at the outer edge of the disk of dyed fluid, fig. 6.15 (*d*). These become more pronounced, fig. 6.15 (*e*), until complete mixing with the lighter fluid on the base plate is accomplished. In spin-up, much the same happens. Heavy fluid is transported up the vertical cylinder in the initial transient stage, and ultimately collapses in the gravitational field. Steady, driven motions may also exhibit instabilities of the same kind.

We will not undertake an analysis of this type of breakdown; the main point here is to emphasize that extreme caution is called for in the analysis of laminar motions of a stratified fluid. Instabilities within the boundary layers may affect the mixing of fluids of different density and, thereby, drastically change the manner in which the motion develops. The validity and applicability of laminar theories for rotating, stratified fluids must then be subjected to close scrutiny and some extra effort is almost mandatory to resolve questions of inherent stability.

6.7. Thermal convection in a rotating annulus

This section should be regarded only as an introduction to a much studied but still current field of activity. Although a complete exposition is not attempted, it is hoped that a foundation, and the incentive, for further study will be provided by even this short descriptive survey of some of the notable results achieved.

The ultimate objective of the research in this area is to understand the manner in which the atmospheric circulation transports the heat provided by the sun, poleward from the equatorial latitudes.

A simple laboratory model for this intricate process consists of a uniformly rotating, fluid filled, cylindrical annulus, fig. 6.16, whose vertical surfaces are maintained at different constant temperatures. The horizontal plates may be thermal insulators. An impressed radial temperature difference, $T_0 - T_1 > 0$, corresponds roughly to conditions at the equatorial and polar latitudes as indicated by the insert in fig. 6.16.

Most of the experimental work utilizing arrangements of this type were initiated by Fultz and Hide, and elaborated upon by their

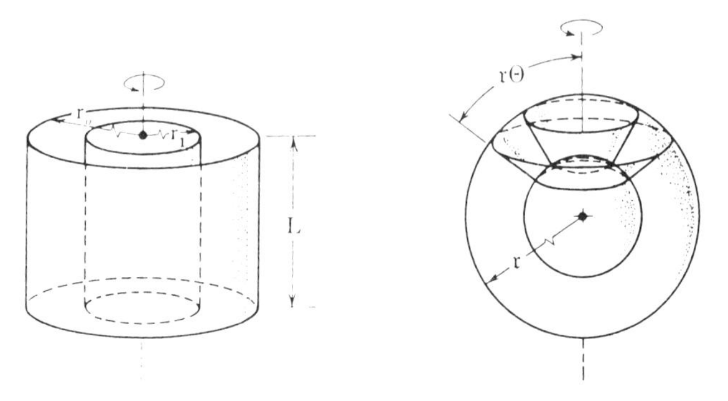

Fig. 6.16. A schematic drawing of the annulus experiments with an insert to show the relevance to the atmosphere. The vertical walls of the annulus are maintained at constant temperatures and the horizontal end plates are insulated. The apparatus rotates uniformly.

students and colleagues; Fultz [58, 59], Hide [80, 81], Bowden and Eden [14], Fowlis and Hide [52], Riehl and Fultz [163, 164].

The usual procedure is either to vary the magnitude of the temperature differential or the rotation rate. Since the configuration and applied boundary conditions are symmetrical about the rotation axis, a symmetrical laminar motion should also exist, at least in a certain range of parameters. This spiral flow is shown in fig. 6.17 (*a*, *b*), photographs by Hide [81]. However, as the rotation rate or temperature difference is increased, this symmetric state becomes unstable and a number of other distinct, non-symmetric flows develop. All of these involve a nearly steady, slowly drifting wave pattern with an associated jet-stream, fig. 6.17 (*c*, *d*, *e*, *f*). The jet is a rather narrow, rapid current possessing a strong span-wise temperature gradient. The number of waves tends to increase as Ω

Fig. 6.17. Photos by Hide [81] of the various types of instability observed in the apparatus of fig. 6.16. Plates (*a*) and (*b*) show regular spiral flow. Plates (*c*), (*d*), (*e*) and (*f*) illustrate the different wave regions; (*g*) shows the state of vacillation and (*h*) is a pattern of irregular eddies.

increases, but eventually a fluctuation sets in, plate g, which in turn evolves into a set of irregular eddies shown in plate h. Four definite regimes are to be identified: symmetrical flow, steady waves, vacillation, irregular flow. Visual evidence and temperature records show that the wave patterns extend almost throughout the entire depth of fluid. The interior horizontal velocity and temperature vary almost linearly with vertical height (see p. 125), and there is, of course, a small circulation induced by the impressed temperature differential. In this connection, it is important to point out that the secondary flow may actually produce a primary temperature field that differs markedly from a purely static distribution. The implication is that non-linear processes, convection in particular, are probably essential to the formation of the basic state.

It appears that the most efficient means of transporting heat depends strongly on the rotation rate and the temperature difference. The process may be one of ordinary conduction and/or convection by means of the high speed current which moves alternately from the hot to the cold surfaces, acquiring and depositing heat. Both of these motions have their analogues in the circulation of the atmosphere. For example, the symmetric flow may be identified with that of the trade winds near the equator, but what is really quite remarkable is the correspondence of the internal jet with the atmospheric jet-stream in the middle and high latitudes. Lorenz [119] observed that these laboratory wave motions were in fact similar to those discussed in connection with the large eddies, or cyclones, in the atmosphere, (see also Charney [32]). (The meteorological terms 'baroclinic waves' and 'baroclinic instability' are used often to designate these disturbances.) Once again, a rather simply conceived experiment is able to reproduce phenomena of global significance.

The immediate objective of experimental and theoretical investigations is the delineation of the various flow regimes and the determination of the conditions underlying their existence. Fortunately, the regions are quite distinct and reproducible. However, the correlation of data from different experiments and investigators is a non-trivial problem because there are many independent dimensionless parameters that can be arbitrarily assigned. It is rather surprising then to find that just two 'universal' parameters are sufficient for this purpose, as long as the media have the same kinematic viscosity and Prandtl number. With this restriction, Fultz [62] has shown that, the results of different experiments (see Fowlis and Hide [52]) define a single neutral stability curve in the plane whose co-ordinates are

$$\frac{gL}{(r_o-r_1)^2}\frac{\Delta\rho}{\rho\Omega^2}, \quad \frac{4\Omega^2(r_o-r_1)^4}{\nu^2}\left(\frac{r_o-r_1}{L}\right)^{\frac{1}{2}}, \quad \text{(see fig. 6.18).}$$

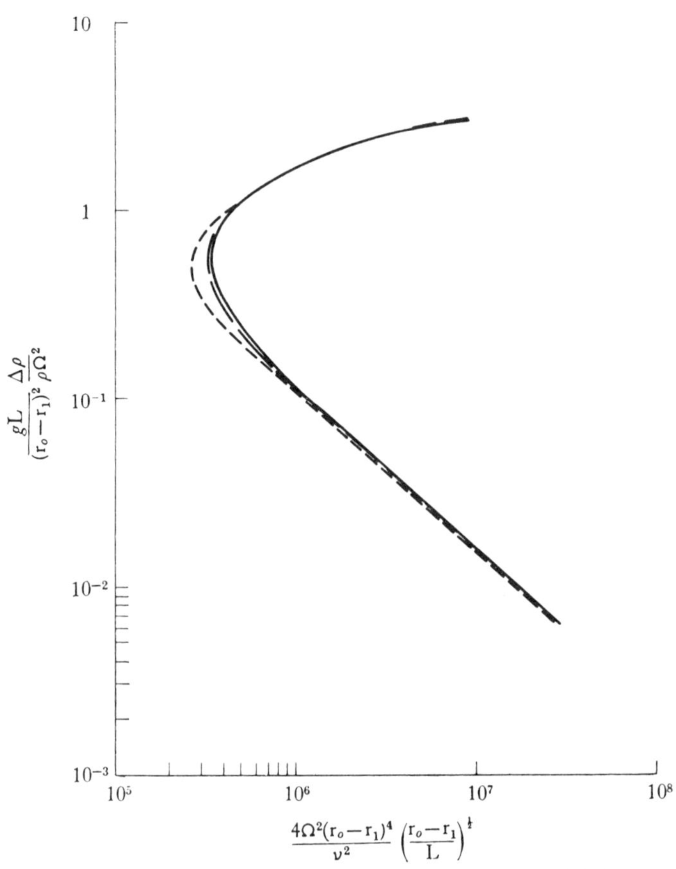

Fig. 6.18. The data from different experiments of Fowlis and Hide[52], define a single neutral stability curve in the plane with co-ordinates, as shown. —— $L = 14$ cm, $r_o - r_1 = 2{\cdot}54$ cm, $\nu = 1{\cdot}01 \times 10^{-2}\,\text{cm}^2\,\text{sec}^{-1}$; - - - $L = 5{\cdot}00$ cm; — — $L = 10$ cm.

(A family of dissimilar curves is generated when ν is varied and a complete flow characterization with just two parameters seems impossible.)

The domain of stability may be divided into upper and lower regions of symmetrical flow. The distinction is based on the mechanism chiefly responsible for the suppression of wave disturbances. A preponderant vertical stability of the primary flow is associated with the former according to Lorenz [119, 120], whereas viscous dissipation is the stabilizing process in the latter.

The delineation of the different wave regimes within the area of instability is a complicated task. The typical spectrum may involve unsteady flows, regions where different wave numbers exist, and hysteresis effects associated with the transition from one wave number to another. Transition curves on the diagram, fig. 6.19, were obtained by Fultz [62] by allowing the radial temperature differential to increase slowly, keeping Ω constant, until a change in wave number occurred. A similar set of transition curves, slightly displaced by hysteresis effects, results when the temperature is allowed to decrease slowly. (The rotation rate can also be changed by small increments, but it seems that varying the temperature is the best method of studying the spectrum.)

A wave number of seven is the most attainable in this particular configuration before irregular motion develops. However, the maximum wave number is definitely related to the dimensions of the tank and Hide [80, 81] deduced from his experimental data that κ_{max} is accurately given as the nearest integer to the value of

$$2{\cdot}05\,\frac{(r_o+r_1)}{(r_o-r_1)}.$$

Moreover, the minimum wave number observed is usually about five integers less,

$$\kappa_{max}-\kappa_{min} \leqslant 5,$$

but this is a rule of thumb rather than a rigorous proposition.

Theoretical work on this problem has the same objectives as the experimental programme. Progress has been hampered by the inability to determine the basic flow state accurately enough to undertake a stability analysis. It is not surprising then to find that current theories provide a reasonably correct qualitative description of the instability mechanism whereas quantitative agreement with experiment is still rather poor.

Most of the experiments deal with flows in which the coupling of the temperature and velocity fields is strongly non-linear. This is a formidable barrier for theoreticians because even the analysis of the strictly linear problem, aspects of which were explored in § 2.21, is far from simple or complete (see Robinson [167] and Hunter [90]).

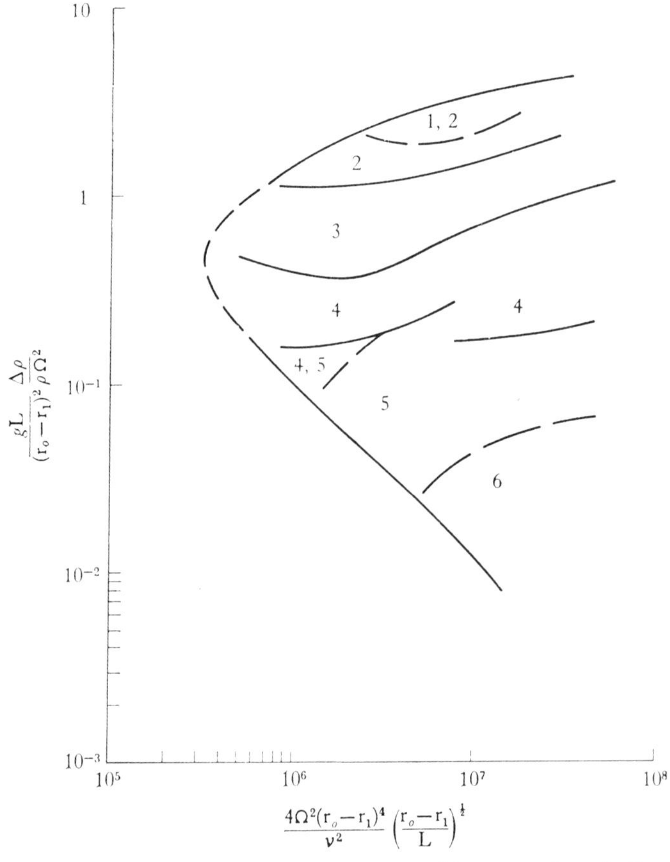

Fig. 6.19. Experimental results of Fultz [62] with an annulus of dimensions, $r_o = 4{\cdot}898$ cm., $r_1 = 2{\cdot}442$ cm., $L = 13{\cdot}00$ cm. The different wave regimes are shown in the region of instability and the number of waves observed is indicated. The spectrum was obtained by decreasing the positive temperature differential.

It is natural then, in view of these difficulties, to base a stability analysis on a relatively crude model of the primary velocity and temperature distributions. The description may be improved, in stages, when necessary. Idealized models of this sort have been advanced by Kuo[101, 102], Brindley[18], Lorenz[120], and Barcilon[4] (see also Eady[40]) and still more sophisticated theoretical models can be expected in the effort to eliminate discrepancies with experiments.

Several important results can be deduced readily by postulating the interior temperature field to vary linearly with both vertical and radial distance. In this case, the velocity is also a linear function of z, as shown by the analysis on p. 125. An examination of the stability of this type of interior flow, reproduces the general qualitative shape of the neutral stability curve displayed in fig. 6.18 and also the formula for the number of waves observed, Barcilon[4].

Potential energy is released as the fluid particles move between the surfaces of constant pressure and temperature. Most of the energy transferred to the disturbance is dissipated in the viscous boundary layers, a process that must be included in any realistic discussion of flow stability. The analysis in [4] incorporates the dissipation produced by the Ekman layers, but there has been, as yet, no successful attempt to evaluate the effects of the vertical shear layers. These may be very important to the total energy balance. However, in this respect, much also depends on the eventual success in relating the vertical gradient of the interior temperature with the impressed horizontal temperature differential.

Davies[37] and Rogers[171, 172] embarked on another tack and obtained exact, jet-like solutions of the inviscid, non-linear equations of motion. These are very informative and provide a means of determining the wave number from the quantity of heat to be transported. The wave form is usually that which will transport the maximum amount of heat. Although solutions of this type should prove indispensable to a study of the non-linear stability problem, much difficult work is required to connect the interior regime with the flow at the side walls.

All of these problems are still in the domain of active research. It seems fitting then to conclude this presentation on an open note and to allow future developments to write their own story.

NOTATION GUIDE

Each of the symbols listed below has a unique definition or interpretation which is employed consistently throughout this book. The symbol is introduced on the page indicated.

Other symbols are defined explicitly in each of the sections where they appear but their meanings in different sections are generally unrelated.

English

a	dimensionless tank radius, p. 82.
B	as a subscript, to denote the evaluation of a function on the bottom surface of the container, e.g. $\hat{\mathbf{n}}_B$.
c_P	specific heat at constant pressure, p. 12.
$\mathscr{C}$	curvature, p. 207.
D	a diffusivity coefficient, p. 15.
$\mathscr{D}$	differential operator, p. 182.
e_*	eccentricity, p. 74.
E	Ekman number, p. 7.
f_R	internal Froude number, p. 15.
f(x, y)	equation of the top surface of the container, p. 38.
F_R	Froude number, p. 13.
$\mathscr{F}$	Fourier transform, p. 207.
g	gravitational acceleration, p. 12.
g(x, y)	equation of the bottom surface of the container, p. 38.
$\mathscr{G}$	differential operator, p. 102.
h_i	metric coefficient, p. 24.
h(x, y)	total height, p. 44.
H	true depth, p. 236.
i	$\sqrt{-1}$.
i	summation index.
$\hat{\mathbf{i}}$	unit vector in the x direction.
$\mathscr{I}m$	imaginary part of.
j	summation index.
$\hat{\mathbf{j}}$	unit vector in the y direction.
k	summation index.
$\hat{\mathbf{k}}$	unit vector in the z direction.

L	length scale, p. 4.
$\mathscr{L}$	Laplace transform, p. 56.
m	summation index.
$\mathscr{M}$	mass flux, p. 107.
n	summation index
$\hat{\mathbf{n}}$	unit normal vector, pp. 10, 24.
$\mathbf{N}$	forcing function, p. 70.
И	an integer.
O	order symbol, p. 26.
p	reduced pressure, pressure, p. 6.
P	pressure, p. 5.
$\mathrm{P}_n^m(\mathrm{x})$	Assoc. Legendre polynomial, p. 64.
$\mathbf{q}$	in all forms, the fluid particle velocity, p. 5.
$\mathbf{q}_0$	geostrophic velocity, p. 40.
$\mathbf{q}_I$	interior velocity, p. 26.
$\tilde{\mathbf{q}}$	boundary layer velocity, p. 26.
$\mathbf{Q}$	velocity amplitude function, p. 10.
(r, θ, z)	cylindrical co-ordinates.
$\hat{\mathbf{r}}$	unit radial vector.
$\mathscr{R}e$	real part of.
$\mathrm{R_E}$	Reynolds number, p. 276.
R_{E}	'Reynolds' number, p. 183.
s, **ds**	arc length, increment of arc length, pp. 19, 48.
$\mathscr{s}$	Laplace transform parameter, p. 56.
$\mathscr{s}_m$	frequency factor, p. 40.
$\mathscr{s}_{m,1}$	decay factor, p. 42.
S(x)	unit step function, p. 102.
$\mathscr{S}$	source distribution, p. 96.
t	time, p. 6.
T	as a subscript, to denote the evaluation of a function on the top surface of the container, e.g. $\hat{\mathbf{n}}_{\mathrm{T}}$.
T	temperature, p. 12.
T	temperature perturbation, p. 14.
$\mathscr{T}$	wall temperature, p. 126.
(u, v, w)	components of velocity in cylindrical co-ordinates (r, θ, z).
$\mathbf{u}$	velocity deviation, p. 176.
$\mathscr{U}$	tangential wall velocity, p. 46.
v	azimuthal velocity component.

V, dV volume, volume element, p. 52.

w velocity component in z direction.

(x, y, z) cartesian co-ordinates.

$\mathfrak{x}$ an unscaled boundary layer co-ordinate, p. 110.

$\boldsymbol{\mathfrak{x}}$ an unscaled boundary layer co-ordinate, p. 103.

$\mathscr{X}(y)$ an arbitrary function, p. 95.

German

$\mathfrak{a}$ dimensionless excitation frequency, p. 70.

$\mathfrak{A}$ force potential, p. 6.

$\boldsymbol{\mathfrak{c}}_g$ group velocity, p. 187.

$\boldsymbol{\mathfrak{c}}_p$ phase velocity, p. 187.

$\mathfrak{C}$ geostrophic contour, p. 44.

$\mathfrak{d}(x, y)$ depth variation, p. 88.

$\mathfrak{D}$ differential operator, p. 216.

$\mathfrak{f}$ Coriolis parameter, p. 236.

$\boldsymbol{\mathfrak{F}}$ body force, p. 6.

$\boldsymbol{\mathfrak{q}}$ depth-averaged horizontal velocity, p. 228.

$\boldsymbol{\mathfrak{Q}}$ depth-averaged horizontal velocity, p. 231.

$\mathfrak{r}$ spherical radial co-odinate.

$\boldsymbol{\mathfrak{r}}$ position vector, p. 6.

$\mathfrak{S}$ salinity, p. 17.

$\boldsymbol{\mathfrak{T}}$ wind-stress curl, p. 228.

$\boldsymbol{\mathfrak{U}}$ uniform free stream velocity, p. 29.

$\mathfrak{V}$ in all forms—vorticity.

$\boldsymbol{\mathfrak{V}}$ relative vorticity, p. 7.

$\boldsymbol{\mathfrak{V}}_a$ absolute vorticity, p. 19.

$\mathfrak{w}$ depth-averaged, vertical velocity component, p. 228.

$\mathfrak{W}$ depth-averaged, vertical velocity component, p. 231.

Greek

α inclination angle, p. 85.

α_* coefficient of thermal expansion, p. 12.

β_* bulk viscosity, p. 11.

β constant approximation for the derivative of the Coriolis parameter, p. 237.

Υ integration contour, p. 19.

Γ circulation or swirl, p. 19.

δ	boundary layer thickness, p. 36.
$\boldsymbol{\delta}(t)$	scaled angular velocity deviation, p. 69.
Δ	as an incremental value, e.g. $\Delta\rho$.
∇	gradient.
$\mathit{\nabla}$	surface gradient, e.g. see pp. 228, 267.
ε	Rossby number, p. 7.
ζ	in all forms, a boundary layer co-ordinate, e.g. see p. 32.
θ	azimuthal angle co-ordinate, (r, θ, z).
$\hat{\boldsymbol{\theta}}$	unit circumferential vector.
Θ	spherical polar angle, p. 64.
κ	wave number.
$\boldsymbol{\kappa}$	wave vector, p. 186.
λ	scaled frequency or eigenvalue, p. 10.
μ_*	dynamic viscosity, p. 11.
ν	kinematic viscosity, p. 4.
ξ_i	orthogonal curvilinear co-ordinates, p. 24.
ρ	fluid density, p. 6.
$\mathit{\rho}$	density deviation, p. 14.
σ_P	Prandtl number, p. 15.
Σ	surface area, p. 19.
$\sum_n$	summation symbol.
τ	a long time variable, p. 40.
Υ	a geometric factor, p. 167.
φ	in all forms, a pressure function.
φ_0	geostrophic pressure, p. 40.
Φ	pressure amplitude, p. 10.
χ	a 'stream' function, p. 31.
ψ	a stream function, e.g. see p. 101.
ω	frequency of excitation or precession, p. 3.
Ω	rotation frequency, p. 3.
$\boldsymbol{\Omega}$	uniform angular velocity, p. 6.
$\mathit{\Omega}(t)$	time-dependent, angular velocity, p. 69.

Russian

и	density ratio, p. 15.
к	thermal conductivity, p. 15.

Л	stress component, p. 155.
Л	surface stress, p. 227.

Special symbols

†	complex conjugate, e.g. $\mathbf{Q}^{\dagger}$, p. 52.
~	to denote a boundary layer function, e.g. $\tilde{\mathfrak{q}}$, p. 26.
~	asymptotic equality, p. 26.
∧	to denote a unit vector, e.g. $\hat{\mathfrak{k}}$.
$\lvert\ \rvert$	absolute value; length of a vector, e.g. $\lvert\mathfrak{r}\rvert = \mathfrak{r}$.
⟦ ⟧	scaling rule, p. 8.
⟨ ⟩	depth-average, p. 54.
≐	approximate equality, p. 26.
≅	approximate equality, p. 26.
$\dfrac{\partial(F, G)}{\partial(x, y)}$	the Jacobian, p. 216.

BIBLIOGRAPHY AND AUTHOR INDEX

Numbered references are cited in the text on the pages ⟨ ⟩ indicated.

[1] Aldridge, K. D. and Toomre, A. (1967). Inertial oscillations of a rotating fluid sphere. Submitted for publication. ⟨pp. 67, 80, 81.⟩

Arons, A. B., Ingersoll, A. P. and Green, T. (1961). Experimentally observed instability of a laminar Ekman flow in a rotating basin. *Tellus*, **13**, 31–9.

Arons, A. B. and Stommel, H. (1956). A β-plane analysis of free periods of the second class in meridional and zonal oceans. *Deep-Sea Res.* **4**, 23–31.

[2] Baker, D. J. and Robinson, A. R. (1966). *A homogeneous β-plane ocean model—experiment and theory.* A report to the International Union of Theoretical and Applied Mechanics Symposium on Rotating Fluid Systems, La Jolla. See [17]. ⟨p. 269.⟩

[3] Banks, W. H. H. (1965). The boundary layer on a rotating sphere. *Quart. J. Mech. Appl. Math.* **18**, 443–54. ⟨p. 150.⟩

[4] Barcilon, V. (1964). Role of the Ekman layers in the stability of the symmetric regime obtained in a rotating annulus. *J. Atmos. Sci.* **21**, 291–9. ⟨p. 299.⟩

Barcilon, V. (1965). Stability of a non-divergent Ekman layer. *Tellus*, **17**, 53–68.

[5] Barcilon, V. (1966). On the motion due to sources and sinks distributed along the vertical boundary of a rotating fluid. *J. Fluid Mech.* **27**, 551–60. ⟨pp. 106, 109, 116.⟩

[6] Barcilon, V. and Pedlosky, J. (1967). Linear theory of rotating stratified fluid motions. *J. Fluid Mech.* **29**, 1–17. ⟨pp. 127, 128.⟩

[7] Batchelor, G. K. (1951). Note on a class of solutions of the Navier–Stokes equations representing steady rotationally-symmetric flow. *Quart. J. Mech. Appl. Math.* **4**, 29–41. ⟨pp. 133, 145.⟩

Benjamin, T. B. and Barnard, B. J. S. (1964). A study of the motion of a cavity in a rotating liquid. *J. Fluid Mech.* **19**, 193–209.

Benton, E. R. (1965). Laminar boundary layer on an impulsively started rotating sphere. *J. Fluid Mech.* **23**, 611–24.

[8] Benton, E. R. (1966). On the flow due to a rotating disk. *J. Fluid Mech.* **24**, 781–800. ⟨pp. 133, 136.⟩

Benton, G. S. and Boyer, D. (1966). Flow through a rapidly rotating conduit of arbitrary cross-section. *J. Fluid Mech.* **26**, 69–80.

[9] Bjerknes, V. and Solberg, H. (1929). Zellulare Trägheitswellen und Turbulence. *Avhandl Norsk Vid. Akad. Mat. Nat. kl.* **7**, 1–16. ⟨p. 82.⟩

[10] Bödewadt, U. T. (1940). Die Drehströmung über festem Grunde. *Z. angew. Math. Mech.* **20**, 241–53. ⟨p. 133.⟩

Bolin, B. (1952). Studies of the general circulation of the atmosphere. *Advances Geophys.* **1**, 87–118.

[11] Bondi, H. and Lyttleton, R. A. (1948). On the dynamical theory of the rotation of the earth. *Proc. Cambridge Phil. Soc.* **44**, 345–59. ⟨pp. 36, 38.⟩

[12] Bondi, H. and Lyttleton, R. A. (1953). On the dynamical theory of the rotation of the earth. *Proc. Cambridge Phil. Soc.* **49**, 498–515. ⟨p. 62.⟩

[13] Bowden, F. P. and Lord, R. G. (1963). Aerodynamic resistance to a sphere rotating at high speed. *Proc. Roy. Soc.* A **271**, 143–53. ⟨p. 150.⟩

[14] Bowden, M. and Eden, H. F. (1964). Thermal convection in a rotating fluid annulus: temperature, heat flow and flow field observations in the upper symmetrical regime. *J. Atmos. Sci.* **22**, 185–95. ⟨p. 294.⟩

[15] Bretherton, F. P. (1964). Low frequency oscillations trapped near the equator. *Tellus*, **16**, 181–5. ⟨pp. 253, 263.⟩

[16] Bretherton, F. P. (1967). The time-dependent motion due to a cylinder moving in an unbounded rotating or stratified fluid. *J. Mech. Fluid* **28**, 545–70. ⟨p. 199.⟩

[17] Bretherton, F. P., Carrier, G. F. and Longuet-Higgins, M. S. (1966). Report of the I.U.T.A.M. symposium on rotating fluid systems. *J. Fluid Mech.* **26**, 393–410. ⟨p. 305.⟩

[18] Brindley, J. (1960). Stability of flow in a rotating viscous incompressible fluid subject to differential heating. *Phil. Trans. Roy. Soc.* A **253**, 1–25. ⟨p. 299.⟩

Brunt, D. (1921). The dynamics of revolving fluid on a rotating earth. *Proc. Roy. Soc.* A **99**, 397–402.

[19] Bryan, G. H. (1889). The waves on a rotating liquid spheroid of finite ellipticity. *Phil. Trans.* A, **180**, 187–219. ⟨p. 64.⟩

[20] Bryan, K. (1963). A numerical investigation of a non-linear model of a wind-driven ocean. *J. Atmos. Sci.* **20**, 594–606. ⟨pp. 240, 255.⟩

[21] Burgers, J. M. (1948). A mathematical model illustrating the theory of turbulence. *Advances in Applied Mechanics*, **1**, 197–9. Academic Press, New York. ⟨p. 181.⟩

[22] Busse, F. H. (1968). Steady fluid flow in a precessing spheroid. *J. Fluid Mech.* **33**, 739–52. ⟨p. 179.⟩

[23] Busse, F. H. (1968). Shear flow instability in rotating systems. *J. Fluid Mech.* **33**, 577–89. ⟨p. 289.⟩

[24] Campbell, G. A. and Foster, R. M. (1948). *Fourier Integrals for Practical Applications*. D. Van Nostrand Co., New York. ⟨p. 31.⟩

[25] Carrier, G. F. (1954). Boundary layer problems in applied mathematics. *Communs. Pure and Appl. Math.* **7**, 11–17. ⟨p. 23.⟩

[26] Carrier, G. F. (1965). Some effects of stratification and geometry in rotating fluids. *J. Fluid Mech.* **23**, 145–72. ⟨pp. 126, 269.⟩

[27] Carrier, G. F. (1966). Phenomena in rotating fluids. *Proc. 11th Int. Congr. Appl. Mech.* (H. Görtler, editor), 69–87. Springer, Berlin. ⟨p. 117.⟩

[28] Carrier, G. F. and Robinson, A. R. (1962). On the theory of the wind-driven ocean circulation. *J. Fluid Mech.* **12**, 49–80. ⟨p. 235.⟩

[29] Cartan, E. (1922). Sur les petites oscillations d'une masse fluid. *Bull. Sci. Math.* **46**, 317–52, 356–69. ⟨pp. 51, 66.⟩

Chandrasekhar, S. (1957). The thermal instability of a rotating fluid sphere heated within. *Phil. Mag.* **2**, 845–58.

[30] Chandrasekhar, S. (1961). *Hydrodynamic and Hydromagnetic Stability*. The Clarendon Press, Oxford. ⟨pp. 185, 271, 293.⟩

[31] Charney, J. G. (1955). The Gulf Stream as an inertial boundary layer. *Proc. Natl. Acad. Sci.* **41**, 731–40. ⟨p. 238.⟩

[32] Charney, J. G. (1959). On the theory of the general circulation of the atmosphere. *The Atmosphere and the Sea in Motion*, (B. Bolin, editor), 178–93. Rockefeller Institute Press, New York. ⟨p. 295.⟩

Charney, J. G. and Drazin, P. G. (1961). Propagation of planetary-scale disturbances from the lower into the upper atmosphere. *J. Geophys. Research*, **66**, 83–109.

[33] Charney, J. G. and Eliassen, A. (1949). A numerical method for predicting the perturbations of the middle latitude westerlies. *Tellus*, **1**, 38–54. ⟨p. 36.⟩

[34] Cochran, W. G. (1934). The flow due to a rotating disc. *Proc. Cambridge Phil. Soc.* **30**, 365–75. ⟨p. 136.⟩

[35] Coles, D. (1965). Transition in circular Couette flow. *J. Fluid Mech.* **21**, 385–425. ⟨p. 272.⟩

[36] Crabtree, L. F., Küchemann, D. and Sowerby, L. (1963). Three-dimensional boundary layers. Chapter 8 of *Laminar Boundary Layers*, (L. Rosenhead editor). Oxford University Press, London. ⟨p. 24.⟩

Crossley, A. F. (1928). On the motion of a rotating circular cylinder filled with viscous fluid. *Proc. Cambridge Phil. Soc.* **24**, 480–8.

Davies, D. R. (1959). On the calculation of eddy viscosity and heat transfer in a turbulent boundary layer near a rapidly rotating disk. *Quart. J. Mech. Appl. Math.* **12**, 211–21.

Davies, T. V. (1948). Rotatory flow on the surface of the earth, Part I. Cyclostrophic motion. *Phil. Mag.* **39**, 482–91.

Davies, T. V. (1956). The forced flow due to heating of a rotating fluid. *Phil. Trans. Roy. Soc.* A **249**, 27–67.

[37] Davies, T. V. (1959). On the forced motion due to heating of a deep rotating liquid in an annulus. *J. Fluid Mech.* **5**, 593–621. ⟨p. 299.⟩

Dean, W. R. (1954). Note on the motion of an infinite cylinder in rotating viscous liquid. *Quart. J. Mech. Appl. Math.* **7**, 257–62.

[38] Donaldson, C. Du P. and Sullivan, R. D. (1960). Behaviour of solutions of the Navier–Stokes equation for a complete class of three-dimensional vortices. *Proc. Heat Transfer and Fluid Mech. Inst.* 16–30. Stanford University Press, Stanford. ⟨p. 181.⟩

[39] Drazin, P. G. and Howard, L. N. (1966). Hydrodynamic stability of parallel flow of inviscid fluid. *Advances in Applied Mechanics*, **9**, 1–85. Academic Press, New York. ⟨pp. 271, 292.⟩

[40] Eady, E. T. (1949). Long waves and cyclone waves. *Tellus*, **1**, 33–52. ⟨p. 299.⟩

[41] Ekman, V. W. (1905). On the influence of the earth's rotation on ocean-currents. *Ark. Mat. Astr. Fys.* **2**, 1–52. ⟨p. 32.⟩

Emslie, A. G., Bonner, F. T. and Peck, L. G. (1958). Flow of a viscous liquid on a rotating disk. *J. Appl. Phys.* **29**, 858–62.

[42] Erdelyi, A. (1953). *Tables of Integral Transforms.* McGraw-Hill, New York. ⟨p. 202.⟩

[43] Ertel, H. (1942). Ein neuer hydrodynamischer Wirbelsatz. *Meteorologische Zeitschrift,* **59**, 277–81. ⟨p. 22.⟩

[44] Ertel, H. and Rossby, C. G. (1949). A new conservation theorem of hydrodynamics. *Geofisica Pura e Appl.* **14**, 189–93. ⟨p. 22.⟩

Fadnis, B. S. (1954). Boundary layer on rotating spheroids. *Z. angew. Math. Phys.* **5**, 156–63.

[45] Faller, A. J. (1960). Further examples of stationary planetary flow patterns in bounded basins. *Tellus,* **12**, 159–71. ⟨p. 117.⟩

[46] Faller, A. J. (1963). An experimental study of the instability of the laminar Ekman boundary layer. *J. Fluid Mech.* **15**, 560–76. ⟨p. 275.⟩

[47] Faller, A. J. and Kaylor, R. E. (1966). A numerical study of the instability of the laminar Ekman boundary layer. *J. Atmos. Sci.* **23**, 466–80. ⟨pp. 275, 276, 287.⟩

[48] Faller, A. J. and Von Arx, W. S. (1959). The modeling of fluid flow on a planetary scale. *Proc. 7th Hydraulics Conf.*, State University of Iowa, 53–70. ⟨p. 238.⟩

[49] Fjortoft, R. (1950). Application of integral theorems in deriving criteria of stability of laminar flow and for the baroclinic circular vortex. *Geofys. Publ.* **17**, 1–52. ⟨p. 285.⟩

[50] Fofonoff, N. P. (1954). Steady flow in a frictionless homogeneous ocean. *J. Mar. Res.* **13**, 254–62. ⟨pp. 244, 258, 268.⟩

[51] Fofonoff, N. P. (1962). Dynamics of ocean currents. Chapter 3 of *The Sea*, I, (M. N. Hill, editor). Interscience, New York. ⟨pp. 235, 239.⟩

[52] Fowlis, W. W. and Hide, R. (1965). Thermal convection in a rotating annulus of liquid: effect of viscosity on the transition between axisymmetric and non-axisymmetric flow regimes. *J. Atmos. Sci.* **22**, 541–58. ⟨pp. 294, 296.⟩

[53] Fox, J. (1964). Boundary layers on rotating spheres and other axisymmetric shapes. *NASA*, TN D-2491. ⟨p. 150.⟩

[54] Fraenkel, L. E. (1956). On the flow of a rotating fluid past bodies in a pipe. *Proc. Roy. Soc.* A **233**, 506–26. ⟨pp. 215, 223.⟩

[55] Frenzen, P. (1955). Westerly flow past an obstacle in a rotating hemispherical shell. *Bull. Am. Meteorol. Soc.* **36**, 204–10. ⟨pp. 254, 265, 266.⟩

[56] Fultz, D. (1950). Experimental studies of a polar vortex. I. *Tellus,* **2**, 137–49. ⟨p. 265.⟩

[57] Fultz, D. (1950). Experimental studies related to atmospheric flow around obstacles. *Proc. 1st Intl. Meeting on Alpine Meteorol.* **17**, 3–4. ⟨p. 265.⟩

[58] Fultz, D. (1951). Experimental analogies to atmospheric motions. *Compendium of Meteorology*, 1235–48, Am. Meteorol. Soc. Boston. ⟨p. 294.⟩

[59] Fultz, D. (1953). A survey of certain thermally and mechanically driven fluid systems of meteorological interest. *Fluid Models in Geophysics. Proc. 1st Symposium on the Use of Models in Geophysical Fluid Dynamics.* Balt., Md. 27–63. ⟨p. 294.⟩

[60] Fultz, D. (1959). A note on overstability and the elastoid-inertia oscillations of Kelvin, Solberg, and Bjerkness. *J. Meteorol.* **16**, 199–208. ⟨pp. 83, 185.⟩

[61] Fultz, D. (1961). Developments in controlled experiments on large-scale geophysical problems. *Advances in Geophys.* **7**, 1–103. ⟨p. 238.⟩

Fultz, D. (1962). An experimental view of some atmospheric and oceanic behavioural problems. *Trans. N.Y. Acad. Sci.* **24**, 421–46.

[62] Fultz, D. (1966). *Spectrum of thermal convection in a rotating annulus.* A report to the International Union of Theoretical and Applied Mechanics. Symposium on Rotating Fluid Systems, La Jolla. See [17]. ⟨pp. 296, 297.⟩

[63] Fultz, D. and Frenzen, P. (1955). A note on certain interesting ageostrophic motions in rotating hemispherical shell. *J. Meteorol.* **12**, 332–8. ⟨pp. 254, 265, 268.⟩

Fultz, D., and Kaylor R. (1959). The propagation of frequency in experimental baroclinic waves in a rotating annular ring. *The Atmosphere and the Sea in Motion*, (B. Bolin, editor), 359–71. Rockefeller Inst. Press, New York.

[64] Fultz, D. and Long, R. R. (1951). Two-dimensional flow around a circular barrier in a rotating spherical shell. *Tellus*, **3**, 61–8. ⟨p. 265.⟩

Fultz, D., Long, R. R., Owens, G. W., Bohan, W., Kaylor, R. and Weil, J. (1959). Studies of thermal convection in a rotating cylinder with some implications for large-scale atmospheric motions. *Meteorol. Monogr.* **4**, 1–104.

Fultz, D. and Murty, T. S. (1964). A three-dimensional vortex instability in rotating fluids. *Proc. of the 11th Int. Congr. Appl. Mech.* (H. Görtler, editor), 1022–29. Springer, Berlin.

Glauert, M. B. (1957). A boundary layer theorem, with applications to rotating cylinders. *J. Fluid Mech.* **2**, 89–99.

[65] Goldstein, S. (editor), (1938). *Modern Developments in Fluid Dynamics.* Clarendon Press, Oxford. ⟨pp. 19, 21.⟩

[66] Görtler, H. (1944). Einige Bemerkungen über Strömungen in rotierenden Flüssigkeiten. *Z. angew. Math. Mech.* **24**, 210–14. ⟨p. 10.⟩

[67] Görtler, H. (1957). On forced oscillations in rotating fluids. *5th Midwestern Conf. on Fluid Mech.* 1–10. ⟨pp. 3, 10, 192, 202.⟩

[68] Grace, S. F. (1922). Free motion of a sphere in a rotating liquid parallel to the axis of rotation. *Proc. Roy. Soc.* A **102**, 89–111. ⟨p. 198.⟩

[69] Grace, S. F. (1923). Free motion of a sphere in a rotating liquid at right angle to the axis of rotation. *Proc. Roy. Soc.* A **104**, 278–301. ⟨p. 198.⟩

Grace, S. F. (1924). A spherical source in a rotating liquid. *Proc. Roy. Soc.* A **105**, 532–43.

[70] Grace, S. F. (1926). On the motion of a sphere on a rotating fluid. *Proc. Roy. Soc.* A **113**, 46–77. ⟨p. 198.⟩

[71] Greenspan, H. P. (1962). A criterion for the existence of inertial boundary layers in oceanic circulation. *Proc. Natl. Acad. Sci.* **48**, 2034–9. ⟨p. 123.⟩

[72] Greenspan, H. P. (1963). A note concerning topography and inertial currents. *J. Mar. Res.* **21**, 147–54. ⟨p. 242.⟩

[73] Greenspan, H. P. (1964). On the transient motion of a contained rotating fluid. *J. Fluid Mech.* **21**, 673–96. ⟨pp. 56, 62, 78.⟩

[74] Greenspan, H. P. (1965). On the general theory of contained rotating fluid motions. *J. Fluid Mech.* **22**, 449–62. ⟨p. 40.⟩

[75] Greenspan, H. P. and Howard, L. N. (1963). On a time dependent motion of a rotating fluid. *J. Fluid Mech.* **17**, 385–404. ⟨pp. 35, 49, 116, 173.⟩

[76] Greenspan, H. P and Weinbaum, S. (1965). On non-linear spin-up of a rotating fluid. *J. Math. and Phys.* **44**, 66–85. ⟨p. 169.⟩

Gregory, N. and Walker, W. S. (1960). Experiments on the effect of suction on the flow due to a rotating disk. *J. Fluid Mech.* **9**, 225–34.

[77] Gregory, N., Stuart, J. T. and Walker, W. S. (1955). On the stability of three-dimensional boundary layers with application to the flow due to a rotating disk. *Phil. Trans. Roy. Soc.* A **248**, 155–99. ⟨pp. 279, 286.⟩

[78] Haurwitz, B. (1940). The motion of atmospheric disturbances. *J. Mar. Res.* **3**, 35–50. ⟨p. 254.⟩

[79] Haurwitz, B. (1940). The motion of atmospheric disturbances on the spherical earth. *J. Mar. Res.* **3**, 254–67. ⟨p. 254.⟩

Hide, R. (1953). Fluid motion in the earth's core and some experiments on thermal convection in a rotating liquid. *Fluid Models in Geophysics, Proc. 1st Symposium on the Use of Models in Geophysical Fluid Dynamics.* Balt., Md. 101–16.

[80] Hide, R. (1953). Some experiments on thermal convection in a rotating liquid. *Quart. J. Roy. Meteorol. Soc.* **79**, 161. ⟨pp. 294, 297.⟩

Hide, R. (1956). The character of the equilibrium of a heavy, viscous, incompressible, rotating fluid of variable density. I and II. *Quart. J. Mech. Appl. Math.* **9**, 22–34, 35–50.

[81] Hide, R. (1958). An experimental study of thermal convection in a rotating liquid. *Phil. Trans. Roy. Soc.* A **250**, 441–78. ⟨pp. 294, 297.⟩

Hide, R. (1964). The viscous boundary layer at the free surface of a rotating baroclinic fluid. *Tellus*, **16**, 523–9.

Hide, R. (1966). Review article on the dynamics of rotating fluids and related topics in geophysical fluid dynamics. *Bull. Am. Meteorol. Soc.* **47**, 873–85.

[82] Hide, R. (1967). Detached shear layers in a rotating fluid. *J. Fluid Mech.* **29**, 39–60. ⟨p. 289.⟩

[83] Hide, R. (1967). On source-sink flows in a rotating fluid. Submitted to *J. Fluid Mech.* ⟨pp. 106, 117.⟩

[84] Hide, R. and Ibbetson, A. (1966). An experimental study of Taylor Columns. *Icarus*, **5**, 279–90. ⟨p. 173.⟩

Hocking, L. M. (1960). The stability of a rigidly rotating column of liquid. *Mathematika*, **7**, 1–9.

Hocking, L. M. (1962). An almost inviscid geostrophic flow. *J. Fluid Mech.* **12**, 129–34.

[85] Hocking, L. M. (1965). On the unsteady motion of a rotating fluid in a cavity. *Mathematika*, **12**, 97–106. ⟨p. 177.⟩

Hocking, L. M. and Michael, D. H. (1959). The stability of a column of rotating liquid. *Mathematika*, **6**, 25–32.

Høiland, E. (1950). On horizontal motion in a rotating fluid. *Geofys. Publik*, **17**, 1–26.

[86] Høiland, E. (1962). Discussion of a hyperbolic equation relating to inertia and gravitational fluid oscillations. *Geofys. Publ.* **24**, 211–27. ⟨p. 59.⟩

Holopainen, E. O. (1961). On the effect of friction in baroclinic waves. *Tellus*, **13**, 363–7.

[87] Holton, J. R. (1965). The influence of viscous boundary layers on transient motions in a stratified rotating fluid, Part I. *J. Atmos. Sci.* **22**, 402–11. ⟨p. 131.⟩

[88] Hough, S. S. (1895). The oscillations of a rotating ellipsoidal shell containing fluid. *Phil. Trans. Roy. Soc.* A **186**, 469–506. ⟨p. 66.⟩

Howard, L. N. (1963). Fundamentals of the theory of rotating fluids. *J. Appl. Mech.* **30**, 481–5.

Howard, L. N. and Drazin, P. G. (1964). On instability of parallel flow of inviscid fluid in a rotating system with variable Coriolis parameter. *J. Math. and Phys.* **18**, 83–99.

[89] Howarth, L. (1951). Note on the boundary layer on a rotating sphere. *Phil. Mag.* **42**, 1308–15. ⟨p. 150.⟩

[90] Hunter, C. (1967). The axisymmetric flow in a rotating annulus due to a horizontally applied temperature gradient. *J. Fluid Mech.* **27**, 753–78. ⟨p. 298.⟩

[91] Ibbetson, A. and Phillips, N. A. (1967). Some laboratory experiments on Rossby waves with application to the ocean. *Tellus*, **19**, 81–8. ⟨p 254.⟩

Jain, M. K. (1961). Flow of a non-Newtonian liquid near a rotating disk. *Appl. Sci. Research*, A **10**, 410–18.

[92] Jacobs, S. J. (1964). The Taylor column problem. *J. Fluid Mech.* **20**, 581–91. ⟨p. 117.⟩

Jacobs, S. J. (1965). On stratified flow over bottom topography. *J. Mar. Res.* **22**, 223–35.

[93] Kármán, Th. v. (1921). Über laminare und turbulente Reibung. *Z. angew. Math. Mech.* **1**, 233–51. ⟨p. 133.⟩

[94] Kármán, Th. v. (1934). Some aspects of the turbulence problem. *Proc. 4th Intl. Cong. Appl. Mech.* Cambridge, 54–9. ⟨pp. 153, 272.⟩

[95] Kelvin, Lord (1880). Vibrations of a columnar vortex. *Phil. Mag.* **10**, 155–68. ⟨pp. 82, 185.⟩

[96] King, W. S. and Lewellen, W. S. (1964). Boundary-layer similarity solutions for rotating flows with and without magnetic interaction. *Phys. Fluids*, **7**, 1674–80. ⟨pp. 142, 153.⟩

[97] Knauss, J. A. (1963). Equatorial current systems. Chapter 10 of *The Sea, II*, (M. N. Hill, editor). Interscience, London. ⟨p. 269.⟩

[98] Knauss, J. A. (1966). Further measurements and observations on the Cromwell current. *J. Mar. Res.* **24**, 205–40. ⟨p. 269.⟩

[99] Kreith, F., Roberts, L. G., Sullivan, J. A. and Sinha, S. N. (1963). Convection heat transfer and flow phenomena of rotating spheres. *Intl. J. Heat and Mass. Transfer*, **6**, 881–95. ⟨p. 150.⟩

Kreith, F., Taylor, J. F. and Chong, J. P. (1959). Heat and mass transfer from a rotating disk. *J. Heat. Transfer*, **81**, 95–105.

[100] Kudlick, M. D. (1966). On transient motions in a contained rotating fluid. Ph.D. Thesis, Math. Dept., M.I.T. ⟨pp. 56, 66, 73, 84, 116.⟩

Kuo, H. L. (1955). On convective instability of a rotating fluid with a horizontal temperature contrast. *J. Mar. Res.*, **14**, 14–32.

[101] Kuo, H. L. (1956). Energy-releasing processes and stability of thermally driven motions in a rotating fluid. *J. Meteorol.* **13**, 82–101. ⟨p. 299.⟩

[102] Kuo, H. L. (1957). Further studies of thermally driven motions in a rotating fluid. *J. Meteorol.* **14**, 553–8. ⟨p. 299.⟩

[103] Lamb, H. (1932). *Hydrodynamics*. Cambridge University Press. ⟨pp. 177, 217, 271.⟩

[104] Lance, G. N. and Rogers, M. H. (1962). The axially symmetric flow of a viscous fluid between two infinite rotating disks. *Proc. Roy. Soc.* A **266**, 109–21. ⟨pp. 133, 145, 147.⟩

[105] Lewellen, W. S. (1962). A solution for three-dimensional vortex flows with strong circulation. *J. Fluid Mech.* **14**, 420–32. ⟨p. 117.⟩

[106] Lewellen, W. S. (1965). Linearized vortex flows. *A.I.A.A. Journ.* **3**, 91–8. ⟨p. 182.⟩

Lewellen, W. S., Ross, D. H. and Rosenzweig, M. L. (1964). Confined vortex flows with boundary-layer interaction. *A.I.A.A. Journ.* **2**, 2127–33.

[107] Lighthill, M. J. (1949). A technique for rendering approximate solutions to physical problems uniformly valid. *London Math. Soc.* **40**, 1179–1201. ⟨p. 171.⟩

Lighthill, M. J. (1954). An elementary reasoned account of the general circulation of the atmosphere. *Phil. Mag.* **45**, 1154–62.

[108] Lighthill, M. J. (1960). Studies on magneto-hydrodynamic waves and other anisotropic wave motions. *Phil. Trans. Roy. Soc.* A **252**, 397–430. ⟨p. 207.⟩

[109] Lighthill, M. J. (1963). Boundary layer theory, Chapter 2 of *Laminar Boundary Layers* (L. Rosenhead, editor). Oxford University Press. ⟨pp. 19, 20.⟩

Lighthill, M. J. (1966). Dynamics of rotating fluids: a survey. *J. Fluid Mech.* **26**, 411–31.

[110] Lilly, D. K. (1966). On the instability of the Ekman boundary layer. *J. Atmos. Sci.* **23**, 481–94. ⟨pp. 284, 286, 287.⟩

[111] Lin, C. C. (1945). On the stability of two-dimensional parallel flows. Parts I, II, III. *Quart. Appl. Math.* **3**, I, 117–42; II, 218–34; III, 277–301. ⟨p. 285.⟩

[112] Lin, C. C. (1955). *The Theory of Hydrodynamic Stability*. Cambridge University Press. ⟨pp. 271, 284.⟩

[113] Long, R. R. (1951). A theoretical and experimental study of the motion and stability of certain atmospheric vortices. *J. Meteorol.* **8**, 207–21. ⟨p. 185.⟩

[114] Long, R. R. (1952). The flow of a liquid past a barrier in a rotating spherical shell. *J. Meteorol.* **9**, 187–99. ⟨pp. 265, 267.⟩

[115] Long, R. R. (1953). Steady motion around a symmetrical obstacle moving along the axis of a rotating liquid. *J. Meteorol.* **10**, 197–202. ⟨pp. 213, 215.⟩

Long, R. R. (1954). Note on Taylor's 'Ink Walls' in a rotating fluid. *J. Meteorol.* **11**, 247–9.

Long, R. R. (1956). Sources and sinks at the axis of a rotating fluid. *Quart. J. Mech. Appl. Math.* **9**, 385–93.

Long, R. R. (1958). Vortex motion in a viscous fluid. *J. Meteorol.* **15**, 108–112.

[116] Longuet-Higgins, M. S. (1964). Planetary waves in a rotating sphere. *Proc. Roy. Soc.* A **279**, 446–73. ⟨pp. 249, 250, 252.⟩

Longuet-Higgins, M. S. (1964). On group velocity and energy flux in planetary wave motions. *Deep-Sea Res.* **11**, 35–42.

[117] Longuet-Higgins, M. S. (1965). The response of a stratified ocean to stationary or moving wind-systems. *Deep-Sea Res.* **12**, 923–73. ⟨p. 252.⟩

[118] Longuet-Higgins, M. S. (1966). Planetary waves on a hemisphere bounded by meridians of longitude. *Phil. Trans. Roy. Soc.* A **260**, 317–50. ⟨p. 252.⟩

[119] Lorenz, E. N. (1953). A proposed explanation for the existence of two regimes of flow in a symmetrically heated cylindrical vessel. *Fluid Models in Geophys.* Proc. 1st Symposium on the Use of Models in Geophysical Fluid Dynamics. Balt., Md. 73–80. ⟨pp. 295, 297.⟩

[120] Lorenz, E. N. (1962). Simplified dynamic equations applied to the rotating basin experiments. *J. Atmos. Sci.* **19**, 39–51. ⟨pp. 297, 299.⟩

[121] Lyttleton, R. A. (1953). *The Stability of Rotating Liquid Masses.* Cambridge University Press. ⟨p. 271.⟩

Mack, L. M. (1960). The compressible viscous heat-conducting vortex. *J. Fluid Mech.* **8**, 284–92.

[122] Mack, L. M. (1962). *The laminar boundary layer on a disk of finite radius in a rotating flow.* Part 1. T.R. 32–224, J.P.L., Cal. Tech. ⟨p. 153.⟩

[123] Mack, L. M. (1963). *The laminar boundary layer on a disk of finite radius in a rotating flow.* Part 2. T.R. 32–366. J.P.L., Cal. Tech. ⟨p. 153.⟩

[124] Malkus, W. V. R. (1967). Hydromagnetic planetary waves. *J. Fluid Mech.* **28**, 793–802. ⟨p. 176.⟩

[125] Malkus, W. V. R. and Busse, F. (1966). *Zonal flows in precessing spheroids.* A report to the International Union Theoretical and Applied Mechanics. Symposium on Rotating Fluid Systems, La Jolla. See [17]. ⟨p. 175.⟩

[126] Margules, M. (1893). Luftbewegungen in einer rotierenden Sphäroidschale (II. Teil). *Sitz. der Math.-Naturwiss. Klasse, Kais. Akad. Wiss., Wien,* **102**, 11–56. ⟨p. 254.⟩

[127] Maxworthy, T. (1964). *The flow between a rotating disk and a coaxial, stationary disk.* Space Programs Summary, **4**, Sec. 327, 37–47. J.P.L., Cal. Tech. ⟨pp. 138, 160.⟩

Mestel, L. (1965). Meridian circulation in stars. Chap. 9 of *Stars and Stellar Systems, VIII.* (G. P. Kuiper, editor). Chicago University Press.

Mihaljan, J. M. (1962). A rigorous exposition of the Boussinesq approximation applicable to a thin layer of fluid. *Astrophys. J.* **136**, 1126–33.

Miles, J. W. (1959). Free surface oscillations in a rotating liquid. *Phys. Fluids*, **2**, 297–305.

Miles, J. W. (1963). The Cauchy–Poisson problem for a rotating liquid. *J. Fluid Mech.* **17**, 75–88.

Miles, J. W. (1964). Free surface oscillations in a slowly rotating liquid. *J. Fluid Mech.* **18**, 187–94.

Miles, J. W. and Ball, F. K. (1963). On free-surface oscillations in a rotating paraboloid. *J. Fluid Mech.* **17**, 257–66.

[128] Moore, D. W. (1963). Rossby waves in ocean circulation. *J. Deep-Sea Res.* **7**, 79–93. ⟨p.259⟩.

[129] Moore, F. K. (1956). Three-dimensional boundary layer theory. *Advances in Appl. Mech.* **4**, 160–244. Academic Press, New York. ⟨p. 23.⟩

[130] Morgan, G. W. (1951). A study of motions in a rotating liquid. *Proc. Roy. Soc.* A **206**, 108–30. ⟨pp. 199, 201.⟩

Morgan, G. W. (1953). Remarks on the problem of slow motions in a rotating fluid. *Proc. Cambridge Phil. Soc.* **49**, 362–64.

[131] Morgan, G. W. (1956). On the wind-driven ocean circulation. *Tellus*, **7**, 301–20. ⟨pp. 235, 238.⟩

[132] Morrison, J. A. and Morgan, G. W. (1956). The slow motion of a disc along the axis of a viscous rotating fluid. *Tech. Rep.* 8, *D.A.M.* Brown University. ⟨pp. 100, 192, 198, 200.⟩

[133] Munk, W. H. (1950). On the wind-driven ocean circulation. *J. Meteorol.* **7**, 79–93. ⟨p. 238.⟩

[134] Munk, W. H. and Carrier, G. F. (1950). The wind-driven circulation in ocean basins of various shapes. *Tellus*, **2**, 158–67. ⟨p. 238.⟩

Munk, W. H. and MacDonald, G. J. F. (1960). *The Rotation of the Earth—A Geophysical Discussion.* Cambridge University Press.

[135] Nigam, S. D. (1953). Growth of boundary layer on a rotating sphere. *Z. angew. Math. Phys.* **4**, 221–3. ⟨p. 152.⟩

[136] Nigam, S. D. (1954). Note on the boundary layer on a rotating sphere. *Z. angew. Math. Phys.* **5**, 151–5. ⟨p. 152.⟩

[137] Nigam, S. D. and Nigam, P. D. (1962). Wave propagation in rotating liquids. *Proc. Roy. Soc.* A **266**, 247–56. ⟨p. 207.⟩

[138] Niiler, P. P. (1966). On the theory of wind-driven ocean circulation. *Deep-Sea Res.* **13**, 597–606. ⟨p. 260.⟩

[139] Oser, H. (1957). Erzwungene Schwingungen in rotierenden Flussigkeiten. *Arch. Ratl. Mech. Anal.* **1**, 81–96. ⟨p. 192.⟩

[140] Oser, H. (1958). Experimentelle Untersuchung uber harmonische Schwingungen in rotierenden Flussigkeiten. *Z. angew. Math. Mech.* **38**, 386–91. ⟨pp. 3, 10, 192.⟩

[141] Payne, R. B. (1958). Calculations of unsteady viscous flow past a circular cylinder. *J. Fluid Mech.* **4**, 81–6. ⟨p. 21.⟩

[142] Pearson, C. E. (1965). A computational method for viscous flow problems. *J. Fluid Mech.* **21**, 611–22. ⟨p. 21.⟩

[143] Pearson, C. E. (1965). Numerical solutions for the time-dependent viscous flow between two rotating coaxial disks. *J. Fluid Mech.* **21**, 623–33. ⟨pp. 21, 133, 136, 145, 147, 172.⟩

[144] Pearson, C. E. (1967). Numerical solutions for the time-dependent viscous flow between two rotating concentric spheres. *J. Fluid Mech.* **28**, 323–36. ⟨p. 153.⟩

[145] Pedlosky, J. (1965). A study of the time-dependent ocean circulation. *J. Atmos. Sci.* **22**, 267–72. ⟨p. 252.⟩

[146] Pedlosky, J. (1965). A note on the western intensification of the oceanic circulation. *J. Mar. Res.* **23**, 207–9. ⟨p. 249.⟩

[147] Pedlosky, J. (1967). Spin-up of a stratified fluid. *J. Fluid Mech.* **28**, 463–80. ⟨p. 131.⟩

[148] Pedlosky, J. and Greenspan, H. P. (1967). A simple laboratory model for the oceanic circulation. *J. Fluid Mech.* **27**, 291–304. ⟨pp. 86, 118.⟩

Peube, J. L. and Kreith, F. (1966). Flow of an incompressible fluid between parallel rotating disks. *J. de Mecanique*, **5**, 261–86.

[149] Phillips, N. A. (1963). Geostrophic motion. *Rev. Geophys.* **1**, 123–71. ⟨p. 254.⟩

[150] Phillips, N. A. (1965). Elementary Rossby waves. *Tellus*, **17**, 295–301. ⟨p. 254.⟩

[151] Phillips, O. M. (1963). Energy transfer in rotating fluids by reflection of inertial waves. *Phys. Fluids*, **6**, 513–20. ⟨pp. 185, 188.⟩

[152] Picha, K. G. and Eckert, E. R. G. (1958). Study of the air flow between coaxial disks rotating with arbitrary velocities in an open or enclosed space. *Proc. 3rd. U.S. Nat'l. Cong. on Appl Mech.* 791–8. ⟨p. 148.⟩

[153] Poincaré, H. (1892). *Les méthodes nouvelles de la méchanique céleste*, **1**, Chapter 3. Gauthier-Villais, Paris. ⟨pp. 42, 171.⟩

[154] Poincaré, H. (1910). Sur la précession des corps déformables. *Bull. Astronomique*, **27**, 321–56. ⟨p. 51.⟩

[155] Proudman, I. (1956). The almost rigid rotation of viscous fluid between concentric spheres. *J. Fluid Mech.* **1**, 505–16. ⟨pp. 100, 105, 152.⟩

[156] Proudman, J. (1916). On the motion of solids in liquids possessing vorticity. *Proc. Roy. Soc.* A **92**, 408–24. ⟨p. 2.⟩

[157] Rattray, M. and Charnell, R. L. (1966). Quasi-geostrophic free oscillations in enclosed basins. *J. Mar. Res.* **24**, 82–102. ⟨p. 254.⟩

[158] Rayleigh, Lord (1880). On the stability, or instablity of certain fluid motions. *Proc. London Math. Soc.* **11**, 57–70. (*Scientific papers*, **1**, 474–87, Cambridge University Press.) ⟨p. 285.⟩

[159] Rayleigh, Lord (1833). Investigation of the character of the equilibrium of an incompressible heavy fluid of variable density. *Proc. London Math. Soc.* **14**, 170–7. (*Scientific papers*, **2**, 200–7, Cambridge University Press.) ⟨p. 293.⟩

[160] Rayleigh, Lord (1917). On the dynamics of revolving fluids. *Proc. Roy. Soc. London*, A **93**, 148–54. (*Scientific papers*, **6**, 447–53, Cambridge University Press.) ⟨p. 272.⟩

[161] Reynolds, A. (1962). Forced oscillations in a rotating liquid (I). *Z. angew. Math. Phys.* **13**, 460–8. ⟨pp. 192, 199, 201.⟩

[162] Reynolds, A. (1962). Forced oscillations in a rotating liquid (II). *Z. angew. Math. Phys.* **13**, 561–72. ⟨pp. 192, 200.⟩

[163] Riehl, H. and Fultz, D. (1957). Jet stream and long waves in a steady rotating dishpan experiment: structure of the circulation. *Quart. J. Roy. Meteorol. Soc.* **83**, 215–31. ⟨p. 294.⟩

[164] Riehl, H. and Fultz, D. (1958). The general circulation in a steady rotating dishpan experiment. *Quart. J. Roy. Meteor. Soc.* **84**, 389–417. ⟨p. 294.⟩

[165] Roberts, P. H. and Stewartson, K. (1963). On the stability of a Maclaurin spheroid of small viscosity. *Astro. J.* **137**, 777–90. ⟨p. 62.⟩

[166] Roberts, P. H. and Stewartson, K. (1965). Motion of a liquid in a spheroidal cavity. *Proc. Cambridge Phil. Soc.* **61**, 279–88. ⟨pp. 75, 78.⟩

[167] Robinson, A. R. (1959). The symmetric state of a rotating fluid differentially heated in the horizontal. *J. Fluid Mech.* **6**, 599–620. ⟨p. 298.⟩

Robinson, A. R. (1960). On two-dimensional inertial flow in a rotating stratified fluid. *J. Fluid Mech.* **9**, 321–32.

[168] Robinson, A. R. (1966). An investigation into the wind as a cause of the equatorial current. *J. Mar. Res.* **24**, 179–204. ⟨p. 269.⟩

Rogers, M. H. (1954). The forced flow of a thin layer of viscous fluid on a rotating sphere. *Proc. Roy. Soc.* A **224**, 192–208.

[169] Rogers, M. H. and Lance, G. N. (1960). The rotationally symmetric flow of a viscous fluid in the presence of an infinite rotating disk. *J. Fluid Mech.* **7**, 617–31. ⟨pp. 133, 136, 139, 145, 147, 164.⟩

[170] Rogers, M. H. and Lance, G. N. (1964). The boundary layer on a disc of finite radius in a rotating fluid. *Quart. J. Mech. Appl. Math.* **17**, 319–30. ⟨p. 139.⟩

[171] Rogers, R. H. (1959). The structure of the jet-stream in a rotating fluid with a horizontal temperature gradient. *J. Fluid Mech.* **5**, 41–59. ⟨p. 299.⟩

[172] Rogers, R. H. (1962). The effect of viscosity near the cylindrical boundaries of a rotating fluid with a horizontal temperature gradient. *J. Fluid Mech.* **14**, 25–41. ⟨p. 299.⟩

[173] Rosenhead, L. (editor) (1963). *Laminar Boundary Layers*. Oxford University Press. ⟨p. 23.⟩

Rosenzweig, M. L., Lewellen, W. S. and Ross, D. H. (1964). Confined vortex flows with boundary layer interaction. *A.I.A.A. Journ.* **2**, 2127–33.

[174] Rossby, C. G. (1939). Relation between variations in the intensity of the zonal circulation of the atmosphere and the displacements of the semi-permanent centers of action. *J. Mar. Res.* **2**, 38–55. ⟨pp. 86, 88, 254.⟩

[175] Rott, N. (1958). On the viscous core of a line vortex. *Z. angew. Math. Phys.* **9**, 543–53. ⟨p. 181.⟩

[176] Rott, N. (1959). On the viscous core of a line vortex, II. *Z. angew. Math. Phys.* **10**, 73–81. ⟨p. 182.⟩

[177] Rott, N. and Lewellen, W. S. (1966). Boundary layers and their interactions in rotating flows. *Progress in Aeronaut. Sci.* **7**, 111–44. ⟨pp. 138, 143, 154, 157.⟩

[178] Schlichting, H. (1955). *Boundary Layer Theory.* McGraw-Hill, New York. ⟨p. 23.⟩

[179] Schultz-Grunow, F. (1935). Der Riebungswiderstand rotierenden scheiben in Gehäusen. *Z. angew. Math. Mech.* **15**, 191–204. ⟨pp. 148, 153.⟩

[180] Smith, N. H. (1947). Exploratory investigation of laminar boundary layer oscillations on a rotating disc. *N.A.C.A., T.N.* 1227. ⟨p. 279.⟩

[181] Squire, H. B. (1956). Rotating fluids. *Surveys in Mechanics*, (G. K. Batchelor, editor), 139–61. Cambridge University Press. ⟨p. 216.⟩

Sobolev, S. L. (1960). Motion of a symmetric top with a cavity filled with fluid. *Zh. Prikl. Mekh.* **3**, 20–55.

[182] Spiegel, E. A. and Veronis, G. (1960). On the Boussinesq approximation for a compressible fluid. *Astrophys. J.* **131**, 442–7. ⟨p. 16.⟩

[183] Stern, M. E. (1960). Instability of Ekman flow at large Taylor number. *Tellus*, **12**, 399–417. ⟨pp. 38, 286.⟩

[184] Stern, M. E. (1963). Trapping of low frequency oscillations in an equatorial 'boundary layer'. *Tellus*, **15**, 246–50. ⟨pp. 253, 263, 265.⟩

[185] Stewartson, K. (1952). On the slow motion of a sphere along the axis of a rotating fluid. *Proc. Cambridge Phil. Soc.* **48**, 169–77. ⟨p. 198.⟩

Stewartson, K. (1953). A weak spherical source in a rotating fluid. *Quart. J. Math. Appl. Mech.* **6**, 45–9.

[186] Stewartson, K. (1953). On the slow motion of an ellipsoid in a rotating fluid. *Quart. J. Math. Appl. Mech.* **6**, 141–62. ⟨p. 173.⟩

[187] Stewartson, K. (1953). On the flow between two rotating coaxial disks. *Proc. Cambridge Phil. Soc.* **49**, 333–41. ⟨pp. 133, 145, 148.⟩

Stewartson, K. (1954). On the free motion of an ellipsoid in a rotating liquid. *Quart. J. Math. Appl. Mech.* **7**, 231–46.

[188] Stewartson, K. (1957). On almost rigid rotations. *J. Fluid Mech.* **3**, 17–26. ⟨pp. 100, 112.⟩

[189] Stewartson, K. (1957). On rotating laminar boundary layers. *Freiburg Symposium Boundary Layer Research*, 59–71. ⟨pp. 133, 139, 150.⟩

[190] Stewartson, K. (1958). On the motion of a sphere along the axis of a rotating fluid. *Quart. J. Math. Appl. Mech.* **11**, 39–51. ⟨pp. 220, 223.⟩

Stewartson, K. (1959). On the stability of a spinning top containing liquid. *J. Fluid Mech.* **5**, 577–92.

[191] Stewartson, K. (1966). On almost rigid rotations, Part 2. *J. Fluid Mech.* **26**, 131–44. ⟨p. 106.⟩

[192] Stewartson, K. and Roberts, P. H. (1963). On the motion of a liquid in a spheroidal cavity of a precessing rigid body. *J. Fluid Mech.* **17**, 1–20. ⟨pp. 75, 78.⟩

[193] Stommel, H. (1948). The westward intensification of wind-driven ocean currents. *Trans. Am. Geoph. Union*, **29**, 202–6. ⟨pp. 238, 257.⟩

[194] Stommel, H. (1957). A survey of ocean current theory. *Deep-Sea Res.* **4**, 149–84. ⟨p. 235.⟩

[195] Stommel, H. (1960). *The Gulf Stream.* University of Calif. Press, Berkeley. ⟨p. 239.⟩

[196] Stommel, H., Arons, A. B. and Faller, A. J. (1958). Some examples of stationary planetary flows. *Tellus*, **10**, 179–87. ⟨pp. 117, 238, 243.⟩

Stuart, J. T. (1954). On the effects of uniform suction on the steady flow due to a rotating disk. *Quart. J. Mech. Appl. Math.* **7**, 446–57.

[197] Stuart, J. T. (1963). Hydrodynamic stability. Chapter 9 in *Laminar Boundary Layers* (L. Rosenhead, editor). Oxford University Press. ⟨p. 271.⟩

[198] Synge, J. L. (1933). The stability of heterogeneous fluids. *Trans. Roy. Soc. Can.* **27**, 1–18. ⟨p. 272.⟩

[199] Tatro, P. R. and Mollö-Christensen, E. L. (1967). Experiments on Ekman layer instability. *J. Fluid Mech.* **28**, 531–44. ⟨p. 275.⟩

Taylor, G. I. (1916). Motion of solids in fluids when the flow is not irrotational. *Proc. Roy. Soc.* A **93**, 99–113.

[200] Taylor, G. I. (1921). Experiments with rotating fluids. *Proc. Cambridge Phil. Soc.* **20**, 326–9. ⟨p. 2.⟩

[201] Taylor, G. I. (1921). Experiments with rotating fluids. *Proc. Roy. Soc.* A **100**, 114–21. ⟨p. 2.⟩

[202] Taylor, G. I. (1922). The motion of a sphere in rotating liquid. *Proc. Roy. Soc.* A **102**, 180–9. ⟨pp. 199, 213, 215.⟩

[203] Taylor, G. I. (1923). Experiments on the motion of solid bodies in rotating fluids. *Proc. Roy. Soc.* A **104**, 213–18. ⟨pp. 2, 213.⟩

[204] Taylor, G. I. (1923). Stability of a viscous liquid contained between two rotating cylinders. *Phil. Trans. Roy. Soc.* A **223**, 289–343. ⟨p. 272.⟩

[205] Taylor, G. I. (1950). The instability of liquid surfaces when accelerated in a direction perpendicular to their planes, I. *Proc. Roy. Soc.* A **201**, 192–6. ⟨p. 293.⟩

[206] Theodorsen, T. and Regier, A. (1947). Experiments on rotating plates, rods and cylinders at high speeds. *N.A.C.A. T.N.* 793. ⟨p. 279.⟩

[207] Thiriot, K. H. (1940) Über die laminare Anlaufströmung einer Flüssigkeit über einem rotierenden Boden bei plötzlicker Änderung des Drehungszustandes. *Z. angew. Math. Mech.* **20**, 1–13. ⟨p. 133.⟩

[208] Titchmarsh, E. C. (1937). *Introduction to the Theory of Fourier Integrals.* Clarendon Press, Oxford. ⟨p. 193⟩.

[209] Truesdell, C. (1954). *The kinematics of vorticity.* Indiana Univ. Press, Indiana. ⟨pp. 19, 181.⟩

[210] Truesdell, C. (1954). Proof that Ertel's vorticity theorem holds in average for any medium suffering no tangential acceleration on the boundary. *Geofisica Pura e Appl.* **19**, 167–9. ⟨p. 22.⟩

[211] Trustrum, K. (1964). Rotating and stratified fluid flow. *J. Fluid Mech.* **19**, 415–32. ⟨pp. 30, 215, 223.⟩

[212] Turner, J. S. (1966). The constraints imposed on tornado-like vortices by the top and bottom boundary conditions. *J. Fluid Mech.* **25**, 377–400. ⟨p. 117.⟩

[213] Van Dyke, M. (1964). *Perturbation Methods in Fluid Mechanics*. Academic Press, New York. ⟨pp. 8, 29.⟩

Veronis, G. (1958). On the transient response of a β-plane ocean. *J. Oceanog. Soc. Japan*, **14**, 1–5.

Veronis, G. (1960). An approximate theoretical analysis of the equatorial undercurrent. *Deep-Sea Res.* **6**, 318–27.

[214] Veronis, G. (1963). On the approximations involved in transforming the equations of motion from a spherical surface to the β-plane. I. Barotropic systems. *J. Mar. Res.* **21**, 110–24. ⟨pp. 236, 252.⟩

[215] Veronis, G. (1963). On the approximations involved in transforming the equations of motion from a spherical surface to the β-plane. II. Baroclinic systems. *J. Mar. Res.* **21**, 199–204. ⟨pp. 236, 252.⟩

[216] Veronis, G. (1966). Wind-driven ocean circulation—Part 1. Linear theory and perturbation analysis. *Deep-Sea Res.* **13**, 17–29. ⟨p. 257.⟩

[217] Veronis, G. (1966). Wind-driven ocean circulation—Part 2. Numerical solutions of the non-linear problem. *Deep-Sea Res.* **13**, 31–55. ⟨pp. 240, 257.⟩

[218] Veronis, G. (1966). Generation of mean ocean circulation by fluctuating winds. *Tellus*, **18**, 67–76. ⟨p. 252.⟩

Veronis, G. (1966). Rossby waves with bottom topography. *J. Mar. Res.* **24**, 338–49.

[219] Veronis, G. and Morgan, G. W. (1955). A study of the time-dependent wind-driven circulation in a homogeneous, rectangular ocean. *Tellus*, **7**, 232–42. ⟨p. 252.⟩

[220] Von Arx, W. S. (1952). A laboratory study of the wind-driven ocean circulation. *Tellus*, **4**, 311–18. ⟨p. 238.⟩

[221] Von Arx, W. S. (1957). An experimental approach to problems in physical oceanography. *Prog. Phys. and Chem. of the Earth*, **2**, 1–29. ⟨p. 238.⟩

Walton, J. (1958). Note on a source in a rotating fluid. *Quart. J. Mech. Appl. Math.* **11**, 208–11.

[222] Warren, B. A. (1963). Topographic influences on the Gulf stream. *Tellus*, **15**, 167–83. ⟨pp. 239, 260.⟩

[223] Wedemeyer, E. H. (1964). The unsteady flow within a spinning cylinder. *J. Fluid Mech.* **20**, 383–99. ⟨pp. 4, 163, 165.⟩

[224] Weske, J. R. and Ranken, T. M. (1963). Generation of secondary motions in the field of a vortex. *Phys. Fluids*, **6**, 1397–403. ⟨p. 291.⟩

[225] Whitham, G. B. (1965). A general approach to linear and non-linear dispersive waves using a Lagrangian. *J. Fluid Mech.* **22**, 273–84. ⟨p. 187.⟩

Whitham, G. B. (1965). Non-linear dispersive waves. *Proc. Roy. Soc.* A **283**, 238–61.

Wood, W. W. (1957). The asymptotic expansions at large Reynolds numbers for steady motion between non-coaxial rotating cylinders. *J. Fluid Mech.* **3**, 159–75.

[226] Wood, W. W. (1966). An oscillatory disturbance of rigidly rotating fluid. *Proc. Roy. Soc.* A **293**, 181–212. ⟨pp. 59, 78.⟩

[227] Yih, C. S. (1965). *Dynamics of Non-Homogeneous Fluids*. Macmillan Company, New York. ⟨pp. 11, 30.⟩

[228] Zeipel, H. V. (1924). The radiative equilibrium of a rotating system of gaseous masses. Monthly notices of *Roy. Ast. Soc.* **84**, 665, 684, 702. ⟨p. 12.⟩

SUBJECT INDEX

SUPPLEMENTARY REFERENCES

Acheson, D. J. (1973). Hydromagnetic wavelike instabilities in a rapidly rotating stratified fluid. *J. Fluid Mech.* **61**, 609-24.

Acheson, D. J. (1975). Forced hydromagnetic oscillations of a rapidly rotating fluid. *Phys. of Fluids* **18**, 961-8.

Acheson, D. J. and Hide, R. (1973). Hydromagnetics of rotating fluids. *Rep. Prog. Phys.* **36**, 159-221.

Adams, M. L. and Szeri, A. Z. (1982). Incompressible flow between finite disks. *J. Appl. Mech.* **49**, 1-9.

Aldridge, K. D. (1975). Inertial waves and the earth's outer core. *Geophys. J. Roy. Astr. Soc.* **42**, 337-45.

Aldridge, K. D. and Toomre, A. (1969). Axisymmetric inertial oscillations of a fluid in a rotating spherical container. *J. Fluid Mech.* **37**, 307-23.

Alfredsson, P. H. and Persson, H. (1989). Instabilities in channel flow with system rotation. *J. Fluid Mech.* **202**, 543-57.

Allen, J. S. (1970). The effect of weak stratification and geometry on the steady motion of a contained rotating fluid. *J. Fluid Mech.* **43**, 128-44.

Allen, J. S. (1971). Some aspects of the initial value problem for the inviscid motion of a contained rotation weakly stratified fluid. *J. Fluid Mech.* **46**, 1-23.

Allen, J. S. (1972). Up-welling of a stratified fluid in a rotating annulus. *J. Fluid Mech.* **52**, 429-45.

Allen, J. S. (1973). Up-welling of a stratified fluid in a rotating annulus: steady state. Part 2. Numerical solutions. *J. Fluid Mech.* **59**, 337-68.

Amberg, G. and Greenspan, H. P. (1986). Boundary layers in a sectioned centrifuge. *J. Fluid Mech.* **181**, 77-97.

Amberg, G., Dahlkild, A. A. et al. (1986). On time-dependent settling of a dilute suspension in a rotating conical channel. *J. Fluid Mech.* **166**, 473-502.

Annamalai, P. and Cole, R. (1986). Particle migration in rotating liquids. *Phys. of Fluids* **29**, 647-9.

Aoki, I. and Tokimoto, T. (1986). Method for separation of particles using a rotating tilted liquid column. *Powder Tech.* **49**, 73-4.

Aoki, I., Shirane, K. et al. (1986). Separation of fine particles using rotating tube with alternate flow. *Rev. Sci. Instrum.* **57**, 2859-61.

Baines, P. G. (1967). Forced oscillations of an enclosed rotating fluid. *J. Fluid Mech.* **30**, 533-46.

Baines, P. G. (1971). The reflexion of internal/inertial waves from bumpy surfaces. *J. Fluid Mech.* **46**, 273-91.

Baines, P. G. (1973). The generation of internal tides by flat-bump topography. *Deep-Sea Res.* **20**, 179-206.

Baker, D. J. (1967). Shear layers in a rotating fluid. *J. Fluid Mech.* **29**, 165-75.

Baker, D. J. (1968). Demonstrations of fluid flow in rotating system II: the spin-up problem. *J. of Phys.* **36**, 980-6.

Baker, D. J. (1968). Shear layers in a rotating fluid. *J. Fluid Mech.* **29**, 165-76.

Baker, D. J. (1971). A source-sink laboratory model of ocean circulation. *Geo. Fluid Dyn.* **2**, 17-30.

Baker, D. J. and Robinson, A. R. (1969). A laboratory model for the general ocean circulation. *Phil. Trans. Roy. Soc.* **265**, 533-66.

Baker, G. R. and Israeli, M. (1981). Spin-up from rest of immiscible fluids. *Stud. Appl. Math.* **65**, 249-68.

Balan, K. C., Rao, A. R. et al. (1975). Unsteady motion of a stratified rotating viscous fluid between two disks. *J. Phys. Soc. Japan* **34**, 1402-7.

Ball, F. K. (1963). Some general theorems concerning the finite motion of a shallow rotating liquid lying on a paraboloid. *J. Fluid Mech.* **17**, 240-56.

Barcilon, A. I. (1967). Vortex decay above a stationary boundary. *J. Fluid Mech.* **27**, 155-75.

Barcilon, A., Lau, J. et al. (1975). Numerical experiments on stratified spin-up. *Geo. Fluid Dyn.* **7**, 29-42.

Barcilon, V. (1968). Stewartson layers in transient rotating fluid flows. *J. Fluid Mech.* **33**, 815-25.

Barcilon, V. (1968). Axisymmetric inertial oscillations of a rotating ring of fluid. *Mathematika* **17**, 93-102.

Barcilon, V. (1970). Some inertial modifications of the linear viscous theory of a steady rotating fluid. *Phys. of Fluids* **13**, 537-44.

Barcilon, V. and Berg, H. C. (1971). Forced axial flow between rotating concentric cylinders. *J. Fluid Mech.* **47**, 469-79.

Barcilon, V. and Bleistein, N. (1969). Scattering of inertial waves in a rotating fluid. *Stud. Appl. Math.* **48**, 91-104.

Barcilon, V. and Pedlosky, J. (1967). A unified linear theory of homogeneous and stratified rotating fluids. *J. Fluid Mech.* **29**, 609-21.

Barcilon, V. and Pedlosky, J. (1967). On the steady motions produced by a stable stratification in a rapidly rotating fluid. *J. Fluid Mech.* **29**, 673-90.

Bark, F. H. and Bark, T. H. (1976). On vertical boundary layers in a rapidly rotating gas. *J. Fluid Mech.* **78**, 749-61.

SUPPLEMENTARY REFERENCES

Bark, F. H. and Meijer, P. S. (1978). Spin-up of rapidly rotating gas. *Phys. of Fluids* **21**, 531-9.

Bark, F. H. and Sundstrom, O. (1981). A note on rotating Hele-Shaw cells. *J. Fluid Mech.* **111**, 271-81.

Bark, F. H., Johansson, A. V. et al. (1984). Axisymmetric stratified two-layer flow in a rotating conical channel. *J. de Mecanique et appliquee* **3**, 861-78.

Barrett, K. E. (1967). On the impulsively started rotating sphere. *J. Fluid Mech.* **27**, 779-88.

Barton, N. G. (1976). An example of the modification of ocean currents by bottom topography. *Tellus* **28**, 261-5.

Beardsley, R. C. (1969). A laboratory model of the wind-driven ocean circulation. *J. Fluid Mech.* **38**, 255-71.

Beardsley, R. C. (1970). An experimental study of inertial waves in a closed cone.. *Stud. Appl. Math.* **49**, 187-96.

Beardsley, R. C. (1971). Integration of the planetary vorticity equation on a parabolic circular grid. *J. Comp. Physics* **7**, 273-88.

Beardsley, R. C. (1973). A numerical model of the wind-driven ocean circulation in a circular basin. *Geo. Fluid Dyn.* **4**, 211-42.

Beardsley, R. C. (1975). The sliced cylinder laboratory model of the wind-driven ocean circulation. *J. Fluid Mech.* **69**, 27-64.

Beardsley, R. C. and Robbins, K. (1974). The sliced-cylinder laboratory model of the wind-driven ocean circulation. *J. Fluid Mech.* **69**, 27-40.

Beardsley, R. C., Saunders, K. D. et al. (1979). An experimental and numerical study of the secular spin-up of a thermally stratified rotating fluid. *J. Fluid Mech.* **93**, 161-84.

Beckett, P. M. (1985). Non-linear oscillations of a spherical particle in a rotating fluid. *Int. J. Non-Linear Mechanics* **20**, 261-71.

Belcher, R. J., Burggraf, O. R. et al. (1972). On generalized vortex boundary layers. *J. Fluid Mech.* **52**, 573-80.

Bellamy-Knights, P. G. (1974). An axisymmetric boundary layer solution for an unsteady vortex above a plane.. *Tellus* **26**, 318-24.

Benjamin, T. B. (1970). Upstream influence. *J. Fluid Mech.* **40**, 49-79.

Bennetts, D. A. and Hocking, L. M. (1973). On non-linear Ekman and Stewartson layers in a rotating fluid. *Proc. Roy. Soc. London A* **333**, 469-89.

Bennetts, D. A. and Hocking, L. M. (1974). Pressure induced flows at low Rossby numbers. *Phys. of Fluids* **17**, 1671-6.

Bennetts, D. A. and Jackson, W. D. N. (1974). Source-sink flows in a rotating annulus: A combined laboratory and numerical study. *J. Fluid Mech.* **66**, 689-705.

Benney, D. J. (1965). The flow induced by a disc oscillating about a state of steady rotation. *Quart. J. Mech. Appl. Math.* **18**, 333-45.
Benton, E. R. (1973). Non-linear hydrodynamic and hydromagnetic spin-up driven by Ekman-Hartmann boundary layers. *J. Fluid Mech.* **57**, 337-60.
Benton, E. R. (1979). Vorticity dynamics in spin-up from rest. *Phys. of Fluids* **22**, 1250-1.
Benton, E. R. and Clark, A. Jr. (1974). Spin-up. *Ann. Rev. of Fluid Mech.* **6**, 257-280.
Benton, E. R. and Loper, D. E. (1969). On the spin-up of an electrically conducting fluid. *J. Fluid Mech.* **39**, 561-86.
Benton, E.R. (1968). A composite Ekman boundary layer problem. *Tellus* **20**, 667-72.
Benton, G. S., Lipps, F. B. et al. (1964). The structure of the Ekman layer for geostrophic flows with lateral shear. *Tellus* **16**, 186-99.
Berman, A. S., Bradford, J. et al. (1978). Two-fluid spin-up in a centrifuge. *J. Fluid Mech.* **84**, 411-31.
Bien, F. and Penner, S. S. (1971). Spin-up and spin-down of rotating flows in a finite cylindrical container. *Phys. of Fluids* **14**, 1305-7.
Blandford, R. (1965). Inertial flow in the Gulf Stream. *Tellus* **17**, 46-52.
Bloor, M. I. G. and Ingham, D. B. (1987). The flow in industrial cyclones. *J. Fluid Mech.* **178**, 507-19.
Blumen, W. and Washington, W. M. (1969). The effect of horizontal shear flow on geostrophic adjustment in a barotropic fluid. *Tellus* **21**, 167-76.
Blumsack, S. and Barcilon, A. (1971). Thermally-driven linear vortex. *J. Fluid Mech.* **48**, 801-14.
Bodonyi, R. J. (1975). On rotationally symmetric flow above an infinite rotating disk. *J. Fluid Mech.* **67**, 657-66.
Bogy, D. B., Fromm, J. E. et al. (1977). Exit region central source flow between finite closely spaced parallel co-rotating disks. *Phys. of Fluids* **20**, 176-86.
Bowden, J. V. (1972). An equatorial boundary layer. *J. Fluid Mech.* **56**, 193-200.
Bowden, M. and Eden, H. F. (1968). Effect of a radial barrier on the convective flow in a rotating fluid annulus. *J. Geophys. Res.* **73**, 6887-95.
Boyer, D. L. (1970). Flow past a right circular cylinder in a rotating frame. *J. Basic Eng.* **92**, 430-6.
Boyer, D. L. (1971). Rotating flow over long shallow ridges. *Geo. Fluid Dyn.* **2**, 164-84.
Boyer, D. L., Davies, P. A. et al. (1984). Rotating flow past disks and cylindrical depressions. *J. Fluid Mech.* **141**, 67-95.

SUPPLEMENTARY REFERENCES

Brady, J. F. and Durlofsky, L. (1987). On rotating disk flow. *J. Fluid Mech.* **175**, 363-94.

Brink, K. H., Veronis, G. et al. (1973). The effect on the ocean circulation of a change in the sign of beta. *Tellus* **25**, 518-22.

Bryan, K. and Cox, M. D. (1967). A numerical investigation of the oceanic general circulation. *Tellus* **19**, 54-81.

Burggraf, O. R., Stewartson, K. et al. (1971). Boundary layer induced by a potential vortex. *Phys. of Fluids* **14**, 1821-33.

Busse, F. H. (1970). Thermal instabilities in rapidly rotating systems. *J. Fluid Mech.* **44**, 441-60.

Busse, F. H. and Carrigan, C. R. (1974). Convection induced by centrifugal buoyancy. *J. Fluid Mech.* **62**, 579-92.

Buzyna, G. and Veronis, G. (1971). Spin-up of a stratified fluid: theory and experiment. *J. Fluid Mech.* **50**, 579-608.

Caldwell, D. R. and Van Atta, C. W. (1970). Characteristics of Ekman boundary layer instabilities. *J. Fluid Mech.* **44**, 79-95.

Caldwell, D. R., Van Atta, C. W. et al. (1972). A laboratory study of the turbulent Ekman layer. *Geo. Fluid Dyn.* **3**, 125-60.

Carrier, G. F. (1971). Swirling flow boundary layers. *J. Fluid Mech.* **49**, 145-58.

Carrier, G. F. and Hammond, A. L. (1971). A model of the mature hurricane. *J. Fluid Mech.* **47**, 145-70.

Cederlof, U. (1988). Free-surface effects on spin-up. *J. Fluid Mech.* **187**, 395-407.

Cerasoli, C. P. (1975). Free shear layer instability due to probes in rotating source-sink flows. *J. Fluid Mech.* **72**, 559-86.

Charney, J. G. (1960). Non-linear theory of a wind-driven homogeneous layer near the equator. *Deep-Sea Res.* **6**, 303-10.

Chawla, S. S. (1972). On hydromagnetic spin-up. *J. Fluid Mech.* **53**, 545-55.

Chawla, S. S. (1976). Non-linear Ekman-Hartmann layers and the flow outside. *J. Fluid Mech.* **76**, 401-13.

Chernousko, F. L. (1964). Motion of a solid body with a cavity containing an ideal fluid and an air bubble. *P.M.M.* **28**, 896-907.

Chernousko, F. L. (1967). Rotational motions of a solid body with a cavity filled with fluid. *P.M.M.* **31**, 416-432.

Chew, J. W., Owen, J. M. et al. (1984). Numerical predictions for laminar source-sink flow in a rotating cylindrical cavity. *J. Fluid Mech.* **143**, 451-66.

Chi, S. W., Ying, S. J. et al. (1969). The ground turbulent boundary layer of a stationary tornado-like vortex. *Tellus* **21**, 693-700.

Childress, S. (1964). The slow motion of a sphere in a rotating, viscous fluid. *J. Fluid Mech.* **20**, 305-14.

Chin, D. T. and Litt, M. (1972). An electrochemical study of flow instability on a rotating disk. *J. Fluid Mech.* **54**, 613-25.

Chun, Ch.- H. and Wuest, W. (1982). Suppression of temperature oscillations of thermal Marangoni convection in a floating zone .... *Acta Astronautica* **9**, 225-30.

Church, C. R., Snow, J. T. et al. (1979). Characteristics of tornado-like vortices as a function of swirl ratio: a laboratory investigation. *Am. Meteor. Soc.* **36**, 1755-76.

Clark, A. Jr. (1973). The linear spin-up of a strongly stratified fluid of small Prandtl number. *J. Fluid Mech.* **60**, 561-80.

Clark, A., Clark. P. A. et al. (1971). Spin-up of a strongly stratified fluid in a sphere. *J. Fluid Mech.* **45**, 131-49.

Collins, R. and Hoath, M. T. (1975). Draining from rapidly spinning tubes. *J. Fluid Mech.* **67**, 763-8.

Conlisk, A. T. and Walker, J. D. A. (1981). Incompressible source-sink flow in a rapidly rotating contained annulus. *Quart. J. Mech. Appl. Math.* **34**, 89-109.

Conlisk, A. T. and Walker, J. D. A. (1982). Forced convection in a rapidly rotating annulus. *J. Fluid Mech.* **122**, 91-108.

Conlisk, A. T., Foster, M. R. et al. (1982). Fluid dynamics and mass transfer in a gas centrifuge. *J. Fluid Mech.* **125**, 283-317.

Cook, L.P. and Ludford, G. S. S. (1975). Higher-order approximation for free shear layers in almost rigid rotations. *J. Fluid Mech.* **69**, 191-6.

Crisalli, A. J. and Walker, J. D. A. (1976). Non-linear effects for the Taylor column for a hemisphere. *Phys. of Fluids* **19**, 1661-8.

Cunzhen, Z. and Conlisk, A. T. (1989). Separation in a gas centrifuge at high feed flow rate. *J. Fluid Mech.* **208**, 355-73.

Davies, P. A. (1972). Experiments on Taylor columns in rotating stratified fluids. *J. Fluid Mech.* **54**, 691-717.

Davies-Jones, R. P. (1973). The dependence of core radius on swirl ratio in a tornado simulator. *J. Atm. Sci.* **30**, 1427-30.

Debnath, L. (1974). On the hydromagnetic spin-up flows in a rotating fluid.. *Letters in Appl. and Eng. Sci.* **1**, 451-63.

Debnath, L. (1974). Resonant oscillations of a porous plate in an electrically conducting rotating viscous fluid.. *Phys. of Fluids* **17**, 1704-6.

Debnath, L. (1974). On the unsteady hydromagnetic boundary layer flow induced by torsional oscillations of a disk. *Plasma Phys.* **16**, 1121-8.

SUPPLEMENTARY REFERENCES

Debnath, L. (1974). On forced oscillations in a rotating stratified fluid. *Tellus* **26**, 652-61.

Debnath, L. (1975). On the unsteady hydromagnetic boundary layer flow induced by torsional oscillations of a disk. *Il Nuovo Cimento* **25**, 711-29.

Debnath, L. and Mukherjee, S. (1973). Unsteady multiple boundary layers on a porous plate in a rotating system. *Phys. of Fluids* **16**, 1418-21.

Debnath, L. and Mukherjee, S. (1974). Inertial oscillations and multiple boundary layers in an unsteady rotating flow. *Phys. of Fluids* **17**, 1372-5.

Devanathan, R. and Rao, R. (1973). Forced oscillations of a contained rotating stratified fluid. *Zeit. angew. Math. Phys.* **53**, 617-23.

Devanathan, R., Rao, A. R. et al. (1974). Note on a point source in a rotating stratified fluid. *Acta Mechanica* **19**, 147-51.

Dietrich, D. (1973). A numerical study of rotating annulus flows using a modified Galerkin method. *Pageoph* **109**, 1826-61.

Dijkstra, D. and van Heijst, G. J. F. (1983). The flow between two finite rotating disks enclosed by a cylinder. *J. Fluid Mech.* **128**, 123-54.

Dolzhanskii, F. V. (1972). A laboratory investigation of the stability of motion of a liquid in an ellipsoidal cavity. *Atm. and Ocean. Phys.* **8**, 661-4.

Dolzhanskii, F. V. (1985). Motion of a fluid between rotating cones. *Fluid Mech. Isv.* **20**, 216-22.

Dolzhanskii, F. V. and Krymov, V. A. (1985). The retardation of a fluid in a cylinder of low height. *Fluid Mech. Isv.* **20**, 15-21.

Donley, H. E. and Ingham, W. H. (1987). Effects of centrifuge shape on the separation of a mixture. *Sep. Sci. and Tech.* **22**, 1691-710.

Duck, P. W. (1986). On the flow between two rotating shrouded discs. *Computers and Fluids* **14**, 183-96.

Dudis, J. J. and Davis, S. H. (1971). Energy stability of the Ekman boundary layer. *J. Fluid Mech.* **47**, 405-13.

Durance, J. A. (1970). Fluid motion in a rotating sliced cylinder. *Proc. Camb. Phil. Soc.* **68**, 203-12.

Durivault, J. and Louvet, P. (1976). Etude de la couche de Stewartson compressible dans une centrifugeuse a contre-courant thermique. *C. R. Acad. Sc. Paris* **283**, 79-82.

Egger, J. and Schmid, S. (1988). Elimination of spurious inertial oscillations in boundary-layer models with time-dependent geostrophic winds. *Boundary-Layer Meteor.* **43**, 393-402.

Enlow, R. (1973). Linearized spin-up in vortex flows.. *J. Appl. Math and Phys.* **24**, 165-80.

Evans, D. J. (1969). The rotationally symmetric flow of a viscous fluid in the presence of an infinite rotating disc with uniform suction. *Quart. J. Mech. Appl. Math.* **22**, 467-85.

Evenson, A. J. and Veronis, G. (1975). Continuous representation of wind stress and wind stress curl over the world ocean. *J. Mar. Res.* **33**, 131-44.

Faller, A. J. and Porter, D. L. (1976). A note on eastern boundary currents in a laboratory analogue of the ocean circulation. *Tellus* **28**, 88-9.

Foster, M. R. (1972). The flow caused by the differential rotation of a right circular cylindrical depression. *J. Fluid Mech. 53*, 647-55.

Foster, M. R. and Duck, P. W. (1982). The inviscid stability of Long's vortex. *Phys. of Fluids* **25**, 1715-9.

Fowlis, W. W. and Martin, P. J. (1975). A rotating laser Doppler velocimeter and some new results on the spin-up experiment. *Geo. Fluid Dyn.* **7**, 67-78 ..

Friedlander, S. (1975). Interaction of vortices in a fluid on the surface of a rotating sphere. *Tellus* **27**, 15-24.

Friedlander, S. (1976). Quasi-steady flow of a rotating stratified fluid in a sphere. *J. Fluid Mech.* **76**, 209-28.

Friedlander, S. and Siegmann, W. (1982). Internal waves in a contained rotating stratified fluid. *J. Fluid Mech.* **114**, 125-56.

Fultz, D. (1965). Some cases of instability in cylindrically symmetric viscous flows. 9th Midwestern Mechanics Conference 37-48.

Fultz, D. and Murty, T. S. (1968). Effects of the radial law of depth on the instability of inertial oscillations in rotating fluids. *J. Atm. Sci.* **25**, 779-88.

Fung, Y. T. and Kurzweg, U. H. (1975). Stability of swirling flows with radius-dependent density. *J. Fluid Mech.* **72**, 243-55.

Gabov, S. A. (1980). On completeness of the system of eigenfunctions of a problem arising in the theory of tidal oscillations. *Soviet Math. Dokl.* **21**, 183-5.

Gabov, S. A. (1980). On the spectrum of a problem of S. L. Sobolev. *Soviet Math. Dokl.* **22**, 104-7.

Gadgil, S. (1970). Structure of jets in rotating systems. *J. Fluid Mech.* **47**, 417-36.

Gans, R. F. (1970). On the precession of a resonant cylinder. *J. Fluid Mech.* **41**, 865-72.

Gans, R. F. (1970). On hydromagnetic precession in a cylinder. *J. Fluid Mech.* **45**, 111-30.

Gans, R. F. (1974). On the Poincaré problem for a compressible medium. *J. Fluid Mech.* **62**, 657-75.

SUPPLEMENTARY REFERENCES

Gans, R. F. (1975). On the gravitationally forced motions of a compressible fluid within a horizontally rotating cylinder. *J. Fluid Mech.* **67**, 611-24.

Gans, R. F. (1975). On the stability of shear flow in a rotating gas. *J. Fluid Mech.* **68**, 403-12.

Gans, R. F. (1976). Poiseuille-like flow in a rotating gas. *Phys. of Fluids* **19**, 1821-2.

Gans, R. F. (1977). On steady flow in partially filled rotating cylinder. *J. Fluid Mech.* **82**, 415-27.

Gans, R. F. (1979). On the flow around a buoyant cylinder within a rapidly rotating horizontal cylindrical container. *J. Fluid Mech.* **93**, 529-48.

Gans, R. F. (1983). Boundary layers on characteristic surfaces for time-dependent rotating flows. *J. Appl. Mech.* **50**, 251-4.

Gans, R. F. and Yalisove, S. M. (1982). Observations and measurements of flow in a partially-filled horizontally rotating cylinder. *J. Fluids Eng.* **104**, 363-6.

Garadzhaev, A. (1987). Spectral theory of a problem concerning small oscillations of a perfect liquid in a rotating elastic container. *Differential Equations* **23**, 38-47.

Gates, W. L. (1968). A numerical study of transient Rossby waves in a wind-driven homogeneous ocean. *J. Atm. Sci.* **25**, 3-22.

Gates, W. L. (1969). The Ekman velocity in an enclosed beta-plane ocean. *J. Mar. Res.* **27**, 99-120.

Gill, A. E. (1971). The equatorial current in a homogeneous ocean. *Deep-Sea Res.* **18**, 421-32.

Gills, A. E. (1974). The stability of planetary waves on an infinite beta-plane. *Geo. Fluid Dyn.* **6**, 29-48.

Gill, A. E. and Bryan, K. (1971). Effects of geometry on the circulation of a three dimensional southern-hemisphere ocean model. *Deep-Sea Res.* **18**, 685-722.

Gilman, P. A. and Benton, E. R. (1968). Influence of an axial magnetic field on the steady linear Ekman boundary layer. *Phys. Fluids* **11**, 2397-401.

Gledzer, Ye. B., Dolzhanskii, F. V. et al. (1975). Experimental and theoretical study of the stability of motion of a liquid in an elliptical cylinder. *Atmospheric and Oceanic Physics, Isv.* **11**, 981-92.

Gledzer, Ye. B., Kovikov, Yu. V. et al. (1974). An investigation of the stability of liquid flows in a three- axis ellipsoid. *Atmospheric and Oceanic Physics, Isv.* **10**, 115-8.

Goller, H. and Ranov, T. (1968). Unsteady rotating flow in a cylinder with a free surface. *J. Basic Eng.* **90**, 445-54.

Gori, F. (1985). Is laminar flow in a cylindrical container with a rotating cover a Batchelor or Stewartson-type solution?. *J. Fluids Eng.* **107**, 436-7.

Graebel, W. P. (1969). On the slow motion of bodies in stratified and rotating fluids. *Quart. J. Mech. Appl. Math.* **22**, 39-54.

Granger, R. A. (1972). A steady axisymmetric vortex flow. *Geo. Fluid Dyn.* **3**, 45-88.

Granger, R. A. (1973). On the decay of a viscous vortex. *Quart. Appl. Math.* **30**, 531-3.

Greenspan, H. P. (1969). On the non-linear interaction of inertial modes. *J. Fluid Mech.* **36**, 257-64.

Greenspan, H. P. (1969). On the inviscid theory of rotating fluids. *Stud. Appl. Math.* **48**, 19-28.

Greenspan, H. P. (1969). A note on the laboratory simulation of planetary flows. *Stud. Appl. Math.* **48**, 147-52.

Greenspan, H. P. (1976). On a rotational flow disturbed by gravity. *J. Fluid Mech.* **74**, 335-51.

Greenspan, H. P. (1980). A note on the spin-up from rest of a stratified fluid. *Geophys. Astrophys. Fluid Dyn.* **15**, 1-5.

Greenspan, H. P. (1983). On centrifugal separation of a mixture. *J. Fluid Mech.* **127**, 91-101.

Greenspan, H. P. (1984). Compressible Ekman layers on curved boundaries. *Stud. Appl. Math.* **70**, 141-50.

Greenspan, H. P. (1988). On the vorticity of a rotating mixture. *J. Fluid Mech.* **191**, 517-28.

Greenspan, H. P. (1989). Stability of the Ekman layer in a mixture. *Stud. Appl. Math.* **81**, 21-32.

Greenspan, H. P. (1990). A note on spin-up effects in a rotating mixture. *Stud. Appl. Math.* **82**, 49-58.

Greenspan, H. P. and Ungarish, M. (1985). On the centrifugal separation of a bulk mixture. *Int. J. Multiphase Flow* **11**, 825-35.

Grimshaw, R. (1975). Internal gravity waves: Critical layer absorption in a rotating fluid. *J. Fluid Mech.* **70**, 287-304.

Grimshaw, R. (1975). Non-linear internal gravity waves in a rotating fluid. *J. Fluid Mech.* **71**, 497-512.

Gunzburger, M. D. and Wood, H. G. (1984). A finite element method for gas centrifuge flow problems. *Siam J. Sci. Stat. Comput.* **5**, 78-94.

Gusev, A. and Bark, F. H. (1980). Stability of rotation-modified plane Poiseuille flow. *Phys. of Fluids* **23**, 2171-7.

Hadlock, R. K. and Hess, S. L. (1968). A laboratory hurricane model incorporating an analog to release of latent heat. *J. Atm. Sci.* **25**, 161-77.

SUPPLEMENTARY REFERENCES

Hall, M. G. (1966). The structure of concentrated vortex cores. *Aero. Sci.* **7**, 53-110.

Hashimoto, K. (1975). A source-sink flow in an incompressible rotating fluid. *J. Phys. Soc. Japan* **38**, 1508-15.

Hashimoto, K. (1976). On the stability of the Stewartson layer. *J. Fluid Mech.* **76**, 289-306.

Hashimoto, K. (1976). On the stability of the Stewartson layer. *J. Phys. Soc. Japan* **38**, 1508-15.

Hatton, L. (1975). Stagnation point flow in a vortex core. *Tellus* **27**, 269-79.

Heikes, K. E. and Maxworthy, T. (1982). Observations of inertial waves in a homogeneous rotating fluid. *J. Fluid Mech.* **125**, 319-45.

van Heijst, G. J. F. (1979). Rotating flow in a cylinder with a circular barrier on the bottom. *J. Eng. Math.* 153-71.

van Heijst, G. J. F. (1984). Source-sink flow in a rotating cylinder. *J. Eng. Math.* **18**, 247-57.

van Heijst, G. J. F. (1986). Fluid flow in a partially-filled rotating cylinder. *J. Eng. Math.* **20**, 233-50.

Heiter, T., Steck, E. et al. (1986). Inviscid and viscous flow through rotating meridional contours. **Finite Approximations in Fluid Mechanics**, Braunschweig 103-117.

Hide, R. (1967). Theory of axisymmetric thermal convection in a rotating fluid annulus. *Phys. of Fluids* **10**, 56-68.

Hide, R. (1971). On geostrophic motion of a non-homogeneous fluid. *J. Fluid Mech.* **49**, 745-51.

Hide, R. and Titman, C. W. (1968). On slow transverse flow past obstacles in a rapidly rotating fluid. *J. Fluid Mech.* **37**, 737-64.

Hide, R., Ibbetson, A. and Lighthill, M.J. (1968). On slow transverse flow past obstacles in a rapidly rotating fluid. *J. Fluid Mech.* **32**, 251-72.

Hide, R., Mason, P. J. et al. (1977). Thermal convection in a rotating fluid subject to a horizontal temperature gradient. *J. Atm. Sci.* **34**, 930-50.

Hinze, J. O. and Milborn, H. (1950). Atomization of liquids by means of a rotating cup. *J. Appl. Mech.* 145-53.

Hocking, L. M. (1970). Radial filling of a rotating container. *Quart. J. Mech. Appl. Math.* **23**, 101-17.

Hocking, L.M., Moore, D.W. et al. (1979). The drag on a sphere moving axially in a long rotating container. *J. Fluid Mech.* **90**, 781-93.

Hogg, N.G. (1973). On the stratified Taylor column. *J. Fluid Mech.* **58**, 517-37.

Holton, J. R. and Stone, P. H. (1968). A note on the spin-up of a stratified fluid. *J. Fluid Mech.* **33**, 127-30.

Homicz, G. F. and Gerber, N. (1987). Numerical model for fluid spin-up from rest in a partially filled cylinder. *J. Fluids Eng.* **109**, 194-7.

Homsy, G. M. and Hudson, J. L. (1968). Transient flow near a rotating disk. *Appl. Sci. Res.* **18**, 384-97.

Homsy, G. M. and Hudson, J. L. (1969). Centrifugally driven thermal convection in a rotating cylinder. *J. Fluid Mech.* **35**, 33-52.

Homsy, G. M. and Hudson, J. L. (1971). Centrifugal convection and its effect on the asymptotic stability of a bounded rotating fluid heated from below. *J. Fluid Mech.* **48**, 605-24.

Hopfinger, E. J. and Linden, P. F. (1990). The effect of background rotation on fluid motions: a report on Euromech 245. *J. Fluid Mech.* **211**, 417-35.

Howard, L. N. and Siegmann, W. L. (1969). On the initial value problem for rotating stratified flow. *Stud. Appl. Math.* **48**, 153-69.

Hsu, C. T. and Fattahi, B. (1976). Mechanism of tornado funnel formation. *Phys. of Fluids* **19**, 1853-7.

Hsu, H. W., Brantley, J. N. et al. (1979). Transport phenomena in zonal centrifuge rotors. *Sep. Sci. and Tech.* **14**, 69-77.

Hsueh, Y. (1969). Buoyant Ekman Layer. *Phys. of Fluids* **12**, 1757-62.

Hsueh, Y. and Legeckis, R. (1973). Western intensification in a rotating water tunnel. *Geo. Fluid Dyn.* **5**, 333-58.

Hunter, C. and Riahi, N. (1975). Non-linear convection in a rotating fluid. *J. Fluid Mech.* **72**, 433-54.

Huppert, H. E. (1975). Some remarks on the initiation of inertial Taylor columns. *J. Fluid Mech.* **67**, 397-412.

Huppert, H. E. and Stern, M. E. (1974). Ageostrophic effects in a rotating stratified flow. *J. Fluid Mech.* **62**, 369-85.

Hyun, J. M. (1983). Axisymmetric flows in spin-up from rest of a stratified fluid in a cylinder. *Geophys. Astrophys. Fluid Dyn.* **23**, 127-41.

Hyun, J. M. (1984). Flows within the Ekman layer during spin-up of a thermally stratified fluid. *Fluid Dynamics* **29**, 65-79.

Hyun, J. M. (1985). Flow in an open tank with a free surface driven by the spinning bottom *J. Fluids Eng.* **107**, 495-9.

Hyun, J. M. (1985). Transient starting flow in a cylinder with counter-rotating endwall disks. *J. Fluids Eng.* **107**, 92-6.

Hyun, J. M., Fowlis, W. W. et al. (1982). Numerical solutions for the spin-up of a stratified fluid. *J. Fluid Mech.* **117**, 71-90.

Hyun, J. M., Leslie, F. et al. (1983). Numerical solutions for spin-up from rest in a cylinder. *J. Fluid Mech.* **127**, 263-81.

Ibbetson, A. and Tritton, D. J. (1975). Experiments on turbulence in a rotating fluid. *J. Fluid Mech.* **68**, 639-72.

Ibrani, S. and Dwyer, H. (1987). Flow interactions during axisymmetric spin-up. *A.I.A.A. Journ.* **25**, 1305-11.

Imawaki, S. and Takano, K. (1974). Planetary flow in a circular basin. *Deep-Sea Res.* **21**, 69-77.

Israeli, M. (1972). On trapped modes of rotating fluids in spherical shells. *Stud. Appl. Math.* **51**, 219-37.

Israeli, M. and Ungarish, M. (1983). Laminar compressible flow between close rotating disks. *Computers and Fluids* **11**, 145-57.

Jacobs, C. (1971). Transient motions produced by disks oscillating torsionally about a state of rigid rotation. *Quart. J. Mech. Appl. Math.* **24**, 221-36.

Jerskey, T. and Penner, S. S. (1973). Velocity profiles in steady and unsteady rotating flows for a finite cylindrical geometry. . *Phys. of Fluids* **16**, 769-74.

Jeyapalan, K. and Bennett, A. F. (1980). Corner effects in free shear layers in rotating flows. *Zeit. angew. Math. Phys.* **31**, 533-5.

Johnson, E. R. (1978). Trapped vortices in rotating flow. *J. Fluid Mech.* **86**, 209-24.

Johnson, E. R. (1982). The effects of obstacle shape and viscosity in deep rotating flow over finite-height topography. *J. Fluid Mech.* **120**, 359-83.

Johnson, E. R. (1984). Starting flow for an obstacle moving transversely in a rapidly rotating fluid. *J. Fluid Mech.* **149**, 71-88.

Johnson, J. A. (1966). The diffusion of a viscous vortex ring in a rotating fluid. *J. Fluid Mech.* **24**, 753-64.

Johnson, J. A., Fandry, C. B. et al. (1971). On the variation of ocean circulation produced by bottom topography. *Tellus* **23**, 113-21.

Joseph, D. D. (1977). Rotating simple fluids. *Arch. Rat. Mech. Anal.* **66**, 311-44.

Jury, S. H. and Locke, W. L. (1957). Continuous centrifugation in a disk centrifuge. *A.I. Ch.E. Journ.* **3**, 480-3.

Kaiser, J. (1969). Rotating deep annulus convection: thermal properties of the upper symmetric regime. *Tellus* **21**, 789-805.

Kakayama, W. and Usui, S. (1974). Flow in rotating cylinder of a gas centrifuge. *J. Nucl. Sci. and Tech.* **11**, 242-62.

Karahalios, G. T. (1986). Singularities on the flow between two rotating surfaces. *Acta Mechanica* **58**, 229-38.

Karanfilian, S. K. and Kotas, T. J. (1981). Motion of a spherical particle in a liquid rotating as a solid body. *Proc. Roy. Soc. Lond.* **376**, 525-44.

Karweit, M. J. (1975). Observation of cellular patterns in a partly filled, horizontal, rotating cylinder. *Phys. of Fluids* **18**, 111-2.

Kasture, D. Y. (1970). Motion of an asymmetric body in a rotating liquid. *J. Math. Phys. Sci.* **4**, 278-85.

Kasture, D. Y. and Rao, Y. S. (1967). Secondary flows in a rotating liquid. *J. Math. and Phys. Sci.* **1**, 65-74.
Katagiri, M. (1974). Flow due to impulsive rotation of infinite disk. *Phys. of Fluids* **17**, 1463-4.
Kheshgi, H. S. and Scriven, L. E. (1985). Viscous flow through a rotating square channel. *Phys. of Fluids* **28**, 2968-79.
Kimura, R. (1976). Barotropic instability of a boundary jet on a sloping bottom. *Geo. Fluid Dyn.* **7**, 205-30.
Kitchens, C. W. Jr. (1980). Navier-Stokes solutions for spin-up in a filled cylinder. *A.I.A.A. Journ.* **18**, 929-34.
Kitchens, C. W. Jr. (1980). Ekman compatibility conditions in Wedemeyer spin-up model. *Phys. of Fluids* **23**, 1062-4.
Kobayashi, R. (1988). Boundary layer transition and separation on spheres rotating in axial flow. *Exp. Thermal and Fluid Sci.* **1**, 99-104.
Kranenburg, C. (1979). Sink flow in a rotating basin. *J. Fluid Mech.* **94**, 65-81.
Kreith, F. and Viviand, H. (1967). Laminar source flow between two parallel coaxial disks rotating at different speeds. *J. Appl. Mech.* **34**, 541-7.
Kroll, J. (1975). The propagation of wind-generated inertial oscillations from the surface into one deep ocean. *J. of Mar. Res.* **33**, 15-36.
Kroll, J. (1977). Uniform flow past an axially uniform viscous vortex. *Stud. Appl. Math.* **57**, 205-23.
Kroll, J. and Veronis, G. (1970). The spin-up of a homogeneous fluid bounded below by a permeable medium. *J. Fluid Mech.* **40**, 225-39.
Krymov, V. A. and Manin, D. Yu. (1986). Spin-down of a fluid in a low cylinder at large Reynolds numbers. *Fluid Mechanics, Isv.* **21**, 39-46.
Krymov, V. A. and Manin, D. Yu. (1986). Spin-down of a fluid between infinite cones. *Fluid Mechanics, Isv.* **21**, 537-43.
Kuiken, H. K. (1972). On the flow between two independently rotating disks of variable distance with blowing. *Philips Res. Repts* **27**, 539-82.
Kuiken, H. K. (1971). The effect of normal blowing on the flow near a rotating disk of infinite extent. *J. Fluid Mech.* **47**, 789-98.
Kuo, H. H. and Veronis, G. (1971). The source-sink flow in oceanic analogy. *J. Fluid Mech.* **45**, 441-64.
Kuo, H. L. (1969). Motions of vortices and circulating cylinder in shear flow with friction. *J. Atm. Sci.* **26**, 390-8.
Kurokawa, J. and Sakuma, M. (1988). Flow in a narrow gap along an enclosed rotating disk with through-flow. *J.S.M.E.* **31**, 243-51.

SUPPLEMENTARY REFERENCES

Lai, C.-Y. Rajagopal, K. R. et al. (1985). Asymmetric flow above a rotating disk. *J. Fluid Mech.* **157**, 471-92.

Landman, M. J. and Saffman, P. G. (1987). The three-dimensional instability of strained vortices in a viscous fluid. *Phys. of Fluids* **30**, 2339-42.

Launder, B. E., Tselepidakis, D. P. et al. (1986). A second-moment closure study of rotating channel flow. *J. Fluid Mech.* **183**, 63-75.

LeBlond, P. H. (1964). Planetary waves in a symmetrical polar basin. *Tellus* **16**, 505-11.

Leibovich, S. (1970). Weakly non-linear waves in rotating fluids. *J. Fluid Mech.* **42**, 803-22.

Lentini, M. and Keller, H. B. (1980). The von Karman swirling flows. *Siam J. Appl. Math.* **38**, 52-64.

Leslie, L. M. (1971). The development of concentrated vortices: a numerical study. *J. Fluid Mech.* **48**, 1-21.

Leslie, L. M. and Smith, R. K. (1970). The surface boundary layer of a hurricane. *Tellus* **22**, 288-97.

Levi, E. (1972). Experiments on unstable vortices. *J. Eng. Mech. Div. A.C.S.E.* **98**, 539-59.

Lezius, D. K. and Johnston, J. P. (1976). Roll-cell instabilities in rotating laminar and turbulent channel flows. *J. Fluid Mech.* **77**, 153-75.

Lighthill, M. J. (1967). On waves generated in dispersive systems by travelling forcing effects with applications to the dynamics of rotating fluids. *J. Fluid Mech.* **27**, 725-52.

Lighthill, M. J. (1970). The theory of trailing Taylor columns. *Proc. Cam. Phil. Soc.* **68**, 485-91.

Lin, Y. Y. (1986). Numerical solutions for flow in a partially filled, rotating cylinder. *Siam J. Stat. Comput.* **7**, 560-70.

Linden, P. F. and van Heijst, G. J. F. (1984). Two-layer spin-up and frontogenesis. *J. Fluid Mech.* **143**, 69-94.

Logan, S. E. (1971). An approach to the dust devil vortex. *A.I.A.A. Journ.* **9**, 660-5.

London, S. D. (1987). Weakly non-linear Rossby waves in a rotating fluid of spheroidal configuration. *Dyn. of Atm. and Oceans* **11**, 39-58.

London, S. and Shen, M. C. (1979). Free oscillation in a rotating spherical shell. *Phys. of Fluids* **22**, 2071-80.

Long, R. R. (1968). Sources and sinks on a beta-earth. *Tellus* **20**, 524-32.

Long, R. R. (1970). Blocking effects in flow over obstacles. *Tellus* **22**, 471-80.

Longuet-Higgins, M. S. (1968). On the trapping of waves along a discontinuity of depth in a rotating ocean. *J. Fluid Mech.* **31**, 417-34.

Longuet-Higgins, M. S. (1972). Topographic Rossby waves. *Memoires Soc. Roy. Sci. De Liege* **2**, 11-16.

Loper, D. (1970). On viscous flow within a rotating spheroidal container. *Quart. J. Mech. Appl. Math.* **23**, 119-25.

Loper, D. E. (1975). A linear theory of rotating thermally stratified, hydromagnetic flow. *J. Fluid Mech.* **72**, 1-16.

Loper, D. and Benton E. R. (1970). On the spin-up electrically conducting fluid. Part 2: Hydromagnetic spin-up between infinite flat insulating plates. *J. Fluid Mech.* **43**, 785-99.

Lots, M. (1973). Kompressible Ekman-Grenzschichten in rotierenden cylinder. *Atom Kernenergie* **22**, 46.

Lugt, H. J. and Abboud, M. (1986). Axisymmetric vortex breakdown with and without temperature effects in a container with a rotating lid. *J. Fluid Mech.* **179**, 179-200.

Lugt, H. J. and Haussling, H. J. (1973). Development of flow circulation in a rotating tank. *Acta Mechanica* **18**, 255-72.

Macey, J. P. and Wellman, E. J. (1969). Instability phenomenon associated with an enclosed rotating disk. *Phys. of Fluids* **12**, 720-2.

Mager, A. (1974). Steady, incompressible, swirling jets and wakes. *A. I. A. A. Journ.* **12**, 1540-7.

Malkus, W. V. R. (1968). Precession of the earth as the cause of geomagnetism. *Science* **169**, 259-64.

Malkus, W. V. R. (1968). Equatorial planetary waves. *Tellus* **20**, 545-7.

Malkus, W. V. R. and Proctor, M. R. E. (1975). The macrodynamics of alpha-effect dynamos in rotating fluids. *J. Fluid Mech.* **67**, 417-43.

Manohar, R. (1967). The boundary layer on a rotating sphere. *Z. angew. Math. Phys.* **18**, 320-30.

Manton, M. J. (1975). The stability of a rotating flow subjected to an axial pulsation. *Quart. J. Math. Appl. Mech.* **28**, 91-105.

Maslennikova, V. N. (1967). Half space potentials for the equations of motion of a rotating fluid. *Soviet Math. Dokl.* **8**, 133-6.

Maslowe, S. A. (1974). Instability of rigidly rotating flows to non-axisymmetric disturbances. *J. Fluid Mech.* **64**, 307-18.

Mason, P. J. (1975). Forces on bodies moving transversely through a rotating fluid. *J. Fluid Mech.* **71**, 577-99.

Mason, P. J. and Sykes, R. I. (1981). A numerical study of rapidly rotating flow over surface-mounted obstacles. *J. Fluid Mech.* **111**, 167-95.

Mathis, D. M. and Neitzel, G. P. (1985). Experiments on impulsive spin-down to rest. *Phys. of Fluids* **28**, 449-54.

SUPPLEMENTARY REFERENCES

Matkowsky, J. and Siegmann, W. L. (1976). The flow between counter-rotating disks at high Reynolds number. *Siam J. Appl. Math.* **30**, 720-7.

Matsson, O. J. and Alfredsson, P. H. (1990). Curvature and rotation induced instabilities in channel flow. *J. Fluid Mech.* **210**, 537-63.

Matsuda, T., Hashimoto, K. et al. (1976). Thermally driven flow in a gas centrifuge with an insulated side wall. *J. Fluid Mech.* **73**, 389-99.

Matsuda, T. and Hashimoto, K. (1976). Thermally, mechanically or externally driven flows in a gas centrifuge with insulated horizontal end plates. *J. Fluid Mech.* **78**, 337-54.

Matsuda, T. and Nakagawa, K. (1983). A new type of boundary layer in a rapidly rotating gas. *J. Fluid Mech.* **126**, 431-42.

Matsuda, T. and Sakurai, T. (1973). Heat-up of rotating stratified fluid. *J. Phys. Soc. Japan* **34**, 1659-66.

Matsuda, T., Sakurai, T. et al. (1975). Source-sink flow in a gas centrifuge. *J. Fluid Mech.* **69**, 197-208.

Maxworthy, T. (1968). The observed motion of a sphere through a short, rotating cylinder of fluid. *J. Fluid Mech.* **31**, 643-55.

Maxworthy, T. (1970). The flow created by a sphere moving along the axis of a rotating slightly viscous fluid. *J. Fluid Mech.* **40**, 453-79.

McCartney, M. S. (1975). Inertial Taylor columns on a beta plane. *J. Fluid Mech.* **68**, 71-95.

McDonald, B. E. and Dicke, R. H. (1967). Solar oblateness and fluid spin-down. *Science* **158**, 1562-4.

McEwan, A. D. (1970). Inertial oscillations in a rotating fluid cylinder. *J. Fluid Mech.* **40**, 603-40.

McIntyre, M. E. (1968). The axisymmetric convective regime for a rigidly bounded rotating annulus. *J. Fluid Mech.* **32**, 625-55.

McIntyre, M. E. (1970). The flow created by a sphere moving along the axis of a rotating, slightly viscous fluid. *J. Fluid Mech.* **40**, 453-79.

McIntyre, M. E. (1972). On Long's hypothesis of no upstream influence in uniformly stratified or rotating flow. *J. Fluid Mech.* **52**, 209-43.

Melander, M. V. (1983). An algorithmic approach to the linear stability of the Ekman layer. *J. Fluid Mech.* **132**, 283-93.

Mellor, G. L., Chapple, P. J. et al. (1968). On the flow between a rotating and a stationary disk. *J. Fluid Mech.* **31**, 95-112.

Mikami, H. (1973). Thermally induced flow in a gas centrifuge. *J. Nucl. Sci. Tech.* **10**, 396-401.

Miles, J. M. (1969). Transient motion of a dipole in a rotating flow. *J. Fluid Mech.* **39**, 433-42.

Miles, J. W. (1971). The Oseenlet as a model for separated flow in a rotating viscous liquid. *J. Fluid Mech.* **42**, 207-17.
Miles, J. W. (1971). Boundary layer separation on a sphere in a rotating flow. *J. Fluid Mech.* **45**, 513-26.
Miles, J. W. (1972). Axisymmetric rotating flow past a circular disk. *J. Fluid Mech.* **52**, 689-700.
Miles, J. W. (1972). Upstream influence of a dipole in rotating flow. *J. Fluid Mech.* **54**, 369-83.
Miles, J. W. (1975). Axisymmetric rotating flow past a prolate spheroid. *J. Fluid Mech.* **72**, 363-71.
Miles, J. W. and Troesch, B. A. (1961). Surface oscillations of a rotating liquid. *J. Appl. Mech.* **28**, 491-6.
Miyasaka, Y. (1974). On the flow of a viscous free boundary jet on a rotating disk. Bull. *J.S.M.E.* **17**, 1461-75.
Moberg, H. and Hultgren, L. S. (1984). Stability of compressible Ekman boundary-layer flow. *J. Fluid Mech.* **147**, 159-68.
Mochizuki, S. and Yang, W. (1986). Flow friction on co-rotating parallel-disk assemblies with forced through-flow. *Exp. in Fluids* **4**, 56-60.
Moffatt, H. K. (1973). Report on the NATO advanced study institute on magnet- hydrodynamic phenomena in rotating fluids. *J. Fluid Mech.* **57**, 625-49.
Moffatt, H. K. (1977). Behaviour of a viscous film on the outer surface of a rotating cylinder. *J. Mecanique* **16**, 651-72.
Moore, D. W. and Niiler, P. P. (1974). A two layer model for the separation of inertial boundary currents. *J. of Mar. Res.* **32**, 457-96.
Moore, D. W. and Saffman, P. G. (1968). The rise of a body through a rotating fluid in a container of finite length. *J. Fluid Mech.* **31**, 635-42.
Moore, D. W. and Saffman, P. G. (1969). The flow induced by the transverse motion of a thin disk in its own plane. *J. Fluid Mech.* **39**, 831-47.
Moore, D. W. and Saffman, P. G. (1969). The structure of free vertical shear layers in a rotating fluid and the motion produced by a slowly rising body. *Phil. Trans. Roy. Soc.* **264**, 597-634.
Morton, B. R. (1966). Geophysical vortices. *Prog. Aero. Sci.* **7**, 145-94.
Morton, B. R. (1969). The strength of vortex and swirling core flows. *J. Fluid Mech.* **38**, 315-33.
Munson, B. R. (1974). Viscous incompressible flow between eccentric coaxial rotating spheres. *Phys. of Fluids* **17**, 528-31.

SUPPLEMENTARY REFERENCES

Munson, B. R. and Menguturk, M. (1975). Viscous incompressible flow between concentric rotating spheres. Part 3. Linear stability and experiments. *J. Fluid Mech.* **69**, 705-20.

Nakayama, W. and Torii, T. (1974). Numerical analysis of separative power of isotope centrifuges. *J. Nucl. Sci. and Tech.* **11**, 495-504.

Nakayama, W. and Usui, S. (1974). Flow in rotating cylinder of a gas centrifuge. *J. Nucl. Sci. and Tech.* **11**, 242-62.

Neitzel, G. P. and Davis, S. H. (1981). Centrifugal instabilities during spin-down to rest in finite cylinders. Numerical experiments. *J. Fluid Mech.* **102**, 329-52.

Newall, A. C. (1969). Rossby wave packet interactions. *J. Fluid Mech.* **35**, 255-71.

Nguyen, N. D. and Ribault, J. P. (1975). Multiple solutions for flow between coaxial disks. *J. Fluid Mech.* **68**, 369-88.

Ogawa, A., Seito, O. et al. (1986). On the spin-up time and the residue time of the turbulent rotational air flows. *J.S.M.E.* 63rd Conf. Tokyo 1698-1703.

Olander, D. R. (1978). The gas centrifuge. *Scientific American* **239**, 37-44.

Owen, J. M., Pincombe, J. R. et al. (1985). Source-sink flow inside a rotating cylindrical cavity. *J. Fluid Mech.* **155**, 233-65.

Page, M. A. (1985). On the low Rossby number flow of a rotating fluid past a circular cylinder. *J. Fluid Mech.* **156**, 205-21.

Page, M. A. (1987). Separation and free-streamline flows in a rotating fluid at low Rossby number. *J. Fluid Mech.* **179**, 155-77.

Pandolfo, J. P. and Brown, P. S. (1967). Inertial oscillations in an Ekman layer containing a horizontal discontinuity surface. *J. Mar. Res.* **25**, 10-28.

Pao, H-P. and Shih, H-H. (1973). Selective withdrawal and blocking wave in rotating fluids. *J. Fluid Mech.* **57**, 459-80.

Pao, H. P. and Kao, T. W. (1969). Sources and sinks at the axis of a viscous rotating fluid. *Phys. of Fluids* **12**, 1536-46.

Pao, S. P. and Siekmann, J. (1968). Oscillations of a vapour cavity in a rotating cylindrical tank. *J. Fluid Mech.* **31**, 249-72.

Paterson, J. A. and Greif, R. (1973). Mass transfer to a rotating disk in a non-Newtonian fluid. *Phys. of Fluids* **16**, 1816-7.

Pedley, T. J. (1968). On the instability of rapidly rotating shear flows to non-axisymmetric disturbances. *J. Fluid Mech.* **31**, 603-7.

Pedley, T. J. (1969). The viscous vortex induced by a sink on the axis of a circulating fluid in the presence of a plane free surface. *J. Fluid Mech.* **36**, 219-38.

Pedlosky, J. (1965). A necessary condition for the existence of an inertial boundary layer in a baroclinic ocean. *J. Mar. Res.* **23**, 69-72.

Pedlosky, J. (1968). An overlooked aspect of the wind-driven oceanic circulation. *J. Fluid Mech.* **32**, 809-21.

Pedlosky, J. (1969). Linear theory of the circulation of a stratified ocean. *J. Fluid Mech.* **35**, 185-205.

Pedlosky, J. (1969). Axially symmetric motion of a stratified, rotating fluid in a spherical annulus of narrow gap. *J. Fluid Mech.* **36**, 401-15.

Pedlosky, J. (1971). A note on the role of the buoyancy layer in a rotating stratified fluid. *J. Fluid Mech.* **48**, 181-2.

Pfeffer, R. L. and Fowlis, W. W. (1968). Wave dispersion in a rotating differentially heated cylindrical annulus of fluid. *J. Atm. Sci.* **25**, 361-71.

Philander, S. G. A. (1971). On the flow properties of a fluid between concentric spheres. *J. Fluid Mech.* **47**, 799-809.

Poe, G. G. and Acrivos, A. (1975). Closed-streamline flows past rotating single cylinders and spheres: Inertial effects.. *J. Fluid Mech.* **72**, 605-23.

Pollard, R. T. (1970). On the generation by winds of inertial waves in the ocean. *Deep-Sea Res.* **17**, 795-812.

Pritchard, W. G. (1969). The motion generated by a body moving along the axis of a uniformly rotating fluid. *J. Fluid Mech.* **39**, 443-64.

Pritchard, W. G. (1970). Solitary waves in rotating fluids. *J. Fluid Mech.* **42**, 61-83.

Puri, P. and Kulshrestha, P. K. (1974). Rotating flow of non-Newtonian fluids. *Applic. Anal.* **4**, 131-40.

Puri, P. and Kulshrestha, P. K. (1977). Unsteady hydromagnetic boundary layer in a rotating medium. *J. Appl. Mech.* **44**, 1-4.

Ralston, J. V. (1973). On stationary modes in inviscid rotating fluids. *J. Math. Anal. and Appl.* **44**, 366-83.

Rao, A. R. (1972). Multipoles in rotating fluids. *Indian J. Phys.* **46**, 451-6.

Rao, A. R. (1973). A note on the application of a radiation condition for a source in a rotating stratified fluid. *J. Fluid Mech.* **58**, 161-4.

Rao, A. R. (1975). On an oscillatory point force in a rotating stratified fluid. *J. Fluid Mech.* **72**, 353-62.

Rao, D. B. and Simons, T. J. (1970). Stability of a sloping interface in a rotating two-fluid system. *Tellus* **22**, 493-503.

Rasmussen, H. (1970). Numerical solutions for steady viscous linearized flow between two finite rotating disks. *Z. angew. Math. Phys.* **21**, 611-9.

Raszillier, H., Durst, F. et al. (1988). Forces of a fluid in a rotating straight pipe. *J. Appl. Math. and Phys.* **39**, 397-403.

SUPPLEMENTARY REFERENCES

Redekopp, L. G. (1971). The boundary layer on a flat plate moving transversely in a rotating stratified fluid. *J. Fluid Mech.* **46**, 769-86.

Redekopp, L. G. (1972). Side-wall boundary layers in rotating axial flow. *J. Fluid Mech.* **55**, 565-76.

Redekopp, L. G. (1975). Wave patterns generated by disturbances travelling horizontally in rotating stratified fluids. *Geo. Fluid Dyn.* **6**, 289-314.

Rhines, P. (1970). Edge, bottom and Rossby waves in a rotating stratified fluid. *Geo. Fluid Dyn.* **1**, 273-302.

Rhines, P. G. (1969). Slow oscillations in an ocean of varying depth. *J. Fluid Mech.* 161-206.

Rhines, P. and Bretherton, F. (1973). Topographic Rossby waves in a rough bottomed ocean. *J. Fluid Mech.* **61**, 583-608.

Ribando, R. J. (1984). A finite-difference solution of Onsager's model for flow in a gas centrifuge. *Computers Fluids* **12**, 235-52.

Ribando, R. J., Palmer, J. L. et al. (1989). Flow in a partially filled, rotating, tapered cylinder. *J. Fluid Mech.* **203**, 541-55.

Ribando, R. J. and Shadday, M. A. (1984). The Ekman matching condition in a partially filled, rapidly rotating cylinder. *J. Comp. Phys.* **53**, 266-88.

Rickard, J. A. (1973). Free oscillations of a rotating fluid contained between two spheroidal surfaces. *Geo. Fluid Dyn.* **5**, 369-84.

Rieutord, M. (1987). Linear theory of rotating fluids using spherical harmonics. *Geophys. Astrophys. Fluid Dyn.* **39**, 163-82.

Roberts, P. H. (1968). On the thermal instability of a rotating-fluid sphere containing heat sources. *Phil. Trans. Roy. Soc.* **263**, 93-117.

Roberts, S. M. and Shipman, J. S. (1976). Computation of the flow between a rotating and stationary disk. *J. Fluid Mech.* **73**, 53-64.

Robinson, A. R. (1970). Boundary layers in ocean circulation models. *Ann. Rev. of Fluid Mech.* **2**, 293-312.

Robinson, A. R., Luyten, J. R. et al. (1975). On the theory of thin rotating jets. *Geo. Fluid Dyn.* **6**, 211-44.

Rosenzweig, M. L., Ross, D. H. et al. (1962). On secondary flows in jet-driven vortex tubes. *J. Aerospace Sci.* **29**, 1142.

Rott, N. and Ohrenberger, J. T. (1968). Transformations of the boundary layer equations for rotating flows. *J. Fluid Mech.* **33**, 113-26.

Rumiantsev, V. V. (1966). On the theory of motion of rigid bodies with fluid-filled cavities. *P. M. M.* **30**, 57-77.

Ruschak, K. J. and Scriven, L. E. (1976). Rimming flow of liquid in a rotating horizontal cylinder. *J. Fluid Mech.* **76**, 113-26.

Ryskin, G. (1988). Brownian motion in a rotating fluid: Diffusivity is a function of the rotation rate. *Phys. Rev. Letters* **61**, 1442-5.

Sakurai, T. (1969). Spin-down problem of rotating stratified fluid in thermally insulated circular cylinders. *J. Fluid Mech.* **37**, 689-99.

Sakurai, T. (1969). Spin-down of Boussinesq fluid in a circular cylinder. *J. Phys. Soc. Japan* **26**, 840-8.

Sakurai, T. (1975). Baroclinic-type instability in a gas centrifuge heated from above. *J. Fluid Mech.* **72**, 321-9.

Sakurai, T. (1976). Non-axisymmetric instability of a rotating sheet of gas in a rotating environment. *J. Fluid Mech.* **75**, 513-24.

Sakurai, T. and Matsuda, T. (1972). A temperature adjustment process in a Boussinesq fluid via a buoyancy-induced meridional circulation. *J. Fluid Mech.* **54**, 417-21.

Sakurai, T. and Matsuda, T. (1974). Gasdynamics of a centrifugal machine. *J. Fluid Mech.* **62**, 727-36.

Sakurai, T., Clark, A. et al. (1971). Spin-down of a Boussinesq fluid of small Prandtl number in a circular cylinder. *J. Fluid Mech.* **49**, 753-73.

Sankara, K. K. and Sarma, L. V. (1985). On the steady flow produced in fluid-particle suspension by an infinite rotating disk with surface suction. *J. Eng. Sci.* **23**, 875-86.

Sarma, L. V. and Sankara, K. K. (1986). On the magneto-hydrodynamic flow past a rotating disk in a fluid-particle suspension with a circular magnetic field. *J. Eng. Sci.* **24**, 649-73.

Saunders, K. D. and Beardsley, R. C. (1975). An experimental study of the spin-up of a thermally stratified rotating fluid. *Geo. Fluid Dyn.* **7**, 1-28.

Savas, O. (1987). Stability of Bodewadt flow. *J. Fluid Mech.* **183**, 77-94.

Scott, W. E. (1975). The large amplitude motion of a liquid-filled gyroscope and the non-interaction of inertial and Rossby waves. *J. Fluid Mech.* **72**, 649-60.

Scott, W. E. (1976). The frequency of inertial waves in a rotating, sectored cylinder. *J. Appl. Mech.* **43**, 571-4.

Scott, W. E. (1976). Refraction of plane Rossby waves. *Phys. of Fluids* **19**, 491-3.

Serotsky, P. and Titman, C. W. (1975). Length of a viscous-limited Taylor column. *Phys. of Fluids* **18**, 748-9.

Serrin, J. (1971). The swirling vortex. *Phil. Trans. Roy. Soc.* **271**, 325-60.

Shadday, M. A., Ribando, R. J. et al. (1983). Flow of an incompressible fluid in a partially filled, rapidly rotating cylinder with a differentially rotating endcap. *J. Fluid Mech.* **130**, 203-18.

Shen, M-C. and Keller, J. B. (1973). Ray method for non-linear wave propagation in a rotating fluid of variable depth. *Phys. of Fluids* **16**, 1565-72.

SUPPLEMENTARY REFERENCES

Shih, H. H. and Pao, H. P. (1971). Selective withdrawal in rotating fluids. *J. Fluid Mech.* 509-27.

Siegmann, W. L. (1971). The spin-down of rotating stratified fluids. *J. Fluid Mech.* **47**, 689-711.

Siegmann, W. L. (1974). Evolution of unstable shear layers in a rotating fluid. *J. Fluid Mech.* **64**, 289-305.

Sirivat, A., Rajagopal, K. R. et al. (1988). An experimental investigation of the flow of non-Newtonian fluids between rotating disks. *J. Fluid Mech.* **186**, 243-56.

Smith, M. K. (1986). Thermocapillary and centrifugal-buoyancy-driven motion in a rapidly rotating liquid cylinder. *J. Fluid Mech.* **166**, 245-64.

Smith, R. (1971). The ray paths of topographic Rossby waves. *Deep-Sea Res.* **18**, 477-84.

Smith, R. K. (1968). The surface boundary layer of a hurricane. *Tellus* **20**, 473-84.

Smith, S. H. (1981). The influence of rotation in slow viscous flows. *Int. J. Multiphase Flow* **7**, 479-92.

Smith, S. H. (1981). The development of vertical layers in a rotating fluid. *Siam J. Appl. Math.* **40**, 78-89.

Smith, S. H. (1983). The development of non-linearities in the E-1/3 Stewartson layer. *Quart. J. Mech. Appl. Math.* **37**, 75-85.

Smith, S. H. (1984). Unsteady flow from a source in a rotating fluid. *J. Eng. Math.* **18**, 235-46.

Smith, S. H. (1985). On the development of characteristics in an oscillating, rotating fluid. *J. Eng. Math.* **19**, 321-7.

Smith, S. H. (1985). Accelerations induced by a radial outflow. *Quart. J. Mech. Appl. Math.* **38**, 66-77.

Smith, S. H. (1986). The formation of Stewartson layers in a rotating fluid. *Quart. J. Mech. Appl. Math.* **40**, 575-94.

Snow, J. T. (1982). A review of recent advances in tornado vortex dynamics. *Rev. Geophys. and Space Phys.* **20**, 953-64.

Souzou, C. (1971). Boundary layer growth on a spinning sphere. *J. Inst. Math. Appl.* **1**, 251-9.

Spall, J. R. and Wood, H. G. (1984). An analysis of the stability of the compressible Ekman boundary layer. *Phys. of Fluids* **27**, 2808-13.

Speziale, C. G. and Thangam, S. (1983). Numerical study of secondary flows and roll-cell instabilities in rotating channel flow. *J. Fluid Mech.* **130**, 377-96.

St.-Maurice, J-P. and Veronis, G. (1975). A multi-scaling analysis of the spin-up problem. *J. Fluid Mech.* **68**, 417-45.

Stephenson, C. J. (1969). Magnetohydrodynamic flow between rotating coaxial disks. *J. Fluid Mech.* **38**, 335-52.

Stergiopoulos, S. and Aldridge, K. D. (1982). Inertial waves in a fluid partially filling a cylindrical cavity during spin-up from rest. *Geophys. Astrophys. Fluid Dyn.* **21**, 89-112.

Stergiopoulos, S. and Aldridge, K. D. (1987). Ringdown of inertial waves during spin-up from rest of a fluid contained in a rotating cylindrical cavity. *Phys. of Fluids* **30**, 302-11.

Stewartson, K. (1967). On slow transverse motion of a sphere through a rotating fluid. *J. Fluid Mech.* **30**, 357-70.

Stewartson, K. (1967). Slow oscillations of fluid in a rotating cavity in the presence of a toroidial magnetic field. *Proc. Roy. Soc.* **299**, 173-87.

Stewartson, K. (1968). On inviscid flow of a rotating fluid past an axially symmetric body using Oseen equations. *Quart. J. Mech. Appl. Math.* **21**, 353-73.

Stewartson, K. (1970). A note on forward wakes in a rotating fluid. *J. Fluid Mech.* **42**, 219-23.

Stewartson, K. (1971). On trapped oscillation of a rotating fluid in a thin spherical shell. Part 1. *Tellus* **23**, 506-10.

Stewartson, K. (1971). On trapped oscillation of a rotating fluid in a thin spherical shell. Part 2. *Tellus* **24**, 283-7.

Stewartson, K. (1972). On trapped oscillations in a slightly viscous rotating fluid. *J. Fluid Mech.* **54**, 749-61.

Stewartson, K. (1972). A spectrum, continuous in discrete intervals for trapped oscillations. *Tellus* **24**, 283-7.

Stewartson, K. and Cheng, H. K. (1979). On the structure of inertial waves produced by an obstacle in a deep, rotating container. *J. Fluid Mech.* **91**, 415-32.

Stewartson, K. and Rickard, J. A. (1969). Pathological oscillations of a rotating fluid. *J. Fluid Mech.* **38**, 759-73.

Stewartson, K. and Walton, I. C. (1976). On inertial oscillations in the oceans. *Tellus* **28**, 71-3.

Stewartson, K., Simpson, C. J. et al. (1982). The unsteady boundary layer on a rotating disk in a counter-rotating fluid. Part 2. *J. Fluid Mech.* **121**, 507-15.

Stone, P. H. and Baker, D. J. (1968). Concerning the existence of Taylor columns in atmospheres. *Quart. J. Roy. Met. Soc.* **94**, 576-80.

Suess, S. T. (1971). Viscous flow in a deformable rotating container. *J. Fluid Mech.* **45**, 189-201.

Szeri, A. Z., Giron, A. et al. (1983). Flow between rotating disks. Part 2. stability. *J. Fluid Mech.* **134**, 133-54.

Szeri, A. Z., Schneider, S. J. et al. (1983). Flow between rotating disks. Part 1. basic flow. *J. Fluid Mech.* **134**, 103-31.

SUPPLEMENTARY REFERENCES

Tam, K. K. (1979). A note on the asymptotic solution of the flow between two oppositely rotating infinite plane disks. *Siam J. Appl. Math.* **17**, 1305-10.

Thompson, R. (1970). Diurnal tides and shear instabilities in a rotating fluid. *J. Fluid Mech.* **40**, 537-52.

Thompson, R. (1973). Stratified Ekman boundary layer models. *Geo. Fluid Dyn.* **5**, 201-10.

Thornley, C. (1968). On Stokes and Rayleigh layers in a rotating system. *Quart. J. Mech. Appl. Math.* **21**, 451-61.

Tiller, F. M. (1987). Theory of batchwise centrifugal filtration. *A.I.Ch.E. Journ.* **33**, 109-20.

Turner, J. S. (1968). The circulation driven by ring convection in a rotating system. *Quart. J. Roy. Meteorol. Soc.* **94**, 589-91.

Ungarish, M. (1988). Two-fluid analysis of centrifugal separation in a finite cylinder. *Int. J. Multiphase Flow* **14**, 233-43.

Ungarish, M. (1988). Numerical investigation of two-phase rotating flow. *Int. J. Multiphase Flow* **14**, 729-47.

Ungarish, M. and Greenspan, H. P. (1984). On the radial filling of a rotating cylinder. *J. Fluid Mech.* **141**, 97-107.

Ungarish, M. and Greenspan, H. P. (1986). On the filling of a rotating cylinder with a mixture. *J. Fluid Mech.* **162**, 117-28.

Ungarish, M. and Israeli, M. (1985). Axisymmetric compressible flow in a rotating cylinder with axial convection. *J. Fluid Mech.* **154**, 121-44.

Uzkan, T. and Lipstein, N. J. (1986). Effects of honeycomb-shaped walls on the flow regime between a rotating disk and a stationary wall. *J. Engin. for Gas Turbines and Power* **108**, 553-61.

Vanyo, J. P. (1960). Centrifugal waves. *J. Fluid Mech.* **7**, 340-52.

Vanyo, J. P. (1973). Cavity radius vs energy dissipation rate in liquid-filled, precessing, spherical cavities. *A.I.A.A. Journ.* **11**, 261-2.

Vanyo, J. P. and Likins, P. W. (1972). Rigid-body approximations to turbulent motion in a liquid filled, precessing, spherical cavity. *J. Appl. Mech.* **39**, 18-24.

Vaziri, A. and Boyer, D. L. (1971). Rotating flow over shallow topographies. *J. Fluid Mech.* **50**, 79-95.

Venezian G. (1969). Non-linear spin-up. *Topics in Ocean Eng.* **II**, 87-96.

Veronis, G. (1967). Analogous behaviour of homogeneous, rotating fluids and stratified, non-rotating fluids. *Tellus* **19**, 326-35.

Veronis, G. (1970). Effect of fluctuating winds on ocean circulation. *Deep-Sea Res.* **17**, 421-34.
Veronis, G. (1973). Model of world ocean circulation. *J. Mar. Res.* **31**, 228-88.
Veronis, G. and Yang, C. C. (1972). Non-linear source-sink flow in a rotating pie shaped basin. *J. Fluid Mech.* **51**, 513-27.
Vladimirov, V. A., Makarenko, V. G. et al. (1987). Experimental investigation of nonaxisymmetric inertia waves in a rotating fluid. *Fluid Dyn. Isv.* **22**, 176-80.
van de Vooren, A. I., Botta, E. F. F. et al. (1987). The boundary layer on a disk at rest in a rotating fluid. *Quart. J. Mech. Appl. Math.* **40**, 15-32.
Waked, A. M. and Munson, B. R. (1977). Laminar-turbulent flow in a spherical annulus. *J. Fluids Eng.* .
Walin, G. (1969). Some aspects of time-dependent motion of a stratified fluid. *J. Fluid Mech.* **36**, 289-307.
Walker, J. D. A. and Stewartson, K. (1974). Separation and the Taylor-column problem for a hemisphere. *J. Fluid Mech.* **66**, 767-90.
Walker, J. S. (1974). Steady flow in rapidly rotating circular expansions. *J. Fluid Mech.* **66**, 657-72.
Walker, J. S. (1975). Steady flow in rapidly rotating variable-area rectangular ducts. *J. Fluid Mech.* **69**, 209-28.
Wang, C. Y. (1970). Cylindrical tank of fluid oscillating about a state of steady rotation. *J. Fluid Mech.* **41**, 581-92.
Wang, C. Y. (1988). Drag of a corrugated plate in a rotating fluid. *J. Fluid Mech.* **195**, 581-6.
Warn-Varnas, A., Fowlis, W. W. et al. (1978). Numerical solutions and laser-Doppler measurements of spin-up. *J. Fluid Mech.* **85**, 609-39.
Watkins, W. B. and Hussey, R. G. (1977). Spin-up from rest in a cylinder. *Phys. of Fluids* **20**, 1596-604.
Weidman, P. D. (1976). On the spin-up and spin-down of a rotating fluid. Part 1. Extending the Wedemeyer model. *J. Fluid Mech.* **77**, 685-708.
Weidman, P. D. (1976). On the spin-up and spin-down of a rotating fluid. Part 2. Measurements and stability. *J. Fluid Mech.* **77**, 709-735.
Weidman, P. D. and Redekopp, L. G. (1976). On the motion of a rotating fluid in the presence of an infinite rotating disk. *Arch. of Mech.* **28**, 5-6.
Welander, P. (1963). Steady plane fronts in a rotating fluid. *Tellus* **15**, 33-43.
Welander, P. (1968). Wind-driven circulation in one-and two-layer oceans of variable depth. *Tellus* **20**, 1-16.

SUPPLEMENTARY REFERENCES

Whitley, S. (1984). Review of the gas centrifuge until 1962. Part I: principles of separation physics. *Rev. Modern Phys.* **56**, 41-97.

Wilcox, D. C. (1972). The motion of a plate in a rotating fluid at an arbitrary angle of attack. *J. Fluid Mech.* **56**, 221-40.

Willey, S. J., Davis, M. M. et al. (1974). Time-dependent response in the flow between eccentric rotating disks. *Trans. Soc. Rheol.* **18**, 515-26.

Williams, E. W. (1976). Non-Newtonian flow caused by an infinite rotating disc. *J. Non-Newtonian Fluid Mech.* **1**, 51-69.

Williams, G. (1967). Thermal convection in a rotating fluid annulus.. *J. Atm. Sci.* **24**, 144-74.

Wimmer, M. (1981). Experiments on the stability of viscous flow between two concentric rotating spheres. *J. Fluid Mech.*, **103**, 117-31.

Wood, H. G. and Sanders, G. (1983). Rotating compressible flows with internal sources and sinks. *J. Fluid Mech.* **127**, 299-313.

Wood, W. W. (1977). Torque-free rotation of a cylinder. *Phys. of Fluids* **20**, 1953-4.

Wood, W. W. (1977). Inertial oscillations in a rigid axisymmetric container. *Proc. Roy. Soc. Lond.* **358**, 17-30.

Wood, W.W. (1981). Inertial modes with large azimuthal wavenumbers in an axisymmetric container. *J. Fluid Mech.* **105**, 427-49.

Ying, S.-J. (1985). Nonsteady sink and source in a rotating fluid. *Int. J. Non-Linear Mechanics* **20**, 145-51.

Zandbergen, P. J. and Dijkstra, D. (1987). von Karman swirling flows. *Ann. Rev. Fluid Mech.* **19**, 465-91.

Zung, L. B. (1969). Flow induced in fluid particle suspension by an infinite rotating disk. *Phys. of Fluids* **12**, 18-23.